Indian Medicinal Plants

Indian Medicinal Plants

Uses and Propagation Aspects

Srinath Rao
Akula Ramakrishna

CRC Press
Taylor & Francis Group
Boca Raton London New York

CRC Press is an imprint of the
Taylor & Francis Group, an **informa** business

First edition published 2021
by CRC Press
6000 Broken Sound Parkway NW, Suite 300, Boca Raton, FL 33487-2742

and by CRC Press
2 Park Square, Milton Park, Abingdon, Oxon, OX14 4RN

CRC Press is an imprint of Taylor & Francis Group, LLC

ISBN: 978-0-367-43214-0 (hbk)
ISBN: 978-1-003-00189-8 (ebk)

Typeset in Times
by Deanta Global Publishing Services, Chennai, India

Contents

Preface

The traditional systems of India, namely the Ayurveda and Siddha systems of medicine, have contributed enormously to the branch of medicinal botany. The earliest records of the use of medicinal plants, primarily herbs, in the prevention and cure of diseases in India are found in Vedic literature, and precisely in the *Rigveda*, which is believed to have been written between 3500 and 1800 BC. In India, several premier institutions, such as the Council of Scientific and Industrial Research (CSIR), New Delhi; the Central Drug Research Institute (CDRI), Lucknow; the Botanical Survey of India (BSI), Kolkata; and the Tropical Botanical Garden and Research Institute (TBGRI), have made considerable efforts to gather information on medicinal plants from different parts of the country. The World Health Organization (WHO) estimated that 80 % of the people in the developing countries of the world rely on traditional medicine for their primary health care needs and about 85 % of traditional medicine involves the use of plants.

The demand for medicinal plants is increasing, and this leads to unscrupulous collection from the wild and adulteration of supplies. Providing high-quality planting material for sustainable use and thereby saving the genetic diversity of plants in the wild is important. In this regard, the in-vitro methods of propagation of some important medicinal plants are provided along with the traditional methods of propagation, which is a highlight of this book. We use the Bentham and Hooker system of classification (proposed between July 1862 and April 1863) in this book. Though this system is not very natural, it is very easy to apply in the field for the identification of plants. Its emphasis on dividing Dicotyledons in to polypetalae, gamopetalae, and monochlamydeae (Incompletae) nature of flowers is very helpful in the identification of plants. It is the best and most accepted system and is still followed in several herbaria globally and in India.

This unique compendium offers the profiles of 274 medicinal plant species with detailed taxonomic classification based on the Bentham and Hooker system of classification. The book also gives the distribution, detailed description and popular nomenclature in Bengali, English, Gujarati, Hindi, Malayalam, Marathi, Sanskrit, Tamil and Telugu. Moreover, it lists the properties, uses, parts used and chemical constituents of the plants. This volume will be of special interest to practitioners of alternative medicine, students of Ayurveda, researchers and industrialists associated with medical botany, pharmacologists, sociologists and medical herbalists.

Prof Srinath Rao
Dr Akula Ramakrishna

Acknowledgements

We are grateful to Renu Upadhyay, assistant commissioning editor, Jyotsna Jangra, editorial assistant and Todd Perry, production editor of the CRC Press/Taylor & Francis Group team, and also Andrew Corrgian, project manager, from Deanta Global for their help in producing this volume in a beautiful and appealing manner. The authors are thankful to Dr Kotresh, Dr Milind Girdhari, Dr Vedavyas Kulkarni, Dr R.S. Modi and Dr Chandralekha CT for providing photographs of medicinal plants for publication in the book. Prof Srinath Rao is thankful to the then vice-chancellor E.T Puttaiaha and registrar S.L. Hiremath of Gulbarga University Kalburgi, Karnataka State, India for sanctioning one year sabbatical leave for the purpose of writing this book. He is also thankful to Prof V.S. Raju professor of Botany (Taxonomy) (retired), Kakatiya University Warangal, Telengana State, India for the identification of plants and for making corrections in the manuscript. The authors are thankful to their families who have encouraged them to take up this task, permitting them to take the time off to complete it. We are also thankful to our friends and academic colleagues for their support in the successful accomplishment of this book.

Prof Srinath Rao
Dr Akula Ramakrishna

Author Bio

Prof Srinath Rao is a retired professor from the Department of Botany, Gulbarga University, Kalburgi, Karnataka, India. For 34 years, he taught post-graduate students of Botany and Biotechnology in the field of in-vitro propagation of medicinal plants and in-vitro enhancement of secondary metabolites. He has successfully guided 25 M.Phil. and 27 Ph.D. students. He has made significant contributions to in-vitro propagation and enhancement of secondary metabolites in some important medicinal plants. He has presented papers in several international conferences held in the USA, UK, Germany, France and Sri Lanka. Apart from this, he has given 50 invited lectures at national and international conferences. He published more than 100 research papers in national and international journals. He has completed 7 major research projects funded by UGC, CSIR and DST.

Dr Akula Ramakrishna is currently a scientist at Bayer Crop Science, Bangalore, India. Dr Ramakrishna holds a master's degree from the Sri Krishna Devaraya University, Anantapur, Andhra Pradesh, India. He obtained his Ph. D in Biochemistry from the University of Mysore, CSIR-CFTRI, Mysuru. He is a senior research fellow of CSIR, New Delhi. He is involved in various research fields such as Plant Biotechnology, Biochemistry, Plant Secondary Metabolites, Food Science and Technology and Vegetable Quality Analytics. He has published several research papers in leading international journals as well as books, reviews and book chapters. He is also serving as editorial board member and reviewer for reputed international journals. Dr Ramakrishna has presented over 20 research papers in symposia in India and abroad, as well as at some invited lectures in India. He attended the Fifth International Symposium on Plant Neurobiology held in 2009 in Florence, Italy. He also attended the Technical Community of Monsanto (TCM) held in 2016 in St Louis, Missouri. He is a member of the Indian Science Congress Association and Society for Biotechnologists (India). He is a fellow of the Society for Applied Biotechnology, India (2012), and a recipient of the Global Vegetable Research Excellence Award (2017), Global Technology Recognition (2014), Rapid Recognition Award (2013) and Special Recognition (2017) from the Monsanto Company.

ABELMOSCHUS MOSCHATUS MEDIK

Synonym: *Hibiscus abelmoschus* L.

Classification:

Kingdom:	Plantae
Subkingdom:	Phanerogamea
Class:	Dicotyledonae
Subclass:	Polypetalae
Series:	Thalamiflorae
Order:	Malvales
Family:	Malvaceae
Genus:	*Abelmoschus*
Species:	*Abelmoschus moschatus Medik*

Common names: Bengali: Kasturidana; **English:** Musk mallow; **Hindi:** Latakasturi, Makadana; **Kannada:** Kadukasturi; **Marathi:** Latakasturi, Kattukasturi; **Sanskrit:** Latakasturika; **Tamil:** Vettrilaikkasturi; **Telugu:** Kasturibendavittu.

Distribution: The plant is native to India and is naturalized in the tropics.

Description: An erect hirsute or hispid annual herb. 60–180 cm in height; leaves alternate, usually palmate, 5–7 lobed, lobes narrow-acute or oblong-lanceolate, crenate, serrate or coarsely toothed, hairy on both surfaces; stipules linear; peduncles 2 in, equaling the petiole. Flowers large, yellow with purple center; calyx oblong, 15-toothed. Corolla yellow with crimson centre. Bracteoles ½–¾ in. Fruits hairy, capsular; seeds many, subreniform, black or grayish brown and scented.

Flowering: September.

Fruiting: January.

Chemical constituents: The leaves and seeds of the plant contain β sitosterol and its glucoside, campesterol, stigmasterol, cholesterol, ergosterol, higher fatty acids. Esters of acetic and amdrettolic acids, ambrettolide (musk-like odour), farnesol, farensylacetate, furfural, linoleic acid, plasmalogen, myricetin and its glucoside, cannabiscitrin, cyaniding-3-sambubioside and cyandin-3-glucoside.

Used in: Ayurveda, Unani system of medicine and also as traditional medicine.

Parts used: Seeds, leaves and roots.

Uses: Seeds – aphrodisiac, antispasmodic, carminative, demulcent, diuretic, stomachic and tonic used as an inhalation to cure hoarseness and dryness of the throat, rubbed to a paste with milk to cure scabies, infusion or decoction useful in nervous disorders and in hysteria.

Leaves and root – efficacious against gonorrhea and venereal diseases; the roots and leaves are administered for gonorrhea.

Methods of propagation: This is propagated by seeds.

Steps to follow:

- Soak the seeds in water for 24 hours. Dibble 2–3 seeds to a depth of 1–2 cm and cover with soil. Sow the seeds at least a foot apart.
- Sow the seeds during early summer or rainy season.
- Seeds take at least 15 days to germinate.

ABROMA AUGUSTA (L.) L.F.

Synonyms: *Abroma fastuosa.*

Classification:

Kingdom:	Plantae
Subkingdom:	Phanerogamea
Class:	Dicotyledonae
Subclass:	Polypetalae
Series:	Thalamiflorae
Order:	Malvales
Family:	Sterculiaceae
Genus:	*Abroma*
Species:	*Abroma augusta*

Common names: Bengali: Olat-kambal; **English:** Devil's cotton, Indian hemp; **Hindi:** Sanukapashi; **Kannada:** Divvahatti; **Marathi:** Ulat kambal; **Sanskrit:** Pisacha karpus, Yonipushpa; **Tamil:** Surapputulti, Sivapputtutti; **Telugu:** Kondagogu.

Distribution: Found in tropical Asia, South and Eastern Africa, and Australia, in both wild and cultivated form. In India it is found throughout the hot and moister parts, from Punjab and Uttar Pradesh eastwards to Arunachal Pradesh, Assam, Meghalaya and Tripura, up to an altitude of 1200 m.

Description: Large, spreading, quick-growing, hairy shrub or small tree. Outer surface of the bark dull brown and longitudinally wrinkled with small warty markings; roots up to 1.5m depth. Leaves polymorphous, about 10–30 cm long and 6–18 cm broad, lower cordate, 3–5-lobed; upper ovate-lanceolate, non-lobed. Flowers 5 cm in diameter, dark red, purple or yellow in colour, occurring on a few flowered cymes. Sepals 2.5 cm, lanceolate, free nearly to the base. Petals scarcely exceeding the sepals, imbricate in bud, deciduous. They fall soon. Stamens present on short staminal tubes, with five staminodes. Capsules are almost 4 cm long, obpyramidal, membranous, pubescent and truncate at the apex. Capsules thrice as long as the persistent calyx. Each carpel has a triangular wing behind it. Many seeds, small, blackish, covered with silky hairs. The leaves and fruits are shown in Pictures 1 and 2 respectively. Capsules membranous, pentagonal, obpyramidal, winged.

Flowering: December–January.

Fruiting: December–January.

Chemical constituents: The root part contains bromine, sterol, friedelin, abromasterol A., choline, betaine, β-sitosterol, stigmasterol and maslinic, α-Protocatechuic, vanillic and caffeic acids. The leaves contain taraxerol and its acetate, β-Sitosterol acetate, lupeol and aliphatic alcohol octacosanol. The stembark of the plant contains β-sitosterol and friedelin. The fatty acid composition of the oil is as follows: palmitic, stearic, oleic, linoleic and hexadecenoic acids. The oil is rich in linoleic acid, which has an important dietary role in the control of arteriosclerosis because of its ability to lower the cholesterol level in blood.

Used in: Ayurveda, homeopathy.

Parts used: Root bark, leaves and seeds.

Uses: Root bark drug, which is considered as a uterine tonic and an emmenagogue. The powdered root acts as an abortifacient and anti-fertility agent. Leaves used to treat uterine disorders, diabetes, rheumatic joint pains, headaches with sinusitis. Cold aqueous infusion of fresh leaves and twigs: efficacious in gonorrhea. Seed oil: rich in linoleic acid, which has an important dietary role in controlling arteriosclerosis.

Methods of propagation: Through seeds. Seeds germinate in 21–30 days. Sometimes root suckers may be used for propagation. Propagation can be done either from seeds or stem-cutting. Since the seeds lose viability fairly quickly and their germination power is poor, fresh seeds soaked in water for an hour should be sown. Insecticidal treatments also improve the germination. The optimum temperature for germination is 32 C°. About 24 kg of seed are required to plant 1 hectare. Sowing is done before the rainy season. A depth of 5 cm is reported to be optimum for germination, and close spacing is recommended (Gupta et al., 2011).

Photo: Prof Srinath Rao

ABRUS PRECATORIUS (L.) W.F.WIGHT

Synonyms: *Abrus abrus* (L.) W. F. Wight.

Classification:

Kingdom:	Plantae
Subkingdom:	Phanerogamea
Class:	Dicotyledonae
Subclass:	Polypetalae
Series:	Calyciflorae
Order:	Rosales (Fabales)
Family:	Leguminosae (Fabaceae)
Subfamily:	Faboideae
Genus:	*Abrus*
Species:	*Abrus precatorius L.*

Common names: Bengali: Kunch, Koonch; **English:** Crab's eye, Jequirity, Indian liquorice; **Gujarati:** Gumchi; **Hindi:** Ghugachi, Chirami, Ratti, Guncaci; **Kannada:** Guluguni, Gulugunchi; **Malayalam:** Kunni; **Marathi:** Dhorgunj; **Tamil:** Kuntumani; **Telugu:** Guruginja.

Distribution: Throughout India. On hedges and bushes in exposed areas.

Description: A perennial twiner. Stems numerous, scarcely woody, slender, glabrous. Leaves 2–4 in long; abruptly pinnate. Petioles thickened and hairy at the base; stipules linear-subulate, pubescent, deciduous; leaflets 10–20 pairs, opposite, very deciduous, thinly membranous, ligulate-oblong, rounded, minutely apiculate, glabrous above, slightly hairy beneath. Petioles very short; flowers small in dense receme; rachis usually thickened in fruit; short pedicels. Calyx glabrous or sparsely silky; teeth very short. Corolla pink or white with a pink tinge. Stamens 9, monadelphous and vexillary stamen is absent, united into a tube slit above; anthers uniform. Ovary subsessile, ovules infinite, style short, incurved, beardless; stigma capitates. Pods oblong, turgid, truncate, silky, pubescent, and thinly septate between seeds. Seeds 3–5 as large as peas, usually bright scarlet with a black spot, sometimes white with a black spot or uniformly white.

Flowering: April–May.

Fruiting: June–July.

Chemical constituents: The toxin abrin is a dimer consisting of two protein subunits, termed A and B. In the seeds, the toxic principle is abrin, a mixture of at least five lectins, abrin A–D, and abrus-agglutinin (Ragasa et al. 2013).

Used in: Siddha, Ayurveda and traditional medicine.

Parts used: Roots, seeds.

Uses: Leaves alexiteric, astringent, diuretic, emetic, sweet, also used for snake bite. The roots are used as substitute for liquorice. They are abortifacient, alexiteric, anthelmintic, antiseptic, astringent diuretic, emetic and sweet. The seeds are abortifacient, acrid, aphrodisiac, astringent, bitter, febrifuge, purgative, toxic and trichogenous. They are also used as weights by jewellers and for making rosaries and ornaments. They contain an acrid poison and have been employed to kill cattle, the seeds being powdered, made into a paste and introduced beneath the skin.

Methods of propagation: The species is easily propagated through seeds.

Steps to follow:

- Soak the seeds in hot water for 5 minutes and then in cold water for 24 hours. This will assure better germination.
- Sow the seeds in polybags.
- In 7–10 days, the seeds begin sprouting, the germination percent being 70 %.
- Transplant the young seedlings in a field.

Photo: Prof Srinath Rao

ABUTILON INDICUM (L.)

Synonyms: *Abutilon asiaticum* W. & A.; *Sida indica* Linn.

Classification:

Kingdom:	Plantae
Subkingdom:	Phanerogamea
Class:	Dicotyledonae
Subclass:	Polypetalae
Series:	Thalamiflorae
Order:	Malvales
Family:	Malvaceae
Genus:	*Abutilon*
Species:	*Abutilon indicum*

Common names: Bengali: Badela; **English:** Country mallow, Flowering maples; **Hindi:** Kanghi; **Kannada:** Shrimudrigida, Mudragida, Turube; **Malayalam:** Uram, Katuvan, Urubam; **Marathi:** Chakrabhendi, Petari, Mudra; **Sanskrit:** Atibala, Bhuribala, Ghanta; **Tamil:** Nallatutti, Paniyarattutti.

Distribution: Warm, temperate regions, as a common weed on road sides and other waste places in plains and hills.

Description: This is a perennial plant with very finely downy branches. It is a hairy herb or an under-shrub, reaching up to a height of 1.0–1.5 m. The leaves of the plant are simple, alternate, stipulated, 2–5 cm long, broadly ovate, cordate at the base, acute, irregularly and coarsely dentate. Petioles are very long, about 3–5 cm. Flowers are regular, bisexual, orange yellow in colour, borne solitary in the axil of the leaf, about 2.5 cm in diameter. Calyx: cup-shaped, with 5 sepals, fused, 5-lobed. Lobes are shallow, apiculate. Petals 5, convolute, adnate to stamen-tube; stamens numerous, fused into a tube. Ovary is superior; carpels 10, style passes through the stamen-tube and divides into 5 branches, each terminating in a capitate stigma. Seeds minutely dotted.

The root system is a tap root. Roots are long, with a number of lateral branches, 1.5–2 cm in diameter, light brown; outer surface smooth with dot-like lenticels; bark thin and easy to peel off.

Flowering: July–September.

Fruiting: December.

Chemical constituents: The leaves contain mucilage, tannin, amino acids, glucose, fructose, galactose, gallic acid, sesquiterpene alkaloids, flavonoids, sterols, triterpenoids, saponins, cardiac glycosides and asparagine (11.5 %). Asparagine has diuretic activities. Roots contain fatty acids such as linoleic, oleic, stearic, palmitic, lauric, myristic, caprylic and capric acids. Flowers contain flavonoid compounds, including quercetin and its glycosides.

Used in: Ayurveda.

Parts used: Fruit, leaves, seeds and roots.

Uses: The plant is used both internally and externally. It possesses potent anti-hemorrhagic, diuretic, demulcent and laxative properties. Its topical application stops bleeding from wounds and heals ulcers, wounds and infections. The seeds are aphrodisiac and also used to treat piles and cough. The leaves are cooked and eaten to cure bleeding piles; the leaf and root juice is used to cure dental problems. A paste made from the leaves and mustered oil is used in rheumatism. A decoction of leaves is employed to cure ulcers, wounds and vaginal infections. The roots of the plant are diuretic.

Methods of propagation: Through seeds.

Photo: Prof Srinath Rao

ACACIA CATECHU (LINN. F.) WILLD.

Synonyms: *Acacia catechu* (Linn. f.) Willd. var. catechuoides (Roxb.) Prain, *Acacia catechuoides* (Roxb.) Benth., *Acacia sundra* (Roxb.) Bedd., *Acacia wallichiana* DC., *Mimosa catechu* Linn. f., *Mimosa catechuoides* (Roxb.)

Classification:

Kingdom:	Plantae
Subkingdom:	Phanerogamea
Class:	Dicotyledonae
Subclass:	Polypetalae
Series:	Calyciflorae
Order:	Rosales (Fabales)
Family:	Leguminosae (Fabaceae)
Subfamily:	Mimosoideae
Genus:	*Acacia*
Species:	*Acacia catechu*

Common names: Bengali: Khera; **English:** Black catechu, Khair gum; **Gujarati:** Kathe, Khair; **Hindi:** Dant-dhavan, Khayar; **Kannada:** Kaachu, Kachinamara, Koggigida; **Malayalam:** Karintaali; **Marathi:** Khayar; **Sanskrit:** Gayatrin, Khadira; **Tamil:** Cenkarungali, Kacu-k-katti; **Telugu:** Khadiramu, Kaviricandra, Nallacandra.

Distribution: Found throughout India, Indonesia, Nepal, Malaysia, Myanmar and Pakistan.

Description: It is a small or medium-sized thorny tree up to 15 m tall; bark dark grey or greyish brown, peeling off in long strips, or sometimes in narrow rectangular plates, brown or red inside; leaves bipinnate, alternate, stipulate; stipular spines paired, 3–10 mm long, straight or hooked, occasionally lacking flowering branchlets; rachis 8–20 cm, slender, pulvinate, downy, grooved above, with a gland near the base of the rachis on the upper side; pinnae 10–20 pairs, 4–8 cm long, opposite, downy, slender, with a gland in between the terminal 6 pairs on the lower side; leaflets 30–50, opposite, paripinnate, sessile; stipels absent; lamina 4–8 × 1–1.5 mm, linear-oblong, base unequally truncate, apex round, obtuse or mucronate, margin entire, pubescent. Flowers white to pale yellow, sessile, in groups of 2–4 axillary spikes; bracts caducous; calyx cupular-campanulate, 1–1.5 × 1–1.5 mm; teeth triangular or deltoid; corolla 2.5–3 mm long, lobes oblong, ovate to linear-lanceolate; stamens many, 4.5–5 mm long; ovary stipitate, 1–1.2 mm long, oblong-ellipsoid; style 4–5 mm long; stigma terminal 5–10 cm long and 1–1.6 cm broad, flat, straight, unlobed or sinuate along the margins, beaked at the apex, brown, narrowed at the base into a stipe; seeds 3–10, orbicular or ovate, flattened.

Flowering: June–August.

Fruiting: September–November.

Chemical constituents: The heartwood contains flavonoids, catechin, epicatechin, epigallocatechin, epicatechin gallate, epigallocatechin gallate, rocatechin, phloroglucinol, protocatechuic and catechu-tannic acids, quercetin, quercitrin. The leaves contain alkaloids, kaempferol, dihydrokaempferol, taxifolin. The bark contains glycosides, poriferasterol, poriferasterol acylglucosides. Tannins: gallic acid and phlobatannins (Monu et al., 2014).

Used in: Ayurveda, traditional medicine.

Parts used: Leaves, flowers, seeds, heartwood and gum.

Uses: The cutch from the heartwood is a powerful anti-diarrheal, astringent, antioxidant, anticancer, anti-diabetic, anti-microbial and hepatoprotective. It is also used as a hemostatic and to cure urinary and vaginal discharge, melancholia, conjunctivitis, catarrh, leprosy and leucoderma. It treats excessive mucous discharges, hemorrhages, conditions of gums, throat and mouth, stomatitis, irritable bowels. The gum of the tree has anti-inflammatory and anti-pyretic activities.

Methods of propagation: Through seeds.

ACACIA FARNESIANA (L.) WILLD.

Synonyms: *Acacia minuta* (M.E. Jones) R.M. Beauch.; *Mimosa farnesiana* L.

Classification:

Kingdom:	Plantae
Subkingdom:	Phanerogamea
Class:	Dicotyledonae
Subclass:	Polypetalae
Series:	Calyciflorae
Order:	Rosales (Fabales)
Family:	Leguminosae (Fabaceae)
Subfamily:	Mimosoideae
Genus:	*Acacia*
Species:	*Acacia farnesiana*

Common names: Bengali: Guya babla, Belatibabla; **English:** Cassie flower, Fragrant acacia, Ironwood, Mimosa bush, Sweet acacia, Sweet wattle; **Gujarati:** Gandhelo babul, Gandhelokhair, Gu baval, Jheribaval; **Hindi:** Wilayati babool, Belati babul, Deobabul; **Kannada:** Banni mara, Karijali, Kari kashturi; **Malayalam:** Kariveelum, Pikharu vil, Pivelam, Pivelum; **Marathi:** Deobabul, Gubabul, Gukikar, Guyababul, Vilayatibabul; **Sanskrit:** Arimeda, Vitkhadira; **Tamil:** Kadivel, Kasthurivel, Kasturi velam; **Telugu:** Kasturi, Kampu tumma, Arimidamu, Murki-tumma.

Distribution: This acacia originates in Central and South America, and is now pantropical, distributed in Pakistan, Nepal, Sri Lanka, Maldives, Myanmar and India. In India, it is found growing in tropical parts, particularly in the sandy soils of riverbeds.

Description: It is a spreading, much-branched thorny shrub or small tree with spines reaching a height of 15 feet or more. Branches are striate, glabrous, curved, with pale-brown lenticels, stipular spines white, straight, long, hard, sharp and divaricate. It has a whitish bark. Leaves contain an alternate compound, bipinnate, rachis 1–2 in long, angular, pubescent, with a small raised gland about the middle of the petiole. Its individual leaflets are small. Leaflets are in 10–20 pairs, linear, acute, glabrous and sessile. The flowers are ball-shaped, bright yellow and sweet-scented. Calyx and corolla glabrous, scented. The fruit pod is, dark brown, nearly calendric, pointed at the ends, broad, glabrous, brown, veined and indehiscent. Seeds in 2 series, embedded in dry, spongy tissue.

Flowering: November–March.

Fruiting: April–September; but in some areas throughout the year.

Chemical constituents: Leaves contain tannins, alkaloids, rutin and apigenin-6, 8-bis-C-glucide, cyanogen, linamarin, lotaustralin. Flowers contain gallic, ellagic and m-digallic acid, isorhamnetin-3, 7-glucorhamnoside, methyl gallate, kaempferol, aromadendrin, naringenin, kaempferol-7-diglucosie, naringenin-7-glucoside and a new glycoside, probably naringenin-7-diglucoside acylate with gallic acid. Pods contain naringenin-7-O-β-D-(6″-O-galloyl) glucopyranoside (purin-O-6″-gallate); rutin and apigenin-6, 8-bis-C-glucopyranoside (0.4 %). Seeds contain the amino acid N-acetyl-L-djenkolic acid.

Used in: Ayurveda, traditional and Unani medicine.

Parts used: Bark, leaves, heartwood, gum, roots.

Uses: Decoctions prepared from the bark are considered to be astringent, anthelmintic, anti-dysenteric, anti-inflammatory, aphrodisiac, antiseptic, demulcent, and are also used to treat arthritis, bronchitis, dental caries, diarrhea, dyspepsia, epilepsy, issues with gums, insanity, nausea, ophthalmia, rabies, skin diseases, snake bite, sores, stomatitis, sterility, swollen ulcers, tuberculosis and urinary disorders. The leaves are used for eye complaints. Tender leaves bruised in a little water are administered orally to treat gonorrhea. The leaf, flower and root are boiled in water and taken orally for a week to treat ringworm infections. The inflorescence was found to be effective in venereal diseases and to treat infertility issues. The roots of the tree are dried and ground to make powder; this is taken in a dose of 1 gm powder with honey twice a day for a month to treat the rickets. A paste made from the heartwood is applied topically to cure swelling and inflammation. The twig or bark is chewed to strengthen teeth and gums.

https://herbs.indianmedicinalplants.info/index.php/a/1238-acacia-farnesiana
(http://tropical.theferns.info/viewtropical.php?id=Vachellia+farnesiana)

Methods of propagation: Through seeds.

Photo: Prof Srinath Rao

ACACIA NILOTICA (L.) WILLD. EX DELILE

Synonyms: *Acacia arabica* Willd. var. *indica* Benth; *Mimosa nilotica* Linn.

Classification:

Kingdom:	Plantae
Subkingdom:	Phanerogamea
Class:	Dicotyledones
Subclass:	Polypetalae
Series:	Calyciflorae
Order:	Rosales (Fabales)
Family:	Leguminosae (Fabaceae)
Subfamily:	Mimosoidae
Genus:	*Acacia*
Species:	*Acacia nilotica*

Common names: Bengali: Babla; **English:** Indian gum arabic tree; **Gujarathi:** Kalobaval; **Hindi:** Babul; **Kannada:** Kari jali; **Malayalam:** Karavelum; **Marathi:** Babhul; **Sanskrit:** Babbula; **Tamil:** Karu velamaram; **Telugu:** Nellatumma.

Distribution: Cultivated or found self-sown throughout the drier parts of India, ascending to an altitude of 900 m. It is indigenous to the plains of Andhra Pradesh and Maharashtra and is generally grown on field boundaries.

Description: A medium to large tree, reaching to a height of 10–12 m. The branches have a tendency to droop downwards. This tree is almost evergreen, with flattish dark brown, deeply grooved, longitudinally fissured bark; branchlets slender. The young twigs are covered with short hairs. Paired, slender, straight spines grow from a single base and sometimes curve backwards; these are up to 80 mm long and white but often reddish brown in colour. Leaves bipinnate, 8 cm long; main rachis downy, often with glands; stipular spines sharp and straight; often pinnae 4–9 pairs. Leaflets subsessile 10–25 pairs, linear-oblong, subobtuse, glabrous or nearly so.

Flowers: Sessile, borne in globose heads, yellow and fragrant; peduncle axillary, in fascicles of 2–6, terete, pubescent. Bracteoles 2. Calyx: campanulate, teeth very short. Corolla: lobes short, triangular. Stamens many and indefinite.

Pods grey, pubescent, distinctly stalked, oblong, moniliform, compressed, constricted at the sutures between the seeds. Leaf shedding commences before May and continues up to the end of May; the new leaves appear in March–May.

Flowering: During the rainy season.

Fruiting: April–June.

Chemical constituents: The plant yields a gum containing arabin as the chief constituent. Other constituents of the gum include tannic and gallic acids, cresol (an antiseptic), methyl salicylate, complex salts of calcium, magnesium and potassium, galactose, l-arabinose, l-rhamnose and 4 aldobiouronic acids. The bark has been found to contain catechin, epicatechin, dicatechin, quercitin, gallic acid and procyanidin. The bark, stems, leaves, roots, pods and heartwood contain several polyphenols, tannins, gallic acid, octacosanol, betulin, flavonoids, α-amyrin and ß-sitosterol. The degraded gum is composed of recurring units comprising 10 galactose residues and 4 glucuronic acid residues. 30 phenolic compounds have been isolated from the tanning extracts of the bark. Epigallocatechin, tannin and polyphenolic phlobaphenes consisting mainly of catechol and pyrogallol have also been isolated from the bark. The flower contains stearic acid, kaempferol-3-glucoside, isoquercitrin.

The roots contain octacosanol, betulin, β-amyrin and β-sitosterol (Bargal and Bargali, 2009;Malviya et al., 2011).

Used in: Ayurveda; Siddha, Unani and traditional medicine; homeopathy.

Parts used: Seeds, leaves, bark and gum.

Uses: It is considered tonic, demulcent, carminative, astringent, antipyretic, antispasmodic, anti-inflammatory, anti-hypertensive and anti-stress. The seeds, leaves and bark of this tree contain chlorogenic acid, garlic acid and catechin for therapeutic use. This plant treats ulcers in animals. A decoction of its pods is used in urinogenital diseases. The tender leaves are used as an astringent and remedy for diarrhea and dysentery. A decoction made from the bark is used for toothache and sore throat. The gum is styptic and astringent.

Methods of propagation:

This is propagated by seeds.

- Seeds can be collected in April.
- They need to be coated with cow dung and kept that way for a month or two.
- Afterwards, they can be directly sown in the field, either in lines or on ridges.

Photo: Dr Kotreh

ACACIA SINUATA (LOUR.) MERR.

Synonyms: *Acacia concinna* (Willd.) DC; *Acacia habbasoides* Bojer; *Mimosa sinuata* Lour.

Classification:

Kingdom:	Plantae
Subkingdom:	Phanerogamea
Class:	Dicotyledonae
Subclass:	Polypetalae
Series:	Calyciflorae
Order:	Rosales (Fabales)
Family:	Leguminosae (Fabaceae)
Subfamily:	Mimosoideae
Genus:	*Acacia*
Species:	*Acacia sinuata*

Common names: English: Soap nut acacia; **Hindi:** Satala, Seekakai Kochi; **Kannada:** Sege; **Malayalam:** Chikaka; **Sanskrit:** Aaavartaki, Saptala; **Tamil:** Cikaikkai, Cityakkai; **Telugu:** Sikaya.

Distribution: It is found throughout tropical jungles, especially in peninsular India.

Description: A perennial, woody, large climbing shrub with brown branches dotted with white growths on big trees. Leaves alternate, bipinnate with sharp prickles on main rachis and a large gland on the petiole. Pinnates, leaflets small, subsessile, unequal-sided, glabrous; flowers with small heads; fruits; short-stalked thin pods with 6–10 seeds per pod, flat and coriaceous.

Flowering: February–March.

Fruiting: April.

Chemical constituents: The bark contains saponins. The sugars identified are glucose, arabinose and rhamnose. It also contains hexacosanol and α-spinasterone. The pods yield saponins which have been reported to contain acacinin-C, D and E and a homologous series of oligosaccharides. The oligosaccharides have been identified as sucrose, raffinose, stachyose and verbascose. The leaves contain alkaloids, nicotine and colycotomine, a triterpentine. The leaves of the Soap Nut contain tannins, rutin, alkaloids and citric acid.

Used in: Folk and Ayurveda medicine.

Parts used: Pods and bark.

Uses: The pods are bitter, astringent, cooling, diuretic, emetic, deobstruent, detergent, depurative and anthelmintic. They are useful in conditions of vitiated pitta, burning sensations, renal constipation and vesical calculi strangury, haemorrhoids, erysipelas, leucoderma, leprosy, prurigo, abscesses and eczema. (https://ayushvedah.com/druginfo.php?drugid=575&info=uses).The powdered pods are the best alternatives to soaps in all cases of skin diseases. Soap nut is a natural hair cleanser that lathers on application. It has hair strengthening and conditioning properties and prevents dandruff.

Methods of propagation:

The plant is propagated through seeds and stem cuttings.

- Sow the seeds directly in the main field. Seeds exhibit a low success rate of about 26 %.
- To propagate it vegetatively, cuttings about 1 m long with a circumference of 4 in are used. These cuttings have to be extracted from the young portion of the stem.
- The cuttings are planted during the first rains.

ACANTHUS ILICIFOLIUS L.

Synonyms: *Acanthus ebracteatus* Vahl var. *xiamenensis* (R. T. Zhang) C. Y. Wu & C.C. Hu, *Acanthus xiamensis* R.T. Zhang.

Classification:

Kingdom:	Plantae
Subkingdom:	Phanerogamea
Class:	Dicotyledonae
Subclass:	Gamopetalae
Series:	Bicarpellatae
Order:	Personiales
Family:	Acanthaceae
Genus:	*Acanthus*
Species:	*Acanthus ilicifolius*

Common names: Bengali: Hargoza, Harkuch Kanta, Kotki; **English:** Holy-leaved acanthus; **Hindi:** Harakata; **Kannada:** Holechulli; **Malayalam:** Chullikantala; **Marathi:** Marandi; **Sanskrit:** Harikusa; **Tamil:** Lakutaimulli; **Telugu:** Alasyakampu.

Distribution: The species is widespread in Southeast Asia, Indochina, Indonesia, the Philippines and northern Australia.

Description: Erect prickly perennial shrub up to 1.5–2 m in height. Leaves: simple, opposite and decussate, lanceolate, narrowed at base, serrate margins with spines, petiole short, 5–6 mm in length, slightly winged with two sharp spurs at the base. They are green in colour when young and yellowish brown when dry; stipules are lacking. Flowers: borne in terminal spike, sessile, about 4 cm long, bisexual, zygomorphic and usually associated with brightly coloured bracts. Fruit: a capsule, ovoid/oblong, about 3 cm long, compressed, brown in colour and shiny. Seed pods are stalked, funiculus, and each seed is modified into a hook-shaped jaculator that flings out the seeds during dehiscence.

Flowering: March.

Fruiting: August.

Chemical constituents: The aerial parts contain alkaloids such as acanthicifoline, trigonelline, flavonoids, as well as quercetin and its glucopyranoside. Triterpenoids, β–amyrin, α–amyrin, lupeol, oleanolic and ursolic acids, β-sitosterol, trigonelline. Fatty acid derivatives, palmitic and octadecanoic acids, stigmasterol octadecanoate, β-sitosterol. The roots contain triterpenoid, saponin, glycosides and apegenin-7-o-glucuronide. A new glycoside, methylapigenin-7-o-β-Dglucuronate, has been isolated from the leaves (Van Kiem et al., 2008).

Used in: Ayurveda.

Parts used: Leaves, roots; the whole plant.

Uses: Various parts of the plant have been used as a crude drug for the treatment of asthma, diabetes, dyspepsia, hepatitis, rheumatoid arthritis and snake bites. An extract prepared from the roots is used to treat high blood pressure. The roots are also used in treating cough and asthma. The leaves help treat rheumatism and neuralgia; they also show antioxidant, hepatoprotective, anti-carcinogenic and anti-tumor properties (Babu et al., 2001 & 2002). A decoction of the plant is given to treat dyspepsia. The plant also possesses anti-inflammatory, antileishmanial, antimicrobial, antioxidant, antiulcer, hepatoprotective and osteoblastic properties. In folklore medicine, it was used for curing leprosy, leucorrhoea and paralysis. An Ayurvedic medicine known as Sahachara, prepared from this plant, is used to treat rheumatic pain.

Methods of propagation: Through seeds.

ACALYPHA INDICA L.

Synonyms: *Acalypha bailloniana* Mull. Arg.; *Acalypha canescens* Benth. nom. nud.; *Acalypha chinensis* Benth.; *Ricinocarpus indicus* (L.) Kuntze.

Classification:

Kingdom:	Phanerogamia
Class:	Dicotyledonae
Subclass:	Monochlamydeae
Series:	Unisexuales
Order:	Malpighiales
Family:	Euphorbiaceae
Genus:	*Acalypha*
Species:	*Acalypha indica*

Common names: Bengali: Muktajhuri, Muktabarsi; **English:** Indian copperleaf, Indian acalypha, Indian nettle; **Gujarati:** Vanchi kanto; **Hindi:** Kuppi, Kuppikhokhali; **Malayalam:** Kuppameni, Kuppi, Kuppikhokhali; **Marathi:** Khokhali; **Sanskrit:** Haritamanjari, Arittamanjarie; **Tamil:** Kuppivaeni, Kuppaimeni; **Telugu:** Kuppichettu, Harita-manjiri.

Distribution: Africa, China, Indian subcontinent, Indonesia, Malaysia, Myanmar, New Guinea, Philippines, Thailand. In India, it occurs as a weed in fields and waste places throughout the plains, in the hotter parts.

Description: An annual, erect herb, 30–80 cm in height. Branches are numerous, long, ascending, angular. Leaves alternate, long, petiolated, 2.5–7.5 cm long, ovate or rhomboid, margins crenate or serrate. Inflorescence terminal or axillary, racemose type. Flowers numerous, unisexual, monoecious. Perianth 1 whorl, green coloured, the males minute, clustered near the summit of the spike, the females scattered, accrescent, 3–5 in number, bracts many-nerved. Ovary hispid. Fruit a capsule, small, hispid, often 1-seeded (Savanur, 2016).

Flowering: June–December.

Fruiting: June–December.

Chemical constituents: The plant contains cyanogenic glucosides, alkaloids such as acalyphine, as well as triacetamine and kaempferol; the leaves and twigs contain acalyphamide and other amides, cyanogenic glycoside, flavonoids, quinine, sterols, hydrocyanic acid, inositol methyl ether, polyphenols, resin, triacetonamine, terpenoids and volatile oils. The other constituents are β-sitosterol, kaempferol, n-octane and quebrachitol (Savanur, 2016).

Used in: Ayurveda, homeopathy.

Parts used: Leaves, roots; the whole plant.

Uses: This plant has emetic, expectorant, antibacterial and anthelmintic properties. The leaf juice is used in the treatment of scabies and bed sores. In congestive headaches, a piece of cotton saturated with the juice from its leaves can be inserted into each nostril. A tincture of the fresh plant is used for treating arterial haemorrhage, tuberculosis of the lungs and pneumonia. The plant is boiled and the extract is used as a purgative. The roots are employed to remove intestinal parasites in children. It is a very good remedy in the treatment of piles (Savanur, 2016).

Methods of propagation: Through seeds.

Photo: Prof Srinath Rao

ACMELLA OLERACEA (L.) R.K. JENSEN

Synonyms: *Spilanthes oleracea* L.; *Spilanthes acmella* (L.) Murr.; *Bidens fixa* Hook.f.; *Bidens fervida* Lam.; *Anacyclus pyrethraria* (L.) Spreng.

Classification:

Kingdom:	Plantae
Subkingdom:	Phanerogamea
Class:	Dicotyledonae
Subclass:	Gamopetalae
Series:	Inferae
Order:	Asterales
Family:	Compositae (Asteraceae)
Genus:	*Acmella*
Species:	*Acmella oleracea*

Common names: Bengali: Akarkara; **English:** Toothache plant, paracress; **Hindi:** Akarkara; **Kannada:** Hemmugalu; **Marathi:** Akkalkada; **Sanskrit:** Akarkara;

Distribution: Native to Brazil, widely distributed in the tropical and subtropical regions of the world, such as Africa, America, Borneo, Sri Lanka and India. In India, they are more common in the Bastar district, Chhattisgarh state.

Description: Annual erect or ascending stout herbs, 20–50 cm high. **Leaves:** opposite, petiolate, broadly ovate and narrowed at the base, acute or obtuse at the apex. Flowers: in axillary and terminal panicles; involucral bracts 2-seriate; ray florets few, inconspicuous; disc florets campanulate. Fruits and achenes dorsally compressed, black.

Flowering: March–April.

Fruiting: March–April.

Chemical constituents: This plant contains spilanthol, which acts as an anaesthetic/analgesic, β-sitosterol, stigmasterol, α and β-amyrin and essential oils such as limonene, β-caryophyllene and β-ocimene. It also contains isovaleric and salicylic acids, asparagine, sterols, flavonoids, bitters, tannins and coumarins. The predominant constituents of its essential oils are sabinene (17.58 %), 1,8-cineole (13.04 %), borneol (12.41 %), bornyl acetate (7.98 %), pinene (6.28 %), pinene (6.26 %), terpinine-4-ol (6.17 %) and chamazulene (5.28 %) (Nadim et al., 2011).

Used in: Ayurveda.

Parts used: Roots, stem, leaves and flower heads.

Uses: The flower heads are chewed to relieve toothaches and other mouth-related troubles, hence the name 'toothache plant'. They are also used to enhance the immune system, improve digestion and fight nausea. **The** leaves are used externally in the treatment of skin diseases. Decoctions made from the roots are purgative, diuretic and lithotriptic. The whole plant is used in the treatment of dysentery. It also shows anti-inflammatory, antibacterial and antifungal properties, as well as being a sialagogue, stimulating the increased flow of saliva and consequently promoting digestion.

Methods of propagation: Through seeds, which readily germinate in 10–15 days. They need direct sunlight to germinate, hence they are not to be buried. The seedlings are raised in beds and later on transferred to fields with a distance of 30 cm between each seedling. The best time to sow the seeds is in early summer.

ACONITUM HETEROPHYLLUM WALL. EX ROYLE

Synonyms: *A. cordatum* Royle, *A. atees* Royle.

Classification:

Kingdom:	Plantae
Subkingdom:	Phanerogamea
Class:	Dicotyledonae
Subclass:	Polypetalae
Series:	Thalamiflorae
Order:	Ranales
Family:	Ranunuculaceae
Genus:	*Aconitum*
Species:	*Aconitum heterophyllum*

Common names: Bengali: Ataicha; **English:** Ativisha; **Hindi:** Atis; **Kannada:** Ativisha; **Malayalam:** Ati vadayam; **Marathi:** Ativish; **Sanskrit:** Ataicha, Ativisa, Shuklakanda; **Tamil:** Ati vadayam.

Distribution: The plant is common in temperate to alpine zones of the Himalayas from Indus to Kumaon, occurring at altitudes between 3000 and 5000 m. It is found growing in the field borders of villages at and around 2400 m altitude.

Description: Erect herb growing up to 1 m in height; stem simple and branched from the base, glabrous below and puberulous above; roots paired, tuberous, whitish or grey; leaves simple, heteromorphic; upper leaves entire; lower leaves with long petioles 5–10 cm in length, ovate or orbicular, cordate, more or less 5-leaved, toothed. Inflorescence: slender raceme or a leafy panicle, flowers bi- or unisexual, regular or irregular, bracteate. Bracts sharply toothed, flowers more than 2.5 cm in length, bright blue or greenish blue with purple veins: sepals 5, rarely persistent, often petaloid, imbricate; petals 5; minutly imbricate; stamens hypogynous, many; anthers adnate opening laterally, carpels numerous, stigma simple follicles. The fruit is a many-seeded follicle. Seeds pyramidal, 3–4 mm long, blackish brown. The dried roots are generally ovoid-conical in form and tapering downwards; 2.5–4 cm long and about 0.5 cm in diameter at the upper end. Externally, they have a greyish or light ash colour, and internally they are milky white. Roots paired, tuberous, greyish white, 2–3 cm long, 0.5–1.5 cm thick.

Flowering: July–September.

Fruiting: July–September.

Chemical constituents: Aconitine is the chief alkaloid of this crop. The root contains non-toxic amorphous alkaloids, atisine, dehydroatisine, geteratisin, hestisine, heterophylline, heterophyllidine and heterophyllisine. It also contains aconitic and tannic acids, pectin and starch. Atisenol, a new entatisene diterpenoid lactone can be found in the roots. Also present are palmitic and stearic acids.

Used in: Ayurveda, homeopathy and Tibetan, Unani and Siddha medicine.

Parts used: Roots.

Uses: The daughter tubers are usually preferred. Shrivelled mother tubers are often sold as an inferior-grade drug. The root cures acute dysentery, chronic enteritis, digestive complaints, piles, vomiting and cough. It is also used in splenic fever and gastric troubles, and as a good astringent, anthelmentic and tonic, but its efficacy as anti-periodic is disputed; it is considered an anti-fertility agent and is used externally for rheumatisms. The ethanolic extract of *Aconitum heterophyllum* acts as an immunomodulator by stimulating the phagocytic function while inhibiting the humoral component of the immune system.

Methods of propagation: Through seeds.

ACONITUM NAPELLUS L.

Synonyms: *Aconitum vulgare* DC; *A. tauricum* Jacq.; *A. angistifolium* Bernh.; *Aconitum carmichaelii* Debeaux.

Classification:

Kingdom:	Plantae
Class:	Dicotyledonae
Subclass:	Polypetalae
Series:	Thalamiflorae
Order:	Ranales
Family:	Ranunculaceae
Genus:	*Aconitum*
Species:	*Aconitum napellus*

Common names: English: Indian aconite, Monk's hood; **Hindi:** Bachnag, Mithazahar; **Kannada:** Vatsanabhi; **Malayalam:** Vatsanabhi; **Sanskrit:** Bachnag, Meetavish, Vatsanabha, Pranahara; **Tamil:** Vasanavi; **Telugu:** Athivasa.

Distribution: Native of the Alps, this species is globally distributed in Western Central Europe, Turkestan, Afghanistan, Pakistan, India and Nepal.

Description: Erect, showy herbaceous perennial, stem often decumbent in small states, glabrous or slightly pubescent. Turnip-like tuberous roots. Leaves: large-lobed, segments linear. Flowers: blue-violet borne in spike-like terminal racemes, bracts entire or trifid. Fruit: a follicle. Follicles: 3–5, bearing many seeds.

Flowering: June–July.

Fruiting: July–August.

Chemical constituents: The roots form the drug aconite. Two types of alkaloids, aconitines and atisine, are found in the roots. Aconitine is the major alkaloid accounting for 30 % of the total yield. Hypoconitine, neopelline, mesaconitine, songaramine, deoxyaconitine and neoline are the other alkaloids present. Trace amounts of ephedrine and sparteine are also reported.

Used in: Ayurveda, homeopathy and Tibetan, Unani and Siddha medicine.

Parts used: Roots, leaves and flowering tops.

Uses: It cures asthma, blood pressure, bradycardia, fever, pneumonia, rheumatism and skin diseases. It prevents heart attacks and hair loss, and promotes hair growth. Aconite is a highly potent and quick-acting poison with drastic cardiac effects. This is stored in cow's urine or subjected to prolonged heating by Ayurvedic physicians. It is then used to cure fever and rheumatism, as an expectorant and to treat general debility. Misuse of the medicinal ingredients contained in this plant can negatively affect the cardiovascular and central nervous systems, resulting in death.

Methods of propagation: This species has a low natural propagation rate under cultivation and is propagated through seeds or by removing the offsets that are generated each year from the root stocks. The seeds can be sown ½ in deep during March–April. Plants flower in 2–3 years.

Source: https://en.wikipedia.org/wiki/Aconitum_napellus

ACORUS CALAMUS L.

Synonyms: *Calamus aromaticus* Garsault.

Classification:

Kingdom:	Plantae
Subkingdom:	Phanerogamea
Class:	Monocotyledonae
Subclass:	Arecide
Series:	Apocarpae (Acorales)
Order:	Acorales
Family:	Aroideae (Acoraceae)
Genus:	*Acorus*
Species:	*Acorus calamus L.*

Common names: English: Sweet flag; **Hindi:** Gorbach, Safed bach; **Kannada:** Baje, Vasa; **Malayalam:** Vaembu, Vashampa; **Marathi:** Vekhanda; **Sanskrit:** Haimavati, Ugragranthi, Vacha; **Tamil:** Pullai valathi, Vasambu; **Telugu:** Vadaja.

Distribution: Its origin is in Europe; however, this species is globally distributed in the temperate Northern Hemisphere and in tropical Asia. Within India, it has been recorded throughout, in marshes, wild or cultivated, ascending to an altitude of 2200 m in the Himalayas.

Description: It is a semi-aquatic perennial plant. Leaves: narrow, up to 80 cm, linear to narrowly ensiform; glossy bright green in colour; acute apex with amplexicaul base. The petioles are sheathing for 20–50 cm. The rhizomes of the plant are long, creeping, jointed and somewhat vertically compressed, with large leaf scars. They are about 1.3–2.5 cm thick and pale to dark brown in colour. They are spongy inside and give off long slender roots below. Flowers: pale yellow green in colour with a sweet aroma. They are arranged compactly on a sessile, cylindrical and stumpy spadix enclosed by a spathe. Fruit: a green berry, 3-celled; each of them contain 1–3 oblong seeds.

Flowering: April–July.

Fruiting: April–July.

Chemical constituents: All parts of the plant contain an essential oil consisting of β-asarone, asaraldehyde, acoradin, phenylindane, calamenol, calamene and sesquiterpene ketones such as acorone, calarene and calacone. The yield of oil is maximum in the rhizomes. In addition, the roots yield choline and acoric acid. Other constituents are α- and β-pinene, myrcene and camphene.

Used in: Ayurveda and traditional medicine.

Parts used: Rhizomes.

Uses: It has numerous medicinal properties. In Ayurveda, the dried rhizomes of the plant constitute the drug Vaca, which is considered to have antispasmodic, carminative and anthelmintic properties. The rhizomes are also considered to be capable of improving memory. They are used to treat insomnia, melancholia, neurosis, remittent fevers, delirium, hysteria, epilepsy and schizophrenia. The oil has carminative, antispasmodic and antibacterial properties. It is also used in rheumatism and eczema. The rhizome powder is a well-known insecticide against flies and bedbugs; it prevents insect pests in rice and other grains and silverfish in libraries. Both roots and leaves have shown antioxidant, antimicrobial and insecticidal activities. This plant is employed in vitiated conditions of *vata* and *kapha*, stomatopathy, hoarseness, colic, flatulence, dyspepsia, amenorrhoea, dismenorrhoea, nephropathy, calculi, cough, bronchitis, depression and other mental disorders.

Methods of propagation: This aquatic herb can be propagated through rhizomes. It prefers moist clayey loams, sandy loams and alluvial soils.

Steps to follow:

- Rhizomes are collected from earlier planting and kept preserved in the soil in constant moisture. After the sprouts emerge, rhizomes are cut into smaller pieces consisting of 2–3 nodes and planted in the nursery.
- These sprouted rhizome pieces may be planted at ½ a finger depth at a distance of 1 foot from each other, preferably in June–August. The success rate is about 90 %.

ADHATODA ZEYLANICA MEDIK

Synonyms: *Adathoda vasica* (L.) Nees.; *Justicia adathoda.*

Classification:

Kingdom:	Plantae
Subkingdom:	Phanerogamea
Class:	Dicotyledonae
Subclass:	Gamopetalae
Series:	Bicapellatae
Order:	Personales
Family:	Acanthaceae
Genus:	*Adhatoda*
Species:	*Adhatoda zeylanica*

Common names: English: Malabar nut tree; **Hindi:** Adusa, Bansa, Vasika; **Kannada:** Adusoge; **Malayalam:** Atalotakam; **Marathi:** Adulsa; **Sanskrit:** Vasaka; **Tamil:** Adadorai, Arathorai; **Telugu:** Adarasamu.

Distribution: This plant occurs throughout the plains and submontane regions of India; it is common near habitations.

Description: A tall, much-branched, dense, evergreen shrub, with large, elliptical lanceolate leaves 10–12 cm in length, mainly pubescent. Flowers in dense, short spikes; stalks of the spikes shorter than the leaves. Bracts, present on the spikes; these are conspicuously veined. The corolla of the flowers is white with a few purplish markings. Fruit is a capsule, 4-seeded.

Flowering: February–May.

Fruiting: February–May.

Chemical constituents: The main alkaloid in the leaves is quinazoline. Compounds such as vasicine, vasicinone and vasicinol are present. Vasicine and vasicinone are well known bronchodilators and respiratory stimulants. Vasicine also exhibits abortifacient properties.

Used in: Ayurveda and Siddha and Tibetan medicine.

Parts used: Leaves, flowers, fruits, seeds, roots, stem and bark.

Uses: The drug Vasaka comprises the fresh or dried leaves of the plant. The chief use of Vasaka is as an expectorant; it is in the form of juice, syrup or decoction. It facilitates the expulsion of sputum and thus gives relief in bronchitis. Its expectorant activity is due to the stimulation of the bronchial glands. Larger doses can, however, cause irritation and vomiting.

Methods of propagation: Propagation is through stem cuttings alone. Soft or semi-hard wood cuttings containing 3–4 nodes are planted in soil, usually during the period April–June. Cuttings will take 15–20 days to sprout and 30 days to root. After rooting, they are transplanted to fields.

Photo: Prof Srinath Rao

AEGLE MARMELOS B. R. ZATE

Synonyms: *Belou marmelos* (L.) Lyons.; *Bilacus marmelos* (L.) Kuntze.; *Cretavia marmelos,* L.; *Feronia pellucida* Roth.

Classification:

Kingdom:	Plantae
Subkingdom:	Phanerogamea
Class:	Dicotyledonae
Subclass:	Polypetalae
Series:	Disciflorae
Order:	Geraniales
Family:	Rutaceae
Genus:	*Aegle*
Species:	*Aegle marmelos*

Common names: Bengali: Bael; **English:** Holy fruit tree, Bengal quince; **Hindi:** Bael; **Kannada:** Bilvapatra; **Malayalam:** Kulakam; **Marathi:** Baela, Kattukasturi; **Sanskrit:** Bilva, Sriphal; **Tamil:** Vilvam Kuvilam; **Telugu:** Bilvapandu.

Distribution: This tree is native to India and is found growing wild from bub-Himalayan tracts to West Bengal, and in central and southern India. It is grown all over the country, especially in the premises of temples and houses. It is a sacred tree associated with the Lord Shiva.

Description: It is a medium-sized deciduous tree growing up to 25–30 feet in height with straight sharp axillary thorns and a yellowish brown, shallowly furrowed corky bark. Leaves are alternate, trifoliate and aromatic. Leaflets ovate crenate, the lateral ones being subsessile and the terminal ones long-petioled. Flowers: greenish white in colour, born on axillary panicles, sweet-scented. Fruit is a berry, globose, woody and pulpy; seeds many, compressed, embedded in pulp, which is orange red.

Flowering: August–September.

Fruiting: October–March.

Chemical constituents: This tree is reported to contain a number of coumarins, alkaloids, sterols and essential oils. Roots and fruits contain coumarins such as scoparone, scopoletin, umbelliferone, marmesin and skimmin. Fruits, in addition, contain xanthotoxol, imperatorin and alloimperatorin as well as alkaloids such as aegeline and marmeline. β- sitosterol and its glycoside are also present in the fruits. Roots contain psoralen, xanthotoxin, 6,7-dimethoxy coumarin, tembamide, mermin and skimmianine. Leaves contain mermesinin, rutin,β-sitosterol and β-D-glucoside.

Used in: Ayurveda.

Parts used: Leaves, stem, bark and roots.

Uses: Root, bark, leaves and fruits are astringent and febrifuge. Root, stem and bark are antidiarrheal and antivenin. Leaves are anti-inflammatory, expectorant, anti-catarrhal, anti-asthmatic, anti-ulcerous, astringent, expectorant, laxative and ophthalmic. Flower is emetic. Unripe fruit is stomachic and demulcent. Ripe fruit is astringent, anti-gonorrheal, cardiotonic, restorative, febrifuge, laxative, antitubercular, anti-dysenteric and antiscorbutic. Seed is anthelmintic and antimicrobial.

Methods of propagation: The plant is propagated mainly through seeds and rarely by root cuttings. Seeds are soaked in water for 6 hours, sown on seed beds and irrigated regularly. Seeds germinate within 15–20 days. 1-month-old seedlings can be transplanted into polybags, which can be planted in a field after 2 months.

Photo: Dr R.S. Modi

AERVA LANATA (L.) JUSS. EX SCHULT

Synonyms: *Aerva pubescens* Mart.; *Acranthus lantana* Roxb.; *Acranthus villosa* Forssk.; *Illecebrum lanatum* L.

Classification:

Kingdom:	Plantae
Subkingdom:	Phanerogamea
Class:	Dicotyledonae
Subclass:	Monochlamydeae
Series:	Curvembryae
Order:	Caryophyllales
Family:	Amaranthaceae
Genus:	*Aerva*
Species:	*Aerva lanata*

Common names: Bengali: Chaya; **English:** Mountain knot grass; **Gujarati:** Kapurimadhvir; **Hindi:** Gorkhabundi, Kapurijadi; **Kannada:** Beeli himdi soppu; **Malayalam:** Cherula; **Marathi:** Kapurmadhura, Kapurimadhuri, Kapurphuti, Kumra; **Sanskrit:** Aadaanpaaki, Gorakshaganjaa, Paashaanabheda; **Tamil:** Ciru pulai, Pulai, Poolaipoo, Koolappoo; **Telugu:** Pindidonda.

Distribution: This plant is found in abundance in drier regions of the Old World tropics and subtropics. It is a common weed found throughout India at up to 900 m elevation.

Description: It is a prostrate herb with woolly branches arising from a woody base. The leaves of the plant are simple, alternate and short-petioled, elliptic-orbicular to obovate. The leaves are about 1.5–3.5 cm long and 1–2.5 cm wide. They become gradually smaller toward the apices, glabrate or finely pubescent above and white woolly beneath. The flowers are borne in axillary spikes; they are minute, sessile, greenish or creamy white in colour, often closely crowded and forming globose clusters; bracteoles 1.25 mm long, membranous, broadly ovate, concave, apiculate. Perianth 1.25–1.5 mm long; sepals oblong, obtuse, sometimes apiculate, silky-hairy on the back, stigmas 2. The fruits of this plant are ovoid, acute. The seeds are smooth, shiny, black and bean-shaped, about 1 mm in diameter.

Flowering: September–March.

Fruiting: September–March.

Chemical constituents: The aerial parts of this plant have been reported to contain aervitrine, feruloyltyramine, narcissin, O-acyl glycosides and syringic, ferulic and vanillic acids (Payal et al., 2015). The whole plant contains β-sitosterol, α-amyrin, betulin, hentriacontane, sitosteryl palmitate, D-glucoside, glycosides, kaempferol-3-galactoside, kaempferol-3-rhamnogalactoside, alkaloids, phenolic compounds and phytosterols (Nagaratna et al., 2014; Payal et al., 2010).

Used in: Ayurveda and folk and tribal medicine Siddha.

Parts used:

Uses: It is used in Ayurveda because of its anthelmintic, anti-inflammatory, demulcent, diuretic, expectorant, hepatoprotective and nephron protective properties (Yamunadevi et al., 2010). The juice from the leaf is an effective remedy to expel kidney stones. The decoction of the plant is used as a diuretic. The flowers are useful for dissolving kidney stones and treating conjunctivitis and gonorrhea. The roots are used to treat strangury. They are demulcent and diuretic and have tonic properties. The ground root mixed with yoghurt is given to treat jaundice (Rajesh et al., 2011; Payal et al., 2015).

Methods of propagation: Through seeds.

Photo: Prof Srinath Rao

AESCULUS INDICA (WALL. EX CAMB.) HOOK

Synonyms: *Pavia indica* Wall. ex Cambess., *Pawia indica* Kuntze.

Classification:

Kingdom:	Plantae
Subkingdom:	Phanerogamea
Class:	Dicotyledonae
Subclass:	Polypetalae
Series:	Disciflorae
Order:	Sapindales
Family:	Sapindaceae
Genus:	*Aesculus*
Species:	*Aesculus indica*

Common names: English: Indian horse chestnut; **Hindi:** Kanor, Bankhor; **Kannada:** Kanor, Kanur;

Distribution: Native to the Himalayas, found at altitudes of 900–3600 m in forests, shady ravines and along water courses; also found in Afghanistan, Pakistan, Nepal.

Description: It is a tall, deciduous, spreading, shady tree, with a straight trunk, and branches in whorls reaching to a height of 50–60 feet; the girth of its trunk is about 3–4 feet. The bark of the tree peels off upwards in narrow strips. The young shoots are minutely velvety, becoming hairless at maturity. The leaves are glossy, palmately compound; they have 7 leaflets arising from the same point on reddish stalks, ovate in shape, margin serrate, undulate. The leaves are highly ornamental and look like tiny umbrellas, about 10–20 cm long, 2–6 cm in width. Inflorescence: a compound raceme, 42 cm long, 12.5 cm broad at the base, bearing 300–350 flowers. The flowers are zygomorphic, bisexual, complete, pedicellate, perigynous, 3–5 mm long, 2–2.5 cm in diameter; calyx gamosepalous, pink, deciduous, tubular, tomentose, having 5 lobes; corolla polypetalous, with 4 petals, clawed, zygomorphic, having a combination of yellow, pink, red and white, each petal 2.5 cm long and 5–8 mm wide; androecium polyandrous, with 7 stamens 3.5 cm long, perigynous, tomentose, narrow and oblong. Ovary with a single style. Fruit, a capsule, 2 cm long, 4.1 cm in diameter, epicarp, brown with rusty spots. Seeds, 1 to 2 per capsule, shining black from outside, and lime white from inside, 3.5 cm in diameter.

Flowering: May–July.

Fruiting: July–October.

Chemical constituents: The seeds contain aescin, aesculuside A and B, also aliphatic esters. The leaves contain aescin, quercetin and β-sitosterol. The stems contain rutin, astragalin and aesculin.

Used in: Traditional medicine.

Parts used: Seeds and bark.

Uses: Antirheumatic, galactogenic, anti-leucorrhoeic. The seeds possess ant-inflammatory properties. They are astringent, acrid and narcotic. The oil from the seeds is applied externally in the treatment of skin diseases and rheumatism. The extract from the seeds is considered to be active against P-388 lymphocytic leukaemia and human epidermoid carcinoma of the nasopharynx. The juice of the bark is also used to treat rheumatism.

Methods of propagation: Through seeds.

AGAVE AMERICANA L.

Synonyms: *Agave altissima* Zumagal; *Agave americana* var. marginata Trel.

Classification:

Kingdom:	Plantae
Subkingdom:	Phanerogamea
Subclass:	Liliidae
Series:	Epigynae
Order:	Liliales
Family:	Amaryllidaceae (Agavaceae)
Genus:	*Agave*
Species:	*Agave americana*

Common names: English: Century plant; Maguey, American aloe, Agave; **Hindi:** Banskeora, Kantala; **Kannada:** Anekatali, Kalanaru; **Malayalam:** Anakyitha; **Sanskrit:** Kalakantala; **Tamil:** Kattaleyi; **Telugu:** Banda-kattla.

Distribution: It is a native of America and was introduced in India in the early 18th century for its leaf fiber. Since then, these plants have become popular throughout the country, usually in the degraded sites near villages and along roads.

Description: It is an erect and stemless plant. Leaves: in rosette, long even up to 6 feet, erect, pointed, with marginal spines and a heavy spike at the tip. They have longitudinal white yellow streaks or bands. Flowers: white in colour, funnel-shaped; their stalks are stout. The plant flowers only once during its lifetime; the flowering stem arises from the centre of the plant like a thick pole. Fruits: capsules, generally oblong and beaked in appearance.

Flowering: The plant lives for a number of years without flowering but dies once it does flower.

Fruiting: Fruiting capsules are large and oblong with small, pointed tips. When mature, fruits split open releasing shiny D-shaped, black seeds, 6–8 mm long.

Chemical constituents: The leaves yield saponins such as agavasaponin, based on hecogenin and chlorogenin. The flowers contain chlorogenin and kaempferol. The seeds contain neotigogenin, hecogenin and kammagenin. Hecogenin extracted from agaves is used in the manufacture of corticosteroids (Bouaziz et al 2014).

Used in: Ayurveda, homeopathy.

Parts used: Leaves; Sap; Seeds and Stem.

Uses: Demulcent, diuretic, emmenagogue, laxative and antiseptic; agave sap is a soothing and restorative remedy for many digestive ailments. It is used to treat ulcers and inflammatory conditions affecting the stomach and intestine. It protects these parts from infection and irritation and encourages healing. Moreover, it was found useful in the treatment of scurvy. Agave is also employed to treat a wide range of other conditions including syphilis, tuberculosis, jaundice and liver diseases. Agave nectar, also called agave syrup, is marketed as a natural sugar substitute with a low glycemic index which is due to its high fructose content (Peana et al 1997).

The roots have diuretic, diaphoretic and anti-syphilitic properties. The juice obtained when cutting the leaves and roots is especially useful in treating syphilis. The juice of the leaves is used as a laxative and emmenagogue, and for treating scurvy, warts, cancerous ulcers and tumors. The core of the plant is used as a febrifuge in treating malaria and other fevers. The leaves yield a strong, durable fiber called pita, used for making ropes, mats, sacking and brushes.

Methods of propagation: Through vegetative methods. It normally produces plenty of suckers during its life and these continue growing for about 10–15 years. Also propagated through seeds. The seed usually germinates in 1–3 months. Seedlings are grown into individual pots of well-drained soil. When they are large enough, they can be grown in a greenhouse until they are at least 20 cm tall. Then they are planed out in late spring or early summer.

Photo: Dr R.S. Modi

AGERATUM CONYZOIDES L.

Synonyms: *A. album* Stend; *A. caeruleum* Hort. ex Poir.; *A. coeruleum* Desf.; *A. cordifolium* Roxb.; *A. hirsutum* Lam.

Classification:

Kingdom:	Plantae
Subkingdom:	Phanerogamea
Class:	Dicotyledonae
Subclass:	Gamopetalae
Series:	Inferae
Order:	Asterales
Family:	Compositae (Asteraceae)
Genus:	*Ageratum*
Species:	*Ageratum conyzoides*

Common names: Bengali: Unchumti; **English:** Goatweed; **Hindi:** Junglipudina; **Kannada:** Naayi tulasi, Helukasa; **Malayalam:** Kattappa; **Marathi:** Ghanera osaadi; **Sanskrit:** Visamusthi, Uchunti, Dochunty; **Tamil:** Appa-k-kodi; **Telugu:** Kampurodda, Gana gaaju.

Distribution: It is native to Central America, Southeast Asia, South China, West Africa and India. Found throughout India up to an altitude of 1800 m.

Description: It is an erect, annual, branched, slender, hairy and aromatic plant. It grows to approximately 1 m in height. Stem is branched and covered with fine white hairs. Leaves are opposite pubescent, broadly ovate or rhomboid, with long petiole, and include glandular trichomes. The inflorescence contains 30–40 pale blue flowers arranged as corymb. Fruit is an achenes, glabrous or thinly hairy, with aristate, serrulate pappus-scales.

Flowering: Almost year-round.

Fruiting: Almost year-round.

Chemical constituents: This plant contains flavonoids, alkaloids, coumarins, essential oils, tannins, terpenoid compounds, precocene I, precocene II and flavone-like ageconyfavones. The species contains alkaloids, mainly belonging to the pyrrolizidinic group. The stem and leaves contain β-caryophyllene, 6-demethoxy-ageratochromene and ageratochromene (Riaz et al., 1995; Nogueira et al., 2010).

Used in: Ayurveda, folk medicine.

Parts used: Leaves, flowers, roots; the whole plant.

Uses: This plant has been known since ancient times for its curative properties and has been utilized for the treatment of various ailments. It is analgesic, antibacterial, antifungal, anti-inflammatory, anti-diarrheal, diuretic, emollient and styptic. It is used in treating kidney stones – the leaf extract is administered twice a day. The leaves are mashed and the juice is used to treat colds and sore throat; a concentrated decoction is used to treat chiggers. To make an infusion, the leaves are kept in water and the liquid is used to cure body itches and 'faja skin'. The root infusion is used to reduce high temperature. The whole plant extract is employed to cure sore throat, colic and issues of the digestive tract, and as a tonic. It is also used to treat burns and wounds, headaches pneumonia, inflammation, asthma, spasmodic and haemostatic effects, stomach ailments, gynaecological diseases, leprosy and other skin diseases (Kamboj and Saluja, 2008). This is the only plant used in the treatment of HIV/AIDS by the Igede people in Nigeria (Igoli, 2005; Amadi et al., 2012).

Methods of propagation: Through Seeds.

Photo: Prof Srinath Rao

ALANGIUM SALVIFOLIUM (L.F.) WANGERIN

Synonyms: *Alangium decapetalum* Lam.; *Alangium acuminatum* Wight ex Steud.; *Alangium lamarkii* Thwaites.; *Alangium salvifolium* subsp. decapetalum (Lam.) Wangerin.; *Grewia salvifolium* L. f.

Classification:

Kingdom:	Plantae
Subkingdom:	Phanerogamea
Class:	Dicotyledonae
Subclass:	Gamopetalae
Series:	Inferae
Order:	Asterales (Cornales)
Family:	Alangiaceae (Cornaceae)
Genus:	*Alangium*
Species:	*Alangium salvifolium*

Common names: Bengali: Akarkanta, Ankhura, Baghankura, Dhalakura; **English:** Sage-leaved alangium, Stone mango; **Gujarati:** Ankol, Ankola, Ankoli; **Hindi:** Akola, Anedhera, Ankol, Dheri; **Kannada:** Ankolemara, Kallu mavu; **Malayalam:** Alinnil, Ankalom, Cemmaram; **Marathi:** Ankol; **Sanskrit:** Kangarola, Gandha puspha, Tamraphala; **Tamil:** Anincil, Arulavam, Atikolam; **Telugu:** Ankolamu, Udugu.

Distribution: This is widely distributed over plains and lower areas throughout most of India. The plants are found near streams in dry deciduous forests, along roadsides or on fallow agricultural lands near villages. *Alangium salvifolium* subsp. hexapetalum is restricted to the wet forests of the Western Ghats within India.

Description: It is a deciduous bushy shrub or small tree growing to 25–30 feet tall, with or without spines. The bark is pale brown, aromatic, rough with shallow cracks, exfoliating in corky scales. Leaves are alternate, narrowly elliptic oblong or ovate in shape, 15 cm long. The flowers are greenish white in colour, fascicled, axillary or on old stems. Style single, with a 4-partite globose stigma rising above the stamens. Fruit is a berry, ellipsoid, ovoid, black, fleshy, crowned by persistent calyx and edible. Seeds are large, enclosed in red, mucilaginous, sweet and astringent pulp (Ratra and Gupta, 2015).

Flowering: February–June.

Fruiting: July–August.

Chemical constituents: Root bark contains alangine A, alangicine, cephaeline, emetine marckine, marckidine, psychotrine and tubulosine. Stem contains cephaeline, deoxytubulosine, emetine, methyl cephaeline and psychotrine. Stem bark contains lamarckinine. Leaves contain alangimarckine, ankorine, campester, chole-chloride, deoxytubulosine, episterol. Seeds contain alangamide, betulinic acid, betulin aldehyde, betulin, lupeol and β-sitosterol (Itho and Nayashiro, 1994; Itho et al., 1995; Ratra and Gupta, 2015).

Used in: Ayurveda and tribal medicine.

Parts used: Fruits, root, bark.

Uses: Each and every part of this plant is used for separate medicinal purposes. The root bark, used in the form of a decoction or powder, is very bitter and credited with antipyretic, astringent, emetic, diaphoretic, pungent and purgative properties. It is a famous drug in Ayurveda, used both externally and internally for rabies and as an antidote for snake bites. It is also used in the form of a paste or decoction to treat diarrhea, syphilis and worms, and applied externally as a cure for leprosy, erysipelas and other skin diseases. A decoction of the root bark is employed for the treatment of dropsy. The oil derived from the root bark is applied externally for the treatment of acute rheumatism. A paste prepared from the root with another herb, *Tragia involucrata* (Euphorbiaceae), is applied as a cure for paralysis. A paste prepared from the root bark is used as a local application to treat snake, cat and dog bites. It is also taken orally to promote the excretion of toxins. The leaves are applied as a poultice to relieve rheumatism and are also considered to be hypoglycemic. The fruits of this plant have anti-phlegmatic, cooling, laxative and tonic properties. The plant is also used for treating emaciation, hemorrhage and strangury, as well as to prevent and cure eye troubles. The seeds are valued in traditional Indian medicine for their cooling and tonic properties and for their use in the treatment of hemorrhage (Ratra and Gupta, 2015).

Methods of propagation: Through seeds.

Photo: Prof Srinath Rao

ALEURITES TRILOBA J.R. FORST & G. FORST

Synonyms: *Aleurites moluccana* (L.) Willd.; *Aleurites ambinux* Pers.; *Aleurites angustifolia* Vieill.; *Aleurites javanica* Gand., *Aleurites lanceolata* Blanco; *Aleurites lobata* Blanco; *Aleurites moluccana* (L.) Wild.; *Jatropha moluccana* L.

Classification:

Kingdom:	Plantae
Subkingdom:	Phanerogamea
Class:	Dicotyledonae
Subclass:	Monochlamidae
Series:	Unisexuales
Order:	Malpighiales
Family:	Euphorbiaceae
Genus:	*Aleurites*
Species:	*Aleurites triloba*

Common names: English: Candle nut, Candleberry, Indian walnut; **Hindi:** Jangli akhrot; **Kannada:** Akroda, Natakrodu, Arkod, Naadu aakrotu; **Malayalam:** Akrottu, Akshotam, Karankolam, Vadam; **Marathi:** Akhod, Japhala, Ranakrot; **Sanskrit:** Akharota, Akhota, Akshota, Asphotaka, Gudashaya; **Tamil:** Nattu akrottu, Woodooga; **Telugu:** Uduga, Natu akrotu.

Distribution: A native of India, China, Myanmar and Sri Lanka.

Description: It is a tree growing to a height of 50–60 feet. When cut, the bark exudes a clear watery sap. The stems, petioles and blades are covered with a whitish starry tomentum. Leaves: simple, spiral and without stipules, petiole long, about 12–15 cm and thin. The blade is lanceolate, rigid, 3–5-lobed, 10–12 cm long and about 12–16 cm broad, and marked at the base with a pair of 1 mm-diameter, disc-shaped glands; margin is recurved. The veins are raised on both surfaces of the blade, which shows 6–7 pairs of secondary veins. The inflorescences are terminal panicles; flowers monoecious. The male flowers surround the female flowers. The pistillate flowers are 13 mm long with 5 white creamy petals, while staminate are longer and thinner. Calyx is 2.5 mm long, 2–3-lobed and covered with a whitish starry tomentum. The corolla is about 1–1.5 cm across.

Flowering: September–October.

Fruiting: Throughout the year.

Chemical constituents: Stems contain scopoletin. The stem bark contains various phytoconstituents such as alkaloids, coumarins, flavonoids, polyphenols, spruceanol, steroids, tannins and triterpenoids. The leaves contain flavonoids, hydrocarbons, sterols, swertisin and triterpenes. The plant also contains acetil aleuritic acid, α- and β-amyrin, campesterol, moretenone, moretenol, moluccanin, swertisin, stigmasterol and β- sitosterol (Shamsuddin, et al., 1988; Meyre Silva et al., 1998; Kaur et al., 2018).

Used in: Ayurveda, folk medicine.

Parts used: Leaves, flowers, seeds, bark.

Uses: It has aperient, cardiotonic, carminative, diaphoretic, expectorant, hematonic and laxative properties. It is used to treat amenorrhea, arthrosis, asthma, bleeding, constipation, edema, gas, gonorrhea, hematochezia, hemorrhoid, hydrophobia, rheumatism, ringworm, sore, sprue, and swelling. A decoction made from the leaves is used in the treatment of hypocholesterolemia, fever, tumor, diarrhea, headache, hernia, pain in chest and cough, while the boiled leaves are used in the treatment of gonorrhea, swollen joints, ulcers and fever. The leaves are also used as poultice for deep contusions, bruises, broken bones, constipation and food poisoning. The flowers and sap are used in maintaining dental hygiene. The seed oil is a laxative. The leaves and bark exhibit antimicrobial properties. The gum is chewed for its aphrodisiac properties. A decoction made from the bark is used to treat infertility and secondary amenorrhea, diarrhea, fever, hypocholesterolemia and tumors (Shamsuddin, et al., 1988; Meyre Silva et al., 1998; Kaur et al., 2018).

Methods of propagation: Through seeds.

Photo source: en. Wikipedia.org

ALLIUM CEPA L.

Synonyms: *Cepa alba* Renault.

Classification:

Kingdom:	Plantae
Subkingdom:	Phanerogamea
Class:	Monocotyledon
Subclass:	Liliidae
Order:	Liliales
Family:	Liliaceae
Genus:	*Allium*
Species:	*Allium cepa*

Common names: English: Bulb onion; **Hindi:** Pyaz; **Kannada:** Irulli, Ullagaddi; **Malayalam:** Ceriyaulli; **Marathi:** Kanda; **Sanskrit:** Dirghapatra, Mahakanda; **Tamil:** Vengayam, Irakayam; **Telugu:** Ullipai.

Distribution: It is a native of Iran, the western Indian subcontinent and Central Asia. About 170 countries cultivate onions for domestic use.

Description: The onion plant has a fan of hollow, bluish green leaves, and its bulb at the base of the plant begins to swell when a certain day length is reached. The bulbs are composed of shortened, compressed, underground stems surrounded by fleshy modified scale (leaves) that envelop a central bud at the tip of the stem. In the autumn (or in spring, in the case of overwintering onions), the foliage dies down and the outer layers of the bulb become dry and brittle. The crop is harvested and dried and the onions are ready for use or storage. The inflorescence takes the form of a globular umbel of white flowers with parts in sixes.

Flowering: June–July.

Fruiting: June–July.

Chemical constituents: Onions contain quercetin, fructose, quercetin-3-glucoside, isorhamnetin-4-glucoside, xylose, galactose, glucose, mannose, organosulfur compounds, allylsulfides, flavonoids, flavonols, S-alk(en)yl cysteine sulfoxides, cycloalliin, selenium, thiosulfinates, and sulfur and seleno-compounds.

Used in: Ayurveda, homeopathy, traditional medicine, Unani.

Parts used: Bulbs and leaves.

Uses: Onion has been an important constituent in foods all over the world. Medicinal uses include: boosting heart health and the immune system, reducing inflammation, fighting cancer and regulating blood sugar.

Methods of cultivation: Onion seeds are short-lived and fresh seed germinate better.

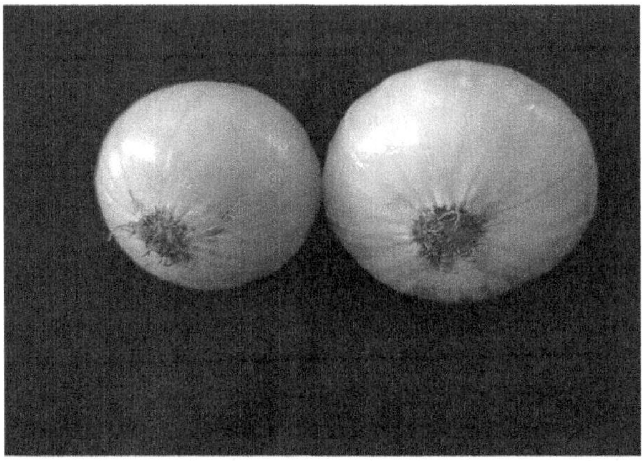

Photo: Prof Srinath Rao

ALLIUM SATIVUM L.

Synonyms: *Allium ophioscorodon* Link; *Allium pekinense* Prokh.; Porrum sativum (L.)

Rchb.

Classification:

Kingdom:	Plantae
Subkingdom:	Phanerogamia
Class:	Monocotyledon
Subclass:	Liliidae
Order:	Liliales
Series:	Coronarieae
Family:	Liliaceae
Genus:	*Allium*
Species:	*Allium sativum*

Common names: Bengali: Rasun; **English:** Garlic; **Gujarati:** Lehasan; **Hindi:** Lahsun; **Kannada**: Belluli, Balloli; **Malayalam:** Vellulli; **Tamil:** Acanam; **Telugu:** Vellipai.

Distribution: Typically it is grown in a temperate climate, similar to those of central Asia. It can be found growing in the Northern and Southern hemispheres.

Description: *Allium sativum* is a bulbous plant, growing up to 1 m in height. It produces hermaphrodite flowers. It is pollinated by bees, butterflies, moths, and other insects. Garlic (*Allium sativum* L.) is a hardy liliaceous sterile plant, the strong-scented, pungent bulb of which is used in cookery and for medicinal purposes.

Chemical constituents: Garlic contains at least 33 sulphur compounds, several enzymes, 17 amino acids, and minerals such as selenium. Garlic's pungent odour and many of its medicinal effects are due to sulphur compounds such as alliin, allicin, ajoene, diallyl trisulfide, S-allyl cysteine, vinyldithiines, allyl propyl disulfide, S-allylmercaptocysteine and others. Approximately 1 % alliin (S-allyl cysteine sulfoxide) is present in dried, powdered garlic. Allicin (diallyl thiosulfinate or diallyl disulfide) is the most biologically active compound in garlic. Enzymes: alliinase, peroxidases, myrosinase and others. Amino acids and their glycosides: arginine and others. Selenium, germanium, tellurium and other trace minerals.

Used in: Folk medicine.

Parts used: Bulbs.

Uses: Garlic has many beneficial effects, such as lipid-lowering, vascular, chemoprotective/antitumoral, immunologic, antimicrobial, anti-viral and glycemic.

Methods of cultivation: It is conventionally propagated by daughter bulbs ("cloves") and bulbils from the flower head.

Photo: Prof Srinath Rao

ALOE BARBADENSIS MILL.

Synonyms: *Aloe vera* (L.) Burm., *Aloe perfoliata* L., *Aloe vulgaris* Lam., *Aloe chinensis* Baker, *Aloe indica* Royle, *Aloe abyssinica* Lam.

Classification:

Kingdom:	Plantae
Class:	Monocotyledonae
Subclass:	Liliidae
Order:	Asparagales
Series:	Coronarieae
Family:	Liliaceae
Genus:	*Aloe*
Species:	*Aloe barbadensis*

Common names: English: Aloe; **Hindi:** Ganwarpatha; **Kannada:** Lolasara; **Malayalam:** Catevala; **Sanskrit:** Kumari; **Tamil:** Akki; **Telugu:** Chinnakalabanda.

Distribution: This species is native of the Canaries, naturalized in the Mediterranean and introduced to warm parts of America. It is found in a semi-wild state, or cultivated, in many parts of India.

Description: It is a succulent perennial with a short stem. Leaves: fleshy, narrow, lanceolate, having dentate margins. Flowers: Yellow, born in racemes on scapes.

Chemical constitutions: The leaves contain mucilage, anthraquinones and chromones. The mucilage on hydrolysis yields glucose, galactose, mannose and galacturonic acid. The quinones present are barbaloin and isobarbaloin and their aglycone, aloe-emodin. The chromones located are aloesone and its C-glycosides-aloesin. Also present are glycoproteins and alocutin A and B.

Used in: Ayurveda, homeopathy and Unani, Tibetan and traditional medicine.

Parts used: Fresh leaves and leaf juice.

Uses: Fresh leaf juice is cathartic and also used for eye troubles, spleen and liver ailments, dermatitis and other skin diseases. The leaves also possess hypocholesterolemic properties. Aloe is also used in menstrual and uterine disorders, stomach pain and also as a tonic after pregnancy. Alocutin A and B are lectins found useful in treating cancer.

Methods of propagation:

- It is propagated through suckers having 4–6 leaves. A healthy mother plant produces about 6 suckers at a time. It can also be propagated through rhizome cuttings. After harvesting the crop, the underground rhizome is dug out and cuttings of 4–6 cm are made with a minimum of 2–3 nodes on them. It is rooted in sand beds and after it has sprouted it can be transplanted. It can be grown on a wide variety of soils; however, waterlogged conditions are not suited for its cultivation.
- Plant the suckers in rows at a spacing of 1 foot. The rows have to be 2 feet apart, formed on a well-prepared and manured loamy soil having plenty of broken bricks and coarse sand.
- With a shallow-root system, the roots do not penetrate deep into the soil. Therefore, whenever the manure is to be applied to the soil, it has to be placed near the base.
- Suckers can be planted during October–December.

Photo: Dr Kotresh

ALPINIA GALANGA (L.) WILLD.

Synonyms: *Amomum galanga* (L.) Lour.

Classification:

Kingdom:	Plantae
Subkingdom:	Phanerogamia
Class:	Monocotyledonae
Family:	Zingiberaceae
Genus:	*Alpinia*
Species:	*Alpinia galanga*

Common names: English: Greater galangal, Javagalangal, Siamese ginger; **Hindi:** Barakalanjar; **Kannada:** Dumparasmi; **Malayalam:** Arattha, Chittaratha; **Marathi:** Kosht-Kulinjan; **Sanskrit:** Sugandhamula, Barakalanjar; **Tamil:** Peraratthei; **Telugu:** Peddaumparashtram.

Distribution: Asia, probably southern China. This species is globally distributed across India and Tropical Asia. Within India, it is mainly distributed in the Eastern Himalayas and in Southwest India, Mysore, Goa, Malabar and Gujrat.

It is very common in West Bengal, Bihar, Assam, Kerala, Karnataka and throughout the

Western Ghats. It is now cultivated also in Indochina, Thailand, Malaysia and Indonesia.

Description: The plant is a perennial clustered herb. It grows up to 2 m. Rhizomes lumpy, with joints, brown-red and with a slightly spicy fragrance. Leaves: oblong or lanceolate, 2 rows, 25–35 cm long, 6–10 cm wide, green above, paler beneath, apex short, acute or acuminate, base gradually narrow, glabrous on both sides or long hairs below, petioles 6 mm long, ligules linear 5 mm long, rachis with hairs, multibranched 2–4 cm long. Flowers: born on panicles 20–30 cm long, acrogenous with bracts and bracteoles. Bracteoles are large and lanceolate and 5–8 mm long. Flowers green-white in colour, refreshing fragrance; calyx tubular, 3-toothed, 0.6–1 cm long, stamens pleurogenous degenerative; corolla tube cylindrical, rarely longer than the calyx. Corolla lobes oblong or linear oblong, the upper ones usually broader and more convex than the lateral ones. Ovary: inferior, 3-celled, glabrous, style slender, stigma slightly inflated, labellum obovoid, 2 cm long, white with red strings, 2-parted. Stamen 1, perfect, filaments 1 cm long and flattened, anthers 7 mm long, diverging at the top; lateral staminodes minute or obsolete. Lip spreading and broadly elliptic, often orbicular. Fruit: Orange-red in colour, capsules, the size of a small cherry, oblong, 1–1.5 cm long, middle part slightly contractive, brown or deep red when mature, smooth or slightly shrivelled with 3–6 seeds.

Flowering: May–August.

Fruiting: September–November.

Chemical constituents: The rhizome contains essential oils consisting of cineol, methyl cinnamate and flavones. Galangin, alpinin, kaempferide, kaemperol, and 3- deoxy 4-methoxy flavones (Al-Adhroey et al., 2010).

Used in: Ayurveda, Siddha, Unani and traditional medicine.

Parts used: Rhizomes.

Uses: The rhizome is an abortifacient. It has carminative, antituberculosis and stimulant properties. Ground rhizome is also used in the treatment of skin infections such as eczema, ringworm, etc. The rhizome has been shown to have weak antimalarial activity in mice. It is used against rheumatism, bronchial catarrh, bad breath, ulcers, whooping colds in children, throat infections and fever. it show promise as anti-fungals, hypotensives, enhancers of sperm count and motility. Anti-tumor and anti-dementia effects have been observed in rodents. *Alpinia galanga* is a stimulating aromatic herb and has been successfully employed to aid the digestive process, preventing fermentation and removing flatus. It is useful in case of dyspepsia, vomiting or sickness of the stomach and facilitates digestion. Homoeopaths use it as a stimulant. It has some reputation as a remedy for perineal relaxation with hemorrhoids and for a lax and pendulous abdomen. It is used as a snuff to treat cold and flu symptoms. Galangal root has also been used as a digestive aid, especially in combating dyspepsia and flatulence. In Ayurveda, '*Rasna-saptak-kwath*' and '*Rasna-adikamath*' are used as anti-inflammatory decoctions. In Unani, it is an ingredient of aphrodisiac preparations '*Majun Mugawivi ma Mumsik*' and '*Majun Samagh*' and antispasmodic nervine tonics '*Majun Chobchine*' and '*Lubab Motadil*'. It is also used in '*Arq Pan*' as a cardiac stimulant and carminative.

Methods of propagation: Through rhizomes. The rhizomes grow best in shaded areas away from direct sunlight. They require well-drained soil for better growth. They are usually planted in spring.

Photo: Prof Srinath Rao

ALSTONIA SCHOLARIS (L.) R. Br.

Synonyms: *Echites scholaris* L.; *Echites pala* Buch.-Ham.; *Pala scholaris* (L.) Robert; *Tabernaemontana alternifolia* Burm.

Classification:

Kingdom:	Plantae
Subkingdom:	Phanerogamea
Class:	Dicotyledonae
Subclass:	Gamopetalae
Series:	Bicarpellatae
Order:	Gentianales
Family:	Apocynaceae
Genus:	*Alstonia*
Species:	*Alstonia scholaris*

Common names: Bengali: Satiani, Chattin, Chatium; **English:** Devil's tree, White Cheesewood, Milk wood; **Gujarati:** Satuparni; **Hindi:** Saitan ka jhad, Chatium, Chitvan; **Malayalam:** Daivapala; **Marathi:** Saptaparni; **Sanskrit:** Saptaparni; **Tamil:** Elalaipalai, Palegaruda; **Telugu:** Edakulaponna.

Distribution: This plant is native to India, Bangladesh, Pakistan, China, Myanmar, Vietnam, Sri Lanka, parts of Indonesia, Malaysia and Australia. It is found throughout India.

Description: Large evergreen tropical tree reaching up to a height of 30–40 m, with bitter milky latex. Trunk furrowed, bark greyish brown, corky, thick, rough, lenticellate. Leaves occur in a whorl of 8–10 and are about 10–12 cm long and 5–7 cm broad, coriaceous, elliptic-oblong, base cuneate, apex usually rounded, upper side of the leaf glossy, pale beneath. Lateral veins are strong, occurring almost at right angles to midveins. Leaf petiole about 8–10 mm long. Inflorescence: cymes, dense and pubescent, 100–120 cm long. Flowers bisexual, small, greenish white, tube-like, 6–8 mm long, greenish, spice-scented. Calyx copular, 5-lobed, ovate, unequal, corolla salver-shaped, 5-lobed, lobes ovate overlapping to the left. Fruit is a follicle, 30–40 cm long. Stamens 5, narrow, carpels 2, free-style filiform stigma, ovules many. Fruit: a follicle, with narrow pendulous mericarps, 30–40 cm long, green. Seeds: papillose, 5–6 mm long, flat, with a tuft of brownish hair at each end.

Flowering: Throughout the year.

Fruiting: Throughout the year.

Chemical constituents: The bark of the tree contains alkaloids such as ditamine, echitenine, echitamine, echitamidin, echicaoutchien and strictamine.

Used in: Ayurveda.

Parts used: Leaf, bark and milky juice.

Uses: The plant is used as a substitute to cinchona and quinine. The roasted leaves are made into poultice and applied on ulcers. The leaf decoction is used to treat beriberi. The fruits are used to treat insanity and epilepsy. The powder of the bark is used to treat various diseases such as diarrhea, indigestion, gastritis and slimy/mucus stools. The paste made from its bark is applied on affected areas to cure joint pain, rheumatism and ulcers. The juice of the bark is used to relieve sores, tooth pain, rheumatism and ulcers. The root powder is used to treat fever, rheumatism and scrotal hernia. The latex is also applied to treat ulcers.

Methods of propagation: This species can be propagated through both generative and vegetative methods.

Photo: Prof Srinath Rao

ALTERNANTHERA PUNGENS KUNTH

Synonyms: *Achyranthes repens* L.; *Alternanthera repens* (L.) Link.; *Alternanthera acranthus* (L.) R. Br. ex Sweet.; *Alternanthera echinata* Smith.

Classification:

Kingdom:	Plantae
Subkingdom:	Phanerogamea
Class:	Dicotyledonae
Subclass:	Monochlamydeae
Series:	Curvembryae
Family:	Amaranthaceae
Genus:	*Alternanthera*
Species:	*Alternanthera pungens*

Common names: English: Khaki joyweed; **Hindi:** Kantevali santi; **Kannada:** Mullu honagonna; **Malayalam:** Ottara mul, Thevadiyaau mul.

Distribution: Native to Central and South America. In India it is a weed of pebbly soils and dry locations.

Description: *Alternanthera pungens* is a herbaceous annual to perennial plant with more or less prostrate, much-branched stems growing from a robust taproot. The plant can grow up to 50 cm tall, the stems are hairy and often have a reddish appearance, forming new roots at the leaf nodes, and often becoming woody at the base. Leaves simple, opposite, entire, ovate; margins that are entire or slightly wavy, 10–60 mm long and 6–30 mm wide, subsessile. The paired leaves usually differ in size, i.e. one leaf of the pair is significantly larger than the other. Flowers inconspicuous, greenish yellow, and grouped together in small globular clusters, borne in the leaf forks; bracts and bracteoles spiny. Each flower has 5 petals (tepals), two of the petals longer than the others, with sharp tips. Some of the flower parts become hardened and form sharp prickles as they mature. Stamens 5, filaments united at the base. One ovary with capitate stigma. Fruit: an utricle, 1.5–2 mm long, apex truncate, hidden in the prickly flower. Seeds are minute, yellowish or orange in colour and shiny.

Flowering: Throughout the year.

Fruiting: Throughout the year.

Chemical constituents: It contains phenols, flavonoids, alkaloids, glycosides, steroids, saponins and tannins.

Used 3: Traditional and folk medicine

Parts used: The whole plant.

Uses: In traditional medicine it is used as a painkiller, for stomachaches, swelling and naso-pharyngeal infections. The extracts of the aerial parts possess antibacterial and antioxidant properties.

Methods of propagation: Through seeds, vegetative, stems developing roots at the nodes.

Photo: Prof Srinath Rao

ALTERNANTHERA SESSILIS (L.) R.Br. ex DC

Synonyms: *A. triandra*; *A. denticulate* R. Brown; *A. repens*; *A. nodiflora* R. Brown; *Gomphrena sessilis* L.

Classification:

Kingdom:	Plantae
Subkingdom:	Phanerogamea
Class:	Dicotyledonae
Subclass:	Monochlamydeae
Order:	Caryophyllales
Series:	Curvembryae
Family:	Amaranthaceae
Genus:	*Alternanthera*
Species:	*Alternanthera sessilis*

Common names: Bengali: Sanchesak, Salincha Sak; **English:** Dwarf copperleaf, Sessile joyweed; **Gujrati:** Jalajambo, Panini nib haji, Paniki bhaji; **Hindi:** Gudari saag; **Kannada:** Honagonne soppu; **Malayalam:** Kozuppa, Ponnankanni, Mukunuwenna; **Marathi:** Kanchari; **Sanskrit:** Matsyagandha, Matsyakshi, Bahli, Matsyaduni; **Tamil:** Ponnanganni keerai; **Telugu:** Ponnaganti kura.

Distribution: It is found growing throughout the warmer parts of India, preferably in wet places, especially around tanks and ponds.

Description: This is a small prostrate rarely ascending herb with several spreading branches. Root is cylindrical, 0.1–0.6 cm diameter, cream to grey, numerous roots arising from the main taproot as lateral root lets. Stem is herbaceous, weak, mostly cylindrical, occasionally sub-quadrangular at the apical region, with spreading branches from the base, yellowish pink to light brown, nodes and internodes distinct, internodes 0.5–5 cm long, often rooting at lower nodes. Leaf about 2–7 cm long, 1–2 cm wide, sometimes reaching 10 cm in length, 2.5 cm wide, sessile, linear-oblong or elliptic, obtuse or subacute, glabrous to sparsely villous, greenish pink, somewhat fleshy. Flower in small axillary, sessile chaffy heads, white, often tinged with pink; bracteoles 1.2 cm long, ovate, scarious; perianth 2.5–3 mm long; sepals ovate, acute, thin; ovary obcordate, compressed; style very short, capitellate. Fruit is an utricle, 1.5 mm long, orbicular, compressed, with thickened margins.

Flowering: July–October.

Fruiting: December.

Chemical constituents: The plant is reported to contain hydrocarbons, enecycloartanol, cycloecucalenol, stigmasterol, campesterol, sitosterol, spinasterol, oleanolic acid rhamnoside, cycloeucarlenol, lupeol and handiano. The leaves also contain a good amount of α- and β-tocopherols. The young shoots contain 5 %protein and 16.7 mg (per 100 g) iron. The antioxidant carotene is found in large amounts in the plant.

Used in: Ayurveda, Siddha and folk medicine.

Parts used: The whole plant.

Uses: The whole plant is used for increasing breast milk, promoting bile flow, inducing abortion and curing fever. A preparation from the fresh plant is helpful in toning up the liver. The juice of the fresh leaves mixed with garlic cloves is a folk remedy for asthma, chronic cough and intermittent fever. The leaf juice mixed with the leaf juice of coriander and wood apple is given to control spermatorrhea. The leaf juice of this plant is also useful against piles. The plant is an excellent coolant and is considered to have immense health benefits for the eye. It is also used to treat gastrointestinal problems such as flatulence, nausea and vomiting, and to get relief from headaches, bronchitis, vertigo, hepatitis and gonorrhea. Moreover, it is employed to treat snake bite. This is a non-toxic plant and can be eaten as a vegetable and in a salad.

Methods of propagation: It can be propagated through vegetative means. Cuttings are made from strong outdoor plants and planted during late summer; divisions of old plants are preferred.

Photo: Prof Srinath Rao

AMARANTHUS SPINOSUS L.

Synonym: *Amaranthus caracasanus* Kunth.; *A. diacanthus* Raf.; *Galliaria spinosa* (L.) Nieuwl.

Classification:

Kingdom:	Plantae
Subkingdom:	Phanerogamea
Class:	Dicotyledonae
Subclass:	Monochlamydeae
Order:	Caryophyllales
Series:	Curvembryae
Family:	Amaranthaceae
Genus:	*Amaranthus*
Species:	*Amaranthus spinosus*

Common names: Bengali: Kanta-notya; **English:** Prickly amaranth; **Hindi:** Kanta chaulai; **Kannada:** Mulluharive soppu, Mulla-dantu; **Malayalam:** Kattumullenkeera, Mullanchira; **Marathi:** Rajgira, Kante bhaji; **Sanskrit:** Tanduliuyah; **Tamil:** Mullikkirai; **Telugu:** Mullutotakura.

Distribution: It is recorded throughout the tropics of the world though it is reported to be a native of Africa. In India it is commonly found in waste places and along roadsides, fields and hedges near villages. It is also cultivated in several places.

Description: It is an erect annual branched herb with axillary spines, red-tinged erect stem, growing about 30–100 cm long. Leaves simple, alternate, membranous, lanceolate-oblong, about 12 cm long and 4–6 cm broad, petiole about 5–6 cm long. The inflorescence is terminal, lax and slender; the upper portion is staminate, the lower portion is pistillate. Flowers: small, sessile, yellowish white or pale green; bracts lanceolate, often spine-like, equalling or slightly shorter than tepals. Tepals: 5, oblong-ovate, equal in length, apex acute or acuminate, incurved against utricle. Fruits: small, orbicular or ovoid, compressed. Seeds: black and shining.

Flowering: Throughout the year.

Fruiting: Throughout the year.

Chemical constituents: Leaves and stems contain α-Spinasterol and hentriacontane. Leaves also contain amino acids with a high content of lysine. Root contains hentriacontane, α-Spinasterol, and saponins such as β-D-glucopyranosyl, β-D-glucopyranosyl, β-D-glucuronopyranosyl and olianiloc acid as well as β-D-glucopyranosyl, β-D-glucopyranosyl, β-D-glucopyranosyl and α-spinasterol. Plant contains sterols such as β-sitosterol, stigmasterol, campesterol and cholesterol.

Used in: Ayurveda.

Parts used: The whole plant: leaves, seeds and roots.

Uses: The plant is astringent, diaphoretic, diuretic, emollient, febrifuge and galactagogue. It is reported to be used in treating burning sensations, hallucinations, leprosy, bronchitis, leucorrhea, haemorrhoids, boils, burns, nausea, flatulence, colic, fever, intermittent fevers and general debility. The roots are used in menorrhagia and leucorrhoea. The boiled leaves and roots are given as a laxative; they are applied as a remedy for snake and scorpion bites. The seeds are used as a poultice for broken bones. The root is emmenagogue and galactagogue.

Methods of propagation: Through seeds; 70–80 % germination is observed. Seeds germinate within 3–4 days. Germination is usually rapid and good if the soil is warm. Cuttings of growing plants can also be used and they root easily.

Photo: Prof Srinath Rao

AMMI MAJUS L.

Synonyms: *Apium ammi-majus* Crantz; *Ammi boeberi* Hell. ex Hoffm.; *Ammi cicutifolium* Willd. ex Schult.; *Ammi intermedium* DC.; *Antheum pinnatum* Ruiz & Pav. ex Urban.; *Atheus ammi* Spreng.; *Carum majus* (L.) Koso-Pol.; *Cuminum aethiopicum* Royle.; *Cuminum regium* Royle.

Classification:

Kingdom:	Plantae
Subkingdom:	Phanerogamea
Class:	Dicotyledonae
Subclass:	Polypetalae
Series:	Calyciflorae
Order:	Umbellales
Family:	Umbelliferae (Apiaceae).
Genus:	*Ammi*
Species:	*Ammi majus*

Common names: English: Bishop's weed, Lady's lace; **Marathi:** Kattukasturi; **Sanskrit:** Kattukasturi.

Distribution: The plant is indigenous to Egypt and grows in the Nile Valley. It is cultivated in

Pakistan. This crop has been grown for its medicinal fruit in several places in Uttar Pradesh, Gujarat, Kashmir and Tamil Nadu.

Description: This is an annual or biennial herb, growing to a height of 80–120 cm. It has a long tap root; the stem is erect, slender, glabrous with fine longitudinal striations; the leaves are alternately arranged, pinnately divided; segments lanceolate to oval. Inflorescence: a compound umbel, the involucral bracts generally divided and about 13 in number; bracteoles of the involucre about 8 in number; primary ray of umbel sometimes 5 cm long, slender; secondary rays 2–5 cm long; flower whitish, actinomorphic or zygomorphic, bisexual, pentamerous and bracteates; calyx teeth obsolete or small; petals obovate with an inflexed point, petals with unequal lobes, the exterior ones frequently longer; stamens epigynous; ovary inferior, 2-celled; stigma capitate; fruit a cremocarp with slender prominent ribs, laterally compressed, oblong, smooth, 1.5–2.0 × 1.0 mm or less in size; mericarps separated by carpophore and bearing a stylopodium at the apex; seed small, pendulous.

Flowering: April–May.

Fruiting: June–October.

Chemical constituents: Furocoumarins such as ammidin, ammoidin, heraclenin, imperatorin, marmesin, majudin and xanthotoxin are present in the seed. The *Ammi majus* fruit contains amorphous glucoside 1 %, tannin 0.45 %, oleoresin 4.76 %, acrid oil 3.2 % and fixed oil 12.92 %. The plant contains bergapten and methoxsalen.

Used in: Ayurveda.

Parts used: Seeds.

Uses: This plant is extensively used in the treatment of leucoderma, vitiligo and psoriasis. It increases the pigmentation of normal skin and induces re-pigmentation in vitiligo. It also has antitumor properties.

Methods of propagation: This plant is propagated through seeds. Its seeds take about 10–12 days to germinate after sowing. The best time for sowing is October. The duration of the crop is 160–170 days. *Ammi* requires cold conditions and grows well under subtropical and temperate conditions. Soils from sandy loam to clay loam are suitable.

ANDROGRAPHIS PANICULATA (BURN.F.)

Synonyms: *Andrographis paniculata* var. glandulosa Trimen; *Justia latebrosa* Russel ex Wall.

Classification:

Kingdom:	Plantae
Subkingdom:	Phanerogamea
Class:	Dicotyledonae
Subclass:	Gamopetalae
Series:	Bicapellatae
Order:	Personales
Family:	Acanthaceae
Genus:	*Andrographis*
Species:	*Andrographis paniculata*

Common names: Bengali: Kalmegh; **English:** Green chirayta; **Gujarati:** Ariyatu; **Hindi:** Hara chirayata Kalmegh; **Kannada:** Nelabevu; **Malayalam:** Kiriyatta, Nilavepp; **Marathi:** Kalamegha, kalpa; **Sanskrit:** Bhunimba, Kalamegha, Mahatikta, Yavatikta; **Tamil:** Nilavempu; **Telugu:** Nelavemu.

Distribution: Throughout India, in agricultural fields.

Description: The plant is native to southern India and Sri Lanka, and is found in China, Pakistan and Thailand. It has also been introduced and cultivated in most other countries.

Flowering: August–September.

Fruiting: September–October.

Chemical constituents: *Andrographis* contains diterpene lactones, glycosides and flavonoids, alkaloids, phenols, catechin, saponins and tannins. In addition, it contains andrographolide, neoandrographolide, tetradecanoic acid, phytol, dioctyl ester, squalene, retinoic acid methyl ester and ß-sitosterol.

Used in: Ayurvedic, traditional and Siddha medicine.

Parts used: All parts.

Uses: This herb is widely used as a treatment for common cold, flu, upper respiratory infections and a variety of infectious diseases. Other folk medicinal uses of the herb are as a treatment for tuberculosis, chicken pox, coughs, headaches, ear infection, inflammation, burns and mumps. It is often used as a herbal treatment for pain associated with conditions such as muscular soreness, arthritis, rheumatism, fibromyalgia and multiple sclerosis. Andrographis is also used as a treatment for malaria, as a replacement for quinine. The herb may be helpful in preventing the condition known as restenosis, or 'reclogging' of arteries following angioplasty. It helps to prevent blood clot formation and is used as a herbal treatment for atherosclerosis and heart attacks. *Andrographis* is believed to be useful in the treatment of liver and gallbladder conditions such as viral hepatitis, jaundice, drug-induced liver damage, liver congestion and sluggish liver. This plant is thought to enhance the immune function, such as the production of white blood cells, to support the release of interferon, and to promote the healthy activity of the lymphatic system. Some scientists believe the herb may help in the fight against HIV and AIDS.

Methods of propagation: Through seeds.

Photo: Prof Srinath Rao

ANETHUM GRAVEOLENS L. (DILL)

Synonyms: *Anethum arvense* Salisb.; *Angelica graveolens* (L.) Steud.; *Ferula graveolens* (L.) Spreng.

Classification:

Kingdom:	Plantae
Subkingdom:	Phanerogamea
Class:	Dicotyledonae
Subclass:	Polypetalae
Series:	Calyciflorae
Order:	Umbellales
Family:	Umbellifera (Apiaceae).
Genus:	*Anethum*
Species:	*Anethum graveolens*

Common names: English: Dill; **Gujarati:** Suva; **Hindi:** Soya; **Malayalam:** Chatakuppa; **Sanskrit:** Satakuppa; **Tamil:** Kattucata kuppai; **Telugu:** Sompa.

Distribution: Believed to be a native of Southwest Asia or Southeast Europe. Distributed in the Mediterranean, Southern USSR and Central Asia, throughout the Indian subcontinent, Malaysian archipelago and Japan.

Description: The plant grows up to 90 cm tall, with slender stems and alternate leaves, finally divided 3–4 times into pinnate sections slightly broader than the similar leaves of fennel. The flower develops into umbels, yellow in colour; the fruit is a schizocarp, oval, compressed, winged, about $\frac{1}{10}$ inch wide, with 3 longitudinal ridges on the back, 3 dark lines between them and 2 on the flat surface.

Flowering: April–July.

Fruiting: July–August.

Chemical constituents: This plant contains volatile oil, dillanoside, xanthone glucoside, coumarins, scopoletin, esculetin, bergapten, umbelliferone, umbelliprenine, kaempferol, 3-gluucuronide, vicenin flavonoids, petroselinic acid triglyceride, β-sitosterol glucoside, phenolic, caffeic, ferulic and chlorogenic acids, carvone, d-limonene, ax-phellandrene, dihydrocarvone, eugenol, β-phellandrene, ax-pinene, anethole, dillapiole, myristicin, carveol, β-caryophyllene, ax-phellandrene, limonene, terpinene, ax-pinene, dillapiole, myristicin, coumaran (http://doctorschar.com/dill-anethum-graveolens/).

Used in: Ayurveda and Unani medicine.

Parts used: Seeds.

Uses: The fruits are acrid, aromatic, antispasmodic, anthelmintic, anodyne, anti-inflammatory, bitter, cardiotonic, carminative, digestive, diuretic, emmenagogue, expectorant, febrifuge, stomachic, stimulant, galactagogue, antispasmodic and anti-dysenteric. This plant is useful in the treatment of flatulence, colic, dyspepsia, intestinal worms, digestive problems, insomnia, inflammation, hiccough, cough, asthma, bronchitis, fever, ulcers, skin diseases, haemorrhoids and cardiac debility. It acts as a stimulant for lactation (https://www.motherherbs.com/anethum-graveolens.html). *Anethum graveolens* is used in more than 56 ayurvedic preparations, which include Dasmoolarishtam, Dhanwanthararishtam, Mrithasanjeevani, Saraswatharishtam, Gugguluthiktaquatham, Maharasnadi kashayam and Dhanwantharam quatham.

Methods of propagation: Through seeds.

Photo: Prof Srinath Rao

ANISOMELES INDICA (L.) KUNTZE

Synonyms: *Anisomeles indica* (Linn.) Kuntze, *Nepeta indica* Linn., *Anisomeles ovata* R. Br.

Classification:

Kingdom:	Plantae
Subkingdom:	Phanerogamea
Class:	Dicotyledonae
Subclass:	Gamopetalae
Series:	Bicarpellatae
Order:	Lamiales
Family:	Labiatae (Lamiaceae)
Genus:	*Anisomeles*
Species:	*Anisomeles indica*

Common names: Bengali: Gobura, Gopali; **English:** English cat mint; **Hindi:** Kalabhangara; **Kannada:** Mangamari soppu; **Malayalam:** Chedayan; **Marathi:** Gopali; **Sanskrit: Tamil: Telugu:** Adabeera.

Distribution: It is found throughout India ascending to 2000 m in the Himalayas. Also found in China and Malaya.

Description: Erect, woody, annual or perennial, pubescent, aromatic herbs or under shrubs, 50–80 cm tall, softly hairy. Stem tetrangular, sparingly hairy or densely pubescent, sometimes almost woolly. Leaves ovate to ovate-lanceolate, 2.0–10.5 × 1.2–7.0 cm, acute at apex, truncate or cordate at base, crenate-serrate at margin, densely pubescent on both surfaces; petioles 1–7 cm long, densely hairy. Inflorescence: long peduncled cymes or spikes. Flower: pale to bluish purple in whorls, bracts linear, 4–5 mm long, pubescent. Calyx: bell-shaped, straight, lobes 5-toothed, 1.2 cm long, ciliate. Corolla: tube short, purple, bilabiate, about 2 cm long; upper lip ciliate, annulate within, upper lip erect. Androecium: stamens 4, exserted, filaments bearded, about 1.2 cm long, anther 2-celled. Gynoecium: ovary superior, bilocular, style bifid. Fruit: nutlets 4, ovoid, 0.1–0.25 cm long, angled brown, compressed. Seeds: 3–4.

Flowering: September–January.

Fruiting: September–January.

Chemical constituents: A macrocyclic diterpene – ovatodiolide, anisofolins B and C isolated from leaves; cerotic, pentacosanic, lignoceric, tricosanoic, behenic, heneicosanic, arachidic, stearic and palmitic acids isolated from roots; ovatodiolide, anisomelic acid, anisomelin, tetracosane, tetracosanol, β-amyrin, friedelin and betulonic acids isolated from roots.

Used in: Folk medicine.

Parts used: Leaves and roots.

Uses: The aerial parts of the plant are antipyretic, astringent, carminative, expectorant and diaphoretic. They can be employed in the treatment of muscle pain, fever and nervous complaints. The roots are also used as an antidote to poisonous bites. The plant extract is used as hair-growth stimulant. The leaves are considered useful in chronic rheumatism, psoriasis and other chronic skin eruptions, as well as in hypertension. The essential oil present in the herb is useful in uterine affections.

Methods of propagation: Through seeds.

Photo: Prof Srinath Rao

ANNONA SQUAMOSA (L.)

Synonyms: *Annona forskahlii* DC.; *Annona cinerea* Dunal; *Annona glabra* Forssk.; *Xylopia frutescens* Sieb. ex Presl.

Classification:

Kingdom:	Plantae
Subkingdom:	Phanerogamea
Class:	Dicotyledonae
Subclass:	Polypetalae
Series:	Thalamiflorae
Order:	Rannales
Family:	Annonaceae
Genus:	*Annona*
Species:	*Annona squamosa*

Common names: Bengali: Ata; **English:** Custardapple; **Hindi:** Seetaphal, Sharifa; **Kannada:** Seetaphal; **Marathi:** Seetaphal; **Sanskrit:** Agrimakhya, Gandha gatram, Krishna beeja; **Tamil:** Cintamaram, Sitapazham; **Telugu:** Sitaphalam.

Distribution: A native of the American continent and West Indies. Cultivated or growing wild in the temperate climate of India.

Description: This is a small–medium-sized tree growing to a height of 15–20 feet. The bark is pale brown in colour. The leaves are simple, alternate, thin, oblong, 10–15 cm long and 4–6 cm wide, rounded at the base and pointed at the tip. The flowers are solitary or in small clusters of 2–4 about

3 cm long and pale green to pale yellow in colour, with 1 inch long drooping stalks. The fruits are heart-shaped, pale green in colour, with white pulp and 30–40 black seeds.

Flowering: May–August.

Fruiting: August–December.

Chemical constituents: The leaves and root contain bullatacinone and squamone, as well as a cardiotonic alkaloid, quinoline. The roots contain anonian, glaucine, iscocorydine, norcorydine and alkaloids. The seeds contain annonacin A and annonastatin. Other compounds present in various parts are anotemoyin-2, cherimolin 1 and 2, motrilin, reticulatin 2 and squamostatin E and D.

Used in: Traditional and folk medicine.

Parts used: Fruit and root.

Uses: A paste made from the dried powder of the fruit is used to treat pus-filled ulcers. A leaf and root paste is applied over fungus-infected areas. The crushed leaves are used to treat hysteria. The leaf powder mixed with honey is taken orally to cure cough and dyspnea. The leaf extract has acaricidal and anti-inflammatory properties. A decoction made from the root is applied to the hair in order to kill head lice. The fruit pulp is a tonic for the cardiac muscle; it is sweet and is used in desserts, jams and syrups. A decoction made from the bark is used to control diabetes. The seeds have an abortifacient action.

Methods of propagation: Through seeds.

Photo: Prof Srinath Rao.

ANTHOCEPHALUS CADAMBA (ROXB.) MIQ.

Synonyms: *Anthocephalus morindifolius* Korth.; *Nauclea cadamba* Roxb.; *Nauclea megaphylla* S. Moore; *Neonauclea megaphylla* S. Moore; *Neolamarckia cadamba* (Roxb.) Bosser; *Samama cadamba* (Roxb.) Kuntze.

Classification:

Kingdom:	Plantae
Subkingdom:	Phanerogamea
Class:	Dicotyledonae
Subclass:	Gamopetalae
Series:	Inferae
Order:	Rubiales
Family:	Rubiaceae
Genus:	*Anthocephalus*
Species:	*Anthocephalus kadamba*

Common names: Bengali: Kadam; **English:** Bur flower, Kadamba; **Hindi:** Kadamb; **Kannada:** Kadam, Kaduavalatige; **Malayalam:** Katampu**; Marathi:** Kadamba, Nhiv; **Sanskrit:** Haripriya; **Tamil:** Katampam, Vellaikkatampu; **Telugu:** Kadambamu, Kadimi, Rudrakskamba.

Distribution: It is an evergreen tropical tree found in different parts of Australia, Bangladesh, Cambodia, Indonesia, Laos, Malaysia, Myanmar, Nepal, Philippines, Sri Lanka and India. Distributed throughout India.

Description: The tree is 40–60 feet tall, with no branches up to 20 feet. The crown is umbrella-shaped; branches arranged tiers; bark grey and longitudinally fissured in old trees. The leaves are simple, opposite, sub-sessile, large, 15–30 cm long, shining, elliptic, oblong. Flowers: yellowish orange, small, in dense globous heads; they appear like solid yellowish balls. Fruits: small capsules packed together to form a yellowish orange-coloured infructescence containing 600–800 seeds. The fruits split into 4 parts releasing the seeds at maturity.

Flowering: June–October.

Fruiting: June–October.

Chemical constituents: This plant contains saponins, indole and quinoline alkaloids, secoiridoids and triterpenes. The stem contains apigenin, β-sitosterol, genkwanin, narigenin, prunetin and sakuranetin. The heartwood contains aromadendrin, chrysin, dihydrotectochrysin, dihydrowogonin, kaempferol, naringenin, pinocembrin, quercetin and taxifolin. The root bark contains glucogenkwanin, prunetinoside and stigmasterol and ursolic acids.

Used in: Ayurveda.

Parts used: Leaves, stem bark.

Uses: The stem bark is considered anti-diuretic, anthelmintic, febrifugal, and hypoglycemic. The fruit is analgesic, anti-catarrhal, blood purifying and cooling. The flowers and roots are abortifacient. The leaves are astringent. A decoction is used to gargle in the treatment of stomatitis and aphthae. A decoction of the leaves is good for ulcers, wounds and metorrhea, and is useful in the treatment of snake bite.

Methods of propagation: Through seeds and stem cuttings. The seeds can be collected in August–September. The germination percentage is poor. A treatment with growth regulators will enhance seed germination.

APIUM GRAVEOLENS L.

Synonym: *Apium dulce* Mill.; *Apium integrilobum* Hayata; *Apium lobatum* Gilib.; *Apium graveolens f. lusitanicum* (Mill.) J. Helm; *Apium vulgare* Bubani, Fl. Pyren.; *Carum graveolens* L. Koso-poljansk.

Classification:

Kingdom:	Plantae
Subkingdom:	Phanerogamea
Class:	Dicotyledonae
Subclass:	Polypetalae
Series:	Calyciflorae
Order:	Umbellales
Family:	Umbellifera (Apiaceae)
Genus:	*Apium*
Species:	*Apium graveolens*

Common names: Bengali: Banyamini, Chiluri; **English:** Celery; **Gujarati:** Bodi ajamoda; **Hindi:** Ajmud, Badi ajamud; **Kannada:** Ajamoda, Guddada sompu, Oma; **Malayalam:** Ayamodakum, Omam; **Marathi:** Ajmoda, Oova; **Sanskrit:** Agani, Ajamoda, Kharashva, Kastumbari parnam,; **Tamil:** Callari; **Telugu:** Naranji vamu.

Distribution: Wide, extending from Sweden to Algeria, Egypt and Ethiopia. In Asia it is found from the Caucasus to Baluchistan and in the mountains of India. Found near ditches, by rivers and in other damp locations, especially near the sea in salt marshes.

Description: Celery is a biennial, which in the first year produces an upright rosette of leaves (40–60 cm in height) with closely appressed succulent leaf stalks. In the second year, it produces a tall flowering stem with terminal and axillary umbels of small, greenish white flowers. Leaves alternate, lower ones pinnate, long-petioled; upper ones trifid, ex-stipulate; a sheath is present; leaflets deltoid, 2–5 cm long and 1–3 cm broad. Flowers bisexual, pentamerous; petals free, ovate to orbicular, 0.5 mm long; apex inflexed. Stamens greenish white, free, alternating with petals. Ovary inferior, 2-celled, styles 2 and divergent. Fruit a schizocarp, ovoid to globous, splitting into 2 1-seeded parts up to 1.5 mm long and ribbed.

Flowering: June–August.

Fruiting: August– September.

Chemical constituents: Contains volatile oils (2–3 %) consisting of d-limonene, d-selinene α and β-eudesmol, santalol and a lactone sedanolide, alkaloids, coumarins, furanocoumarins such as apigravin, bergapten, celerin, coumarine glycosides and gravebioside A and B. Celery oil is dominated by terpenes, mostly limonene (70–80 %) and the sesquiterpenes β-selinene (10 %) and humulene. It also contains flavonoids such as apigenin, apiin, isoquercetin and furanocoumarins, as well as fatty acids such as linoleic, myristic, oleic, palmitic and steric acids.

Used in: Ayurveda, Siddha and Unani medicine.

Parts used: Leaves, fruits (seeds) and roots.

Uses: Wild celery has a long history of medicinal and food use. It is an aromatic bitter tonic herb that reduces blood pressure, relieves indigestion, stimulates the uterus and is anti-inflammatory. The ripe seeds, herb and root are aperient, carminative, diuretic, emmenagogue, galactagogue, nervine, stimulant and tonic. Wild celery is said to be useful in cases of hysteria, promoting restfulness and sleep and diffusing through the system a mild sustaining influence. The root is harvested in the autumn and can be used fresh or dried. The whole plant is harvested when fruiting and is usually liquidized to extract the juice. The seeds are harvested as they ripen and are dried for later use. An essential oil obtained from the plant has a calming effect on the central nervous system. Some of its constituents have antispasmodic, sedative and anticonvulsant properties. This herb has been shown to be of value in treating high blood pressure. A homeopathic remedy is made from it. It is used in treating rheumatism and kidney complaints. It is antibacterial, analgesic, antihepatotoxic, anti-fertility, anti-inflammatory and antihypercholesterolemic. The alkaloids from the seeds have a depressant and tranquillizing effect on the central nervous system (CNS), and are used to treat psychiatric and epilepsy-like disorders. The roots are used for treating paralysis and palsy. They are diuretic. Apigenin has vasodilatory properties. Fatty oils are useful as antispasmodic and helpful for curing urinary calculi.

Methods of propagation: Through seeds. Sow in spring or autumn.

ARGEMONE MEXICANA (L.)

Classification:

Kingdom:	Plantae
Subkingdom:	Phanerogamea
Class:	Dicotyledonae
Subclass:	Polypetalae
Series:	Thalamiflorae
Order:	Parietales
Family:	Papaveraceae
Genus:	*Argemone*
Species:	*Argemone mexicana*

Common names: Bengali: Siyal kanta; **English:** Mexican poppy, Prickly popp; **Gujarati:** Darudi; **Hindi:** Satyanashi; **Kannada:** Datturigida; **Malayalam:** Ponnummattu; **Marathi:** Fringi dhotra; **Sanskrit:** Swaranakishirini; **Tamil:** Kudiyotti; **Telugu:** Bramhadandi.

Distribution: Native to America; naturalized throughout India up to an altitude of 1500 m.

A genus of annual, biennial or rarely perennial prickly herbs. Some species are cultivated for their showy flowers. Three species are recorded from India.

Description: An erect, prickly annual herb, up to 1.2 m in height. Leaves sessile, semi-amplexicaule, sinuately pinnatifid, spiny on margins, mid-rid and veins beneath; flowers prickly, rarely unarmed; seeds small, round, blackish brown, deeply reticulate-scrobiculate.

Flowering: April–May.

Fruiting: May–October.

Chemical constituents: Whole plant: berberine, protopine, sanguinarine, chelerytherine, pancorine, (+)-argenaxine, angoline, aronttianamide, dihydrocheilantifoline, allocryptopine, coptisine, jatrorrhizine, columbamine, oxyberberine, N-demethyloxysanguinarine (Chang et al., 2003). Seed oil: myristic, palmitic, oleic, linoleic acids. Yellow juice: berberine. Leaves: mexicanol, mexicanic acid. Seeds: dihydropalmitine hydroxide; berberine, protopine, ferulic acid, tannic acid, caffeic acid, benzoic acid, cinnamic acid (Rajvaidhya et al., 2012). Leaves: protomexicine, mexitin, 8-methoxydihydrosanguinarine, 13-oxoprotopine, rutin, quercitrin, eriodictyol (Koumari et al., 2013).

Parts used: Leaves, seeds, flowers, mid-rib.

Uses: Anti-insect, antifeedant, antifungal and antibacterial properties. It is also used in the treatment of cough, phlegm in the throat, dysentery, rheumatism. In America, the latex of the plant was used for the treatment of syphilis and skin diseases. The seeds are expectorant and sedative; the seed oil is purgative. The root is diuretic, alterative, anodyne, hypnotic and used in chronic skin diseases.

Photo: Prof Srinath Rao

ARGYREIA NERVOSA (BURM. F.) BOJER

Synonyms: *Argyreia speciosa* Sweet; *Convolvulus speciosa* L. f.; *Convolvulus nervosa* Burm. f.; *Rivea nervosa* (Burm. f.) Hallier f.

Classification:

Kingdom:	Plantae
Subkingdom:	Phanerogamea
Class:	Dicotyledonae
Subclass:	Gamopetalae
Series:	Bicarpellatae
Order:	Polymoniales
Family:	Convolvulaceae
Genus:	*Argyreia*
Species:	*Argyreia nervosa*

Common names: Bengali: Bichtaraka, goguli; **English:** Elephant creeper, Hookers woodrose; **Hindi:** Adhogoda, Vridhadaraka, Vidhara; **Kannada:** Samudravalli; **Malayalam:** Samudrappacha; **Marathi:** Gayri; **Sanskrit:** Samudraphalaka, Murva; **Tamil:** Katrapalai; **Telugu:** Chandrapoda.

Distribution: Its origin is India. It is found throughout the country in tropical and sub-tropical climates. It is also distributed in Africa, the Caribbean Islands and Hawaii.

Description: It is a perennial climbing vine up to 10 m long, found in the moist deciduous forests of the Western Ghats. It has grooved, smooth stems. Broad, ovate leaves with a heart-shaped base, 15–20 cm, with long stalks of 10–12 cm. Flowers trumpet-shaped, 5–7 cm, pink in colour, arranged in a few flowered cymes on long stalks of 15–20 cm.

Flowering: December–January.

Fruiting: January.

Chemical constituents: Alkaloids, β-sitosterol, ergometrine, ergometrinine, lysergic acid-α-hydroxy ethyl amide and iso-lysergic acid-α-hydroxy ethyl amide, agroclavine, chanoclavine I and II.

Used in: Ayurveda and Unani medicine.

Parts used: Leaves, bark of the stem, roots and seeds.

Uses: It is highly effective as an antioxidant, adaptogenic, antifungal, antiphlogistic, antidiabetic, antiaging, anti-arthritic, anti-viral, emollient and tonic. A poultice of the whole plant is applied to cure skin infections; the plant extract taken orally (10 ml) cures diabetes; a seed decoction prepared in water enhances male sexual activity. The root is acrid, bitter, alterative, antigonorrhoeic, anti-inflammatory, antirheumatic, aperient, aphrodisiac, digestive, diuretic, nervine, purgative and intellect-promoting. The root is also useful in anorexia, ascites, colic, chronic ulcer, dyspepsia, hemiplegia, neuralgic pains and general debility. This plant is very effective against gangrene: the affected area is washed with juice made from the leaves. This juice is also used in treating bed sores. Moreover, the leaves are used to treat tumors: the upper surface of the leaves is warmed, simmered with butter and tied locally on the tumor. The tumor (cyst) will burst and healing will start. If the leaves are tied on its lower surface, the tumor will be suppressed inside the body. The root powder is used to treat dog bite.

Methods of propagation: Through seeds, soft wood cuttings.

Photo: Prof Srinath Rao

ARTEMISIA VULGARIS L.

Synonyms: *Artemisia selengensis* Turcz. ex Besser; *Artemisia tilesii* Ledeb. var. aleutica (Hultén) S. L. Welsh; *Artemisia unalaskensis* Rydb.

Classification:

Kingdom:	Plantae
Subkingdom:	Phanerogamea
Class:	Dicotyledonae
Subclass:	Gamopetalae
Series:	Inferae
Order:	Asterales
Family:	Compositae (Asteraceae)
Genus:	*Artemisia*
Species:	*Artemisia vulgaris*

Common names: Bengali: Nagdoni; **English:** Indian wormwood, Mugwort; **Gujarati:** Damoro, Nagadavano **Hindi:** Dona, Majpatri, Mastaru, Nagdon, **Kannada**: Panchapatre; **Malayalam:** Makkippoovu, Tiru-nitripachcha; **Marathi:** Davana, Surband, Surpan; **Sanskrit:** Barhikusum, Chauhar, Dhavana, Granthika; **Tamil:** Machipattari; **Telugu:** Davanamu, Machipatri.

Distribution: *Artemisia vulgaris* is native to Asia and North America and is found throughout the hilly regions of India up to 3600 m in the Western Himalaya, Sikkim and Khasi hills.

Description: It is a perennial, shrubby, aromatic plant. The plant is long-stemmed, about 70–150 cm in height. The shoots are slightly, pubescent, often red-tinged and have an unpleasant smell. The rhizome is creeping, branched and many-headed. Leaves 5–10 cm long, ashy and tomentose on the lower side, lobes toothed. The inflorescence is a heavily branched panicle with numerous bracts; the bracts are whitish with a green midrib, lanceolate/acuminate. The flower is numerous, short-stemmed, erect or slightly drooping. The flower heads are ovoid, 3–4 mm long.

Flowering: July–September.

Fruiting: December.

Chemical constituents: Camphor, linalool, vulgarin, pilostachyin, pilostachyin-c, lipophilic flavonoids, polyenes and hydroxycoumarins.

Used in: Ayurveda, folk medicine.

Parts used: Leaves, roots, flowering tops; the whole plant.

Uses: The plant is anthelmintic, antidepressant, antiseptic, antispasmodic, anodyne, carminative, emmenagogue and vermifuge. It is also used as appetizer, diaphoretic, diuretic, expectorant, purgative and as a nervine tonic. All parts of mugwort stimulate the secretion of gastric juices, thereby improving digestion. It also has a sedative, antidepressant and calming action on the mind. Mugwort roots are used as general tonic and to boost physical and mental energy, as well as to treat anxiety, hypochondria, insomnia, irritability neurasthenia and psychoneurosis.

Methods of cultivation: Through seeds.

ASCLEPIAS CURASSAVICA (L.)

Synonyms: *Asclepias aurantiaca* Salisb.; *Asclepias bicolor* Moench; *Asclepias nivea* var. curassavica.

Classification:

Kingdom:	Plantae
Subkingdom:	Phanerogamea
Class:	Dicotyledonae
Subclass:	Gamopetalae
Series:	Bicarpellatae
Order:	Gentianales
Family:	Apocynaceae
Genus:	*Asclepias*
Species:	*Asclepias curassavica*

Common names: English: Scarlet milkweed, Swallow-wort; **Hindi:** Kakatundi; **Kannada**: Chikka kaage thonde; **Malayalam:** Chemullichedi; **Sanskrit:** Kakanasika; **Tamil:** Ariyaman; **Telugu:** Jilledu mandaara.

Distribution: It is native to South America but has become a naturalized weed in tropical and subtropical pastures, fields and disturbed areas throughout the world, including central and southern Florida.

Description: It is an evergreen perennial shrub that grows up to 1 m tall and has pale gray stems. The leaves are arranged oppositely on the stems and are lanceolate-shaped, ending in acuminate or acute tips. Like in other members of the genus, the sap is milky. The flowers are in cymes with 10–20 flowers each. They have purple or red corollas and corona lobes that are yellow or orange. Flowering occurs nearly year round. Its blossoms are red and orange, less than 1 inch across, and appear in clusters at the top of 2–4 feet stalks. The 5–10 cm-long, fusiform-shaped fruits are called follicles.

Flowering: Throughout the year.

Fruiting: Throughout the year.

Chemical constituents: The plant contains tannin, phenol, terpenoids and xanthoprotein, among others. It also contains a glucoside, asclepiadin. The physiological action of unaltered asclepiadin was reported to resemble that of emetine. The roots are reported to contain vincetoxin, also resembling emetine in its action. The root contains an active principle, asclepiadin and curassavine, identical in therapeutical value to digitalin. The plant also contains cardenolide glycosides, α-asclepiadin and β-asclepiadin; β-sitosterol and seed oil (53 % linoleic acid, 1 % linolenic acid). The latex contains cardenolides and esterified triterpenes (Kalidas et al., 2009; Warashina and Noro, 2008).

Used in: Ayurveda, Siddha and traditional Chinese medicine.

Parts used: Roots, flowers, stems, leaves and latex.

Uses: As a remedy for cancers, warts and similar growths. The extract from the root is used in traditional medicine as an emetic and laxative. Other uses are against warts, fever, vomiting and as an expectorant. The root is febrifuge. A decoction is used in the treatment of dysentery and as an eyewash for infected eyes. The roots are also used as an emetic and purgative. A paste made from the crushed leaves, combined with salt, vegetable oil and bread, is used for treating skin ulcers. The white latex is applied to warts and corns in order to kill them.

Methods of cultivation: Scarlet milkweed can be started from cuttings, and it grows quickly from seed.

Photo: Prof Srinath Rao

ASPARAGUS RECEMOSUS WILLD.

Synonym: *Asparagus rigidulus* Nakai; *Protasparagus racemosus* (Willd.) Oberm.

Classification:

Kingdom:	Plantae
Subkingdom:	Phanerogamea
Class:	Monocotyledonae
Series:	Coronarieae
Family:	Liliaceae
Genus:	*Asparagus*
Species:	*Asparagus recemosus*

Common names: English: Wild asparagus; **Hindi:** Chatwal, Satawar; **Kannada:** Aheru balli; **Malayalam:** Catavali; **Sanskrit:** Satamuli, Satavari; **Tamil:** Aikkuncu, Ammaikodi; **Telugu:** Challa gadda, pilligadalu.

Distribution: This species is globally distributed. Within India, it is found growing wild in tropical and sub-tropical parts, including the Andaman Islands; and ascending in the Himalayas up to an altitude of 1500 m.

Description: This plant is a spiny and climbing shrub with branched, woody stems, whitish grey or brown in colour. The spines of this plant are usually 0.5–1.3 cm long. Flowers: white in colour. These have a sweet fragrance and appear in a bunch; they are reduced to minute chaffy scales and spines. They are in simple or branched racemes in the axils of thorns or on the nodes of the main shoots. The fruits or berries of this medicinal plant are globose in structure or obscurely 3-lobed, purplish black when ripe. The diameter of this fruit varies from 0.4–0.6 cm in diameter. Rootstocks are short and tuberous and bear numerous succulent tuberous roots, 30–100 cm long.

Flowering: December–January.

Fruiting: December–January.

Chemical constituents: The tubers contain saponins named shatavarins I–IV, glycoside, tridecaacetyl shatavarin-I. Also present are sarasapogenin and β-sitosterol. The other components of the tubers are carbohydrates, mucilage, polyfructosans, free sugars and insoluble polysaccharides. Alcoholic extracts yield asparagine, an anti-cancer drug. The leaves contain rutin, diosgenin and quercetin. The flowers contain quercetin, hyprosid and rutin.

Used in: Ayurveda, folk, Tibetan, Unani and Siddha medicine.

Part used: Tuberous roots and aerial parts.

Uses: The root of this plant is used in the form of juice, paste, decoction and powder to treat intrinsic haemorrhage, diarrhea, piles, hoarseness of voice, cough, arthritis, poisoning, diseases of the female genital tract, erysipelas and fever; as an aphrodisiac and as a rejuvenating product. It is also used in the treatment of kidney and liver disorders, gleet and gonorrhea. Ghee cooked with paste and decoction of satavari and mixed with sugar is used to rejuvenate. Satavari pounded and taken with milk increases the flow of breast milk. It is useful in vitiated conditions of *vata* and *pitta*, burning sensations, hyperdipsia and cardiac and general debility.

Methods of propagation:

This species is propagated either through seeds or vegetatively by tillers. It grows on sandy loam well-drained soil. Seeds are sown in June–July. As a pre-treatment, the seeds are soaked in water for 2 days or treated with gibberellic acid to obtain higher and early germination. Germination takes place in 15–20 days and about 70–80 % of seeds germinate. The seedlings are then transferred to a field.

Photo: Vedavyas Kulkarni

ASCLEPIAS CURASSAVICA (L.)

Synonyms: *Asclepias aurantiaca* Salisb.; *Asclepias bicolor* Moench; *Asclepias nivea* var. curassavica.

Classification:

Kingdom:	Plantae
Subkingdom:	Phanerogamea
Class:	Dicotyledonae
Subclass:	Gamopetalae
Series:	Bicarpellatae
Order:	Gentianales
Family:	Apocynaceae
Genus:	*Asclepias*
Species:	*Curassavica*

Common names: English: Scarlet milkweed, Swallow-wort; **Hindi:** Kakatundi; **Kannada:** Chikka kaage thonde; **Malayalam:** Chemullichedi; **Sanskrit:** Kakanasika; **Tamil:** Ariyaman; **Telugu:** Jilledu mandaara.

Distribution: It is native to South America but has become a naturalized weed in tropical and subtropical pastures, fields and disturbed throughout the world, including central and southern Florida.

Description: It is an evergreen perennial shrub that grows up to 1 m tall and has pale gray stems. The leaves are arranged oppositely on the stems and are lanceolate-shaped, ending in acuminate or acute tips. Like in other members of the genus, the sap is milky. The flowers are in cymes with 10–20 flowers each. They have purple or red corollas and corona lobes that are yellow or orange. Flowering occurs nearly year round. Its blossoms are red and orange, less than 1 inch across, and appear in clusters at the top of 2–4 feet stalks. The 5–10 cm-long, fusiform-shaped fruits are called follicles.

Flowering: Throughout the year.

Fruiting: Throughout the year.

Chemical constituents: The plant contains tannin, phenol, terpenoids and xanthoprotein, among others, as well as a glucoside, asclepiadin. The physiological action of unaltered asclepiadin was reported to resemble that of emetin. The roots are reported to contain vincetoxin, resembling emetine in its action, as well as an active principle, asclepiadin and curassavine, identical in therapeutical value to digitalin. The plant also contains a cardenolide glycosides, α-asclepiadin and β-asclepiadin, β-sitosterol and seed oil (53 % linoleic acid, 1 % linolenic acid). The latex contains cardenolides and esterified triterpenes.

Used in: Ayurveda, Siddha, traditional and Chinese medicine.

Parts used: Roots, flowers, stems, leaves and latex.

Uses: Used as a remedy for cancers, warts and similar growths. The extract made from the root is used in traditional medicine as an emetic and laxative. Other uses are against warts, fever, vomiting and as an expectorant. The root is febrifuge. A decoction is used in the treatment of dysentery and as an eyewash for infected eyes. The roots are also used as an emetic and purgative. A paste made from the crushed leaves, combined with salt, vegetable oil and bread, is used for treating skin ulcers. The white latex is applied to warts and corns in order to kill them.

Methods of cultivation: Scarlet milkweed can be started from cuttings, and it grows quickly from seed.

Photo: Prof Srinath Rao

ASTERACANTHA LONGIFOLIA NEES.

Synonyms: *Barleria longifolia*; *Hygrophila spinosa* T. Anders; *Hygrophila auriculata* (Schumach.) Heine.

Classification:

Kingdom:	Plantae
Subkingdom:	Phanerogamea
Class:	Dicotyledonae
Subclass:	Gamopetalae
Series:	Bicarpellatae
Order:	Personiales
Family:	Acanthaceae
Genus:	*Asteracantha*
Species:	*Asteracantha latifolia*

Common names: **Bengali:** Kandekshu, Kantakalika; **English:** Asteracantha; **Hindi:** Gokhulajanum; **Malayalam:** Bhelshulli, Vajrakantaka; **Marathi:** Kolshinda; **Sanskrit:** Vajrakantaka; **Tamil:** Mirmalli, Neermulli; **Telugu:** Kokilaksha.

Distribution: Throughout India and Sri Lanka, found in hedges, margins of tanks, plains, ponds, rice fields, road sides, swampy places, waste land and wet places.

Description: Stout thorny herb, stem quadrangular or slightly circular, hairy, hispidulous below nodes. Leaves opposite, simple, sessile, 3 pairs of each node, lamellate, entire, acute, hispid, up to 4–11 × 0.5–1.4 cm; thorns brownish yellow, straight, 3.5 cm long. Four pairs of purplish-blue two lipped flowers whorts of 8 flowers each; bracts 2.5 cm long, hispid, yellowish, hairy, lanceolate, ciliate, bracteolate, smaller than bracts,1.8 cm long. Calyx 4, partite, larger, up to 1.5 cm long, linear-lanceolate, hairy on the back, ciliate. Corolla purple, blipped, tube 1.3 cm long; lobes oblong, stamens 4, didynamous. Other cells parallel, glabrous, appendaged.

Flowering: February–March.

Fruiting: August–September.

Chemical constituents: The seeds contain 23 % of yellow, semidrying oil. They also contain diastase, lipase, protease, alkaloids, glucosides, steroids and flavonoids. The seeds contain sitosterol I, II, III and IV, asteracanthin, asteracanthicin, large amounts of mucilage and potassium salts. The whole plant contains palmitic, stearic, oleic and linoleic acids.

Used in: Ayurveda and Unani medicine.

Parts used: Roots, leaves, whole plant.

Methods of propagation: Through cuttings.

Photo: Prof Srinath Rao

AVERRHOA CARAMBOLA L.

Synonyms: *Averrhoa acutangula* Stokes; *Averrhoa pentandra* Blanco; *Connaropsis philippica* Villar.

Classification:

Kingdom:	Plantae
Subkingdom:	Phanerogamea
Class:	Dicotyledons
Subclass:	Polypetalae
Series:	Disciflorae
Order:	Geraniales
Family:	Oxalidaceae
Genus:	*Averrhoa*
Species:	*Averrhoa carambola*

Common names: Bengali: Kamarana; **English:** Karambole, Gooseberry, Star fruit; **Gujarati:** Kamarakh; **Hindi:** Dantasatha, Kamaraka; **Kannada:** Kamaraakshi; **Malayalam:** Aarampuli: **Marathi:** Kamarakha, Karamala; **Sanskrit:** Kamarakha**,** Karamaranga; **Tamil:** Kantacatkam, Tamarattai; **Telugu:** Ambanapukaya, Karamalangaum.

Distribution: This plant is believed to have originated in Ceylon and the Moluccas, but it has been cultivated in Southeast Asia and Malaysia for hundreds of years. It is cultivated in Australia, Brazil, China, Florida, Israel, Taiwan and Thailand. In India it is commonly grown in warmer parts.

Description: It is a small tree 15–20 feet in height, spreading to 20–25 feet in diameter. It is branched, with a bushy shape. Leaves: compound, alternate, spirally arranged, ovate/oblong-ovate, 15–20 cm long, soft, medium, green, smooth on the upper surface, whitish below. Flowers: purple in colour, about 6 mm wide, pedicellate, petals 5 with curved ends produced in the axil of the leaves, arranged in clusters, attached to the tree with a red stalk. Sepals 5 in number. The fruits are fleshy, oblong, longitudinally 5–6 angled, 10–15 cm long and up to 9 cm wide. They are green in colour when unripe, but turn to yellow when ripe. The fruits are crunchy and when cut in cross section star-shaped. They are edible and their taste vary from sour to mildly sweetish. Seeds: up to 12, thin, about 5 mm long, brown in colour, enclosed in a gelatinous aril.

Flowering: In India, flowering was observed in three flushes: April–May, July–August, September–October.

Fruiting: December–January.

Chemical constituents: The fruits contain saponins, alkaloids, flavonoids and tannins. The main phytochemicals from the fruits are epicatechin, gallic acid and L-ascorbic acid; the major steroids present are β-sitosterol, campesterol, lupeol and isofucosterol. Fatty acids such as palmitic, oleic, linoleic and linolenic acids are also present. Other components, such as quercetin-3-0-β-D-glycoside, β-amirin and rutin, have also been identified.

Used in: Ayurveda and Unani medicine.

Parts used: Leaves, fruits, flowers, stems and root.

Uses: The leaves and shoots are applied externally to treat ringworm infections, scabies and chickenpox. The flower is vermicidal. The fruit syrup is an excellent antiscorbutic; it is also used to treat bilious fever and cracks in the feet. The fruit is laxative, anti-dysenteric, antiphlogistic, febrifuge, anti-inflammatory, antispasmodic (used in hepatic colic, bleeding piles), hypoglycemic. The seeds are galactogenic and in large doses act as emmenagogue and cause abortion. The fruit and leaves are appetite-stimulant, antidiarrheal, febrifugal, and used to pacify impaired *kapha*, *pitta*, pruritis, skin diseases, hemorrhoids and general debility (Chau et al., 2004; Moresco et al., 2012). The stem extract has shown activity against brain tumor cells, while the extract from the leaves was effective against liver carcinoma cells (Tadros and Sleem, 2004).

Methods of propagation: Through seeds

Photo: C.T. Chandralekha

AZADIRACHTA INDICA A. JUSS.

Synonyms: *Antelae azadirachta* (L.) Adelb.; *Antelae javanica* Gaertn.

Classification:

Kingdom:	Plantae
Subkingdom:	Phanerogamea
Class:	Dicotyledonae
Subclass:	Polypetalae
Series:	Disciflorae
Order:	Geraniales
Family:	Meliaceae
Genus:	*Azadirachta*
Species:	*Azadiractha indica*

Common names: Bengali: Nim; **English:** Indian lilac, Margosa tree, Neem tree; **Gujarati:** Limbado; **Hindi:** Nim, Nimb; **Kannada:** Bevinamara; **Malayalam:** Veppa; **Marathi:** Limba; **Sanskrit:** Arishta, Nimba; **Tamil:** Vembu, Veppam; **Telugu:** Veepachettu, Yapachettu.

Distribution: Native to India, Pakistan and Bangladesh, growing in tropical and semi-tropical regions.

Description: A large, evergreen tree, 12–18 m in height and 1.8–2.4 m in girth, with a straight bole and long spreading branches forming a broad crown, commonly found throughout the greater part of India, and often cultivated. Bark grey or dark grey, rough, reddish brown inside, with numerous oblique furrows and scattered tubercles; leaves imparipinnate, alternate, 20–38 cm long; leaflets 8–19, alternate or opposite, ovate-lanceolate, oblique or sub-falcate, falcate-lanceolate, glossy, bluntly serrate; flowers white or pale-yellow, small, scented, numerous, in long, slender, very lax, axillary panicles; drupes green, turning yellow on ripening, aromatic, oblong or ovoid-oblong, smooth, 1.3–1.8 cm long, with a single ex albuminous seed.

Flowering: February–June.

Fruiting: February–June.

Chemical constituents: Three bitter compounds from neem oil, named nimbin, nimbinin and nimbidin. Azadirachtin, meliacin, gedunin, salanin and vasallin are the other compounds present.

Used in: Ayurvedic and Unani medicine.

Parts used: Seed oil and leaves.

Uses: Products made from neem trees have been used in India for over 2 millennia for their medicinal properties. Many benefits of neem are listed in ancient documents such as *Charak-Samhita* and *Susruta-Samhita*. Neem products are believed by Ayurvedic practitioners to be anthelmintic, antifungal, anti-diabetic, antibacterial, anti-viral, contraceptive and sedative. Neem is considered a major component in Ayurvedic and Unani medicine and is particularly prescribed for skin diseases. Neem oil is also used for healthy hair; to improve liver function, detoxify the blood and balance blood sugar levels. Neem leaves have also been used to treat skin diseases such as eczema and psoriasis.

Methods of propagation: Through seeds.

Photo: Prof Srinath Rao

AZIMA TETRACANTHA LAM.

Synonyms: *Azima spinosissima* Engl.

Classification:

Kingdom:	Plantae
Subkingdom:	Phanerogamea
Class:	Dicotyledonae
Subclass:	Gamopetalae
Series:	Bicarpellatae
Order:	Gentianales
Family:	Salvadoraceae
Genus:	*Azima*
Species:	*Azima tetracantha*

Common names: Bengali: Trikanta jati; **English:** Bee sting bush, Mistletoe; **Hindi:** Kanta-gurkamai; **Kannada:** Ganji mullu; **Malayalam:** Sankukuppi; **Marathi:** Sukkapat; **Sanskrit**: Kundali; **Tamil:** Mulsangu, Sungam-chedi; **Telugu:** Tella-upi.

Distribution: South Africa to India; hot, dry riverine scrub in most parts of south India, Ceylon, Philippines and Burma. It occurs naturally in Central, Eastern and Southern Africa. Found in bush, scrub and forest, along rivers and the coast, up to 1100 m altitude.

Description: Spinous shrub belonging to the family Salvadoraceae. The presence of quadrangular spines at the nodes is considered to be a unique character. Dioecious, erect shrub up to 90 cm tall with 1–2 spines 0.5–5 cm long in each leaf axil, sometimes scandent with stems up to 8 m long; branchlets terete or quadrangular, glabrous to densely hairy. Leaves decussately opposite, simple and entire; stipules absent or rudimentary; petiole short; blade elliptical-oblong to ovate-oblong or orbicular, 1.5–5.5 cm × 0.5–4.5 cm, base rounded or somewhat narrowed, apex mucronate, pinnately veined with one pair of lateral veins from near the base. Inflorescence an axillary, sometimes terminal spike or cyme up to 3 cm long or flowers solitary; bracts ovate, often with long and spinous mucro. Flowers unisexual, regular, 4-merous, usually sessile; calyx campanulate, 2–4 mm long, with triangular lobes; petals linear-oblong to oblong, greenish to yellowish, the upper part reflexed over the calyx, 2–5 mm long; male flowers with stamens inserted at the base of the rudimentary ovary, exserted; female flowers with staminodes and superior ovary, up to 4.5 mm long with a broad sessile stigma. Fruit a globose berry, 0.5–1 cm in diameter, 1–2-seeded, green turning white, with persistent stigma. Seeds disk-like, brown to black.

Flowering: August–September.

Fruiting: August–September.

Chemical constituents: The dimeric piperidine alkaloids azimine, azcarpine and carpaine have been isolated from all plant parts. Terpenoids are present in the roots and the leaves, while the seeds contain a complex mixture of about 25 flavonoids, predominantly glycosides and acyl-glycosides, the most important being quercetin, isorhamnetin, rhamnetin and rhamnazin. All parts contain glucosinolates. The seed and roots of *Azima tetracantha* contain high concentrations of N-methoxy-3-indolylmethyl-glucosinolate, whereas the stem and young leaves contain lower concentrations. The seed oil contains the fatty acids myristic acid 0.2 %, palmitic acid 5 %, stearic acid 15 %, arachidic acid 7 %, behenic acid 2 %, oleic acid 32 %, linoleic acid 18 % and eicosenoic acid 21 %, indicating that the oil could be suitable for use as an edible oil.

Used in: Folk medicine.

Parts used: The whole plant.

Uses: It has been reported for its antimicrobial, analgesic, anti-inflammatory and wound-healing properties. The roots and leaves were used as stimulant and tonic. It is considered diuretic and is also used to treat dropsy, dyspepsia and chronic diarrhea. In Western India, the juice of the leaves is applied as eardrops against earache and the crushed leaves are placed on sore teeth. The leaves are also used as an appetizer and against colds (Gowthami et al., 2012).

Methods of cultivation: Through seeds. Multiplication through cuttings is possible.

Photo: Prof Srinath Rao

BACOPA MONNIERI (L.) WETT.

Synonym: *Herpestis monniera* (L.) H. B. and K.; *Monniera cuneifolia* Michx.

Classification:

Kingdom:	Plantae
Subkingdom:	Phanerogamea
Class:	Dicotyledonae
Subclass:	Gamopetalae
Series:	Inferae
Order:	Personales
Family:	Scrophulariaceae
Genus:	*Bacopa*
Species:	*Bacopa monnieri*

Common names: English: Indian pennywrot; **Hindi:** Brahmi, Safed chamni; **Kannada:** Brahmi; **Malayalam:** Barna, Nirbhrahmi; **Marathi:** Ghola; **Sanskrit:** Brahmi, Sarasvati; **Tamil:** Sambra ne elaye; **Telugu:** Sambra ne akhlu.

Distribution: This herb is found in moist or wet places, such as on borders of water channels, wells, irrigated fields, etc., in all parts of India.

Description: This herb spreads on the ground and its stems and small leaves are succulent, i.e. fleshy. Roots also arise on the nodes of the stem. Leaves: simple, opposite, obovate, small, sessile. Flowers arise in the axils of the leaves and are borne on short pedicels, bluish white. Calyx: one of the 5 sepals is larger than the others. Corolla: bluish white in colour and about 1 cm across. Fruit: ovoid, acute, 2-celled, 2-valved capsules, tipped with style base, seeds minute, numerous.

Flowering: January–May.

Fruiting: January–May.

Chemical constituents: The entire plant contains alkaloids and saponins. The alkaloids found are brahmine, herpestine and nicotine. The saponins present are monnierin, hersaponin and bacoside A, A_3 and B.

Used in: Ayurvedic, Siddha, Unani and traditional medicine; homeopathy.

Parts used: The whole plant.

Uses: The drug Brahmi consists of the whole plant. Famed in Ayurvedic medicine, Brahmi has antioxidant properties. It has been reported to reduce the oxidation of fats in the blood stream, which is a risk factor for cardiovascular diseases. It has been used for centuries to help cure epilepsy, improve memory capacity, increase concentration, and reduce stress-induced anxiety. It is listed as a nootropic, a drug that enhances cognitive ability. It is bitter, pungent, heating, emetic, laxative and useful in bad ulcers, tumors, ascites, enlargement of the spleen, indigestion, inflammations, leprosy, anaemia, biliousness, etc. It is also good in scabies, leucoderma, syphilis, etc. It is a blood purifier and useful in diarrhea and fevers.

Methods of propagation:

The plant is propagated by stem cuttings. The whole plant is cut into small units with 4–5 internodes and planted in the field.

Photo: Prof Srinath Rao

BALANITES AEGYPTIACA (L.) DEL.

Synonyms: *Ximenia aegyptiaca* L.; *Balanites roxburghii* Planch.; *Agialida senegalensis* van.

Kingdom:	Plantae
Subkingdom:	Phanerogamea
Class:	Dicotyledonae
Subclass:	Polypetalae
Series:	Disciflorae
Order:	Geraniales
Family:	Zygophyllaceae
Genus:	*Balanites*
Species:	*Balanites aegyptiaca*

Common names: English: Desert date; **Hindi:** Hingot, Hingan; **Kannada:** Ingalarade; **Malayalam:** Namunta; **Marathi:** Hingam; **Sanskrit:** Tapasadruma, Angaravrksa; **Tamil:** Nanchundan; **Telugu:** Gara-chettu.

Distribution: Found in dry land areas of Africa and South Asia. In India, it is found in Rajasthan, Gujarat, Madhya Pradesh and Deccan.

Description: It is a thorny tree growing up to 30 feet tall. The branches are thorny. Leaves: compound dark green. Leaflets: 2, leathery, variable in size and shape, entire, generally up to 6 cm long, 4 cm broad. Flowers: small, inconspicuous, hermaphroditic, stalkless and borne in few-flowered fascicles. Flower buds: ovoid and velvety. Individual flowers are 8–14 mm in diameter and generally greenish yellow. Flower stalks: densely grayish, velvety, less than 1 cm in length. Sepals: 5 in number (free), ovate, 3 mm long. Petals: 5 in number (2 free), oblong-obovate, longer than the sepals. Stamens: 10 in number, filaments glabrous, anthers dorsifixed. Ovary: ovoid, silky, 5-celled; ovules solitary in each cell, their style short and conical. Fruit: ellipsoid, up to 4 cm long. Young fruits are green and tormentose, turning yellow and glabrous when mature, with a brittle coat enclosing a brown sticky pulp and a hard-stone seed (Anon, 1986, Bukar et al., 2004; Hamid et al., 2001; Ojo et al., 2006).

Flowering: January–February.

Fruiting: March.

Chemical constituents: Mesocarp of fruit contains 3-rutinoside and 3-rhamnogalactoside, diosgenin, balanitoside, 6-methyldiosgenin, balanitin-3, balanitin-6 and -7 and diosgenyl saponins.

Used in: Folk medicine.

Parts used: Fruits, seed oil.

Uses: It is traditionally used in treatment of various ailments, i.e. jaundice, intestinal worm infections, wounds, malaria, syphilis, epilepsy, dysentery, constipation, diarrhea, haemorrhoids, stomachaches, asthma and fever. The latex of the plant is used in epilepsy. The bark of the plant is useful in curing mental diseases, epilepsy, yellow fever, jaundice and syphilis and can also act as a fumigant to heal circumcision wounds. The boiled root of the plant can be used as a soup against stomach pain, anthrax, and the infusion of root also acts as an antidote to snake bite. The infusion of root bark has been used in diarrhea and haemorrhoids. The paste of the shoots has been used for dressing wounds and as tooth brushes. The thorns are used in the treatment of leprosy. The leaves are used in curing anthrax, for their anthelmintic properties and to clean malignant wounds. The fruit can cure mouth ulcer, whooping cough, sleeping sickness and skin diseases. The fruit kernel acts as a mild laxative and an antidote to arrow poison, and also as a vermifuge. The kernel oil helps in curing skin diseases. The seeds are useful as ointments and to cure cough and colic pain (Anon, 1986; Hamid et al., 221; Bukar et al., 2004; Ojo, 2006).

Methods of propagation: Through seeds.

Photo: Prof Srinath Rao

BASELLA ALBA (L.)

Synonym: *Basella rubra* L.

Classification:

Kingdom:	Plantae
Subkingdom:	Phanerogamea
Class:	Dicotyledonae
Subclass:	Monochlamydeae
Series:	Curvembryae
Order	Centrospermae
Family:	Chenopodiaceae
Genus:	*Basella*
Species:	*Basella alba*

Common names: Bengali: Pui shak; **English:** Malabar spinach; **Hindi:** Poi saag; **Kannada:** Basale soppu; **Malayalam:** Mayalu; **Marathi:** Poi saag; **Tamil:** Bachhali; **Telugu:** Bachhali.

Distribution: It is found in tropical Asia and Africa. It is native to the Indian Subcontinent, Southeast Asia and New Guinea.

Description: *Basella alba* is a fast-growing, soft-stemmed vine, with green or red stem reaching 10 m in length. Leaves: thick, semi-succulent, heart-shaped, with a mild flavour and a mucilaginous texture. The stem of the cultivar *Basella alba* 'rubra' is reddish purple.

Flowering: May–September.

Fruiting: July–August.

Chemical constituents: The leaves contain proteins (~3 %) carbohydrate, betacyanins, carotenoids; flavonoids such as acacetin, 7,4-diOMe kaempferol and 4'OMe isovitexin and phenolic acids such as vanillic, syringic and ferulic acids. The fruits contain gomphrenin I, II and III. The plant contains specific pigments: red-violet betacyanins and yellow betaxanthins, β-carotene and vitamin A.

Used in: Ayurveda, siddha, folk medicine.

Parts used: Leaves, stem.

Uses: The leaves are used in digestive disorders, fistulae, pustules, inflammatory tumours and syphilitic ulcers in the nose. The leaf juice has laxative and diuretic properties and is useful against gonorrhea. The plant is said to exhibit anti-viral properties.

Methods of propagation: Vegetative by stem cutting.

Photo: Prof Srinath Rao

BENINCASA HISPIDA (THUNG.)

Synonyms: *Benincasa cylindrica* Ser; *Benincasa pruriens* f. *hispida* (Thunb.) W. J. de Wilde & Duyfjes; *Cucurbita alba* Roxb. ex Wight & Arn.; *Cucurbita littoralis* Hassk.; *Lagenaria leucantha* var. hispida (Thunb.) Nakai.

Classification:

Kingdom:	Plantae
Subkingdom:	Phanerogamea
Class:	Dicotyledonae
Subclass:	Polypetalae
Series:	Calyciflorae
Order:	Passiflorales
Family:	Cucurbitaceae
Genus:	*Benincasa*
Species:	*Benincasa hispida*

Common names: Bengali: Chalkumra; **English:** Ash gourd; **Hindi:** Pethaa kaddu, Safed kaddu; **Kannada:** Bude kumbalakayi; **Malayalam:** Kumbalangal; **Marathi:** Kohlal; **Sanskrit:** Kushmanda; **Tamil:** Neer poosanikai; **Telugu:** Boodidae gummadikaaya.

Distribution: Native to Japan and Java, now cultivated more or less throughout India and warm countries.

Description: It is a climber which is cultivated mainly in warm countries. The plant has a musky smell. The stems are angular, 3–4 mm in diameter and hispid. The leaves are simple, alternate and without stipules, kidney-shaped, large, 5–7-lobed. The petiole is 6–8 cm × 2 mm and hispid. The blade is palmately 5-lobed, very thin, and hispid underneath. The blade is serrate, and shows 5–6 pairs of secondary nerves. The tendrils are axillary and slender. The flowers are simple, axillary, unisexual and showy. The male flowers are 5–15 cm and the female flowers are 2–4 cm long. The flower pedicel is 5–10 cm long and hispid. The calyx consists of 5 linear, 8 mm-long hispid sepals. The corolla consists of 5 yellow, orbicular, very thin and showily nerved petals. Stamens: anthers trilobed, filaments of the stamens inflated and hairy at the base. The fruits are massive, 40 cm long, ovate, covered with a dense chalky white powder.

Flowering: July–August.

Fruiting: August–October.

Chemical constituents: It is found to contain free amino acids such as γ-aminobutric acid and serine sugars such as mannitol, triacontanol, lupeol, β-sitosterol, trigonelline and histidine. The major constituents of the fruits are volatile oils, flavonoids, glycosides, saccharides, proteins, carotenes, vitamins, minerals, ß-sitosterol and uronic acid. Linoleic and oleic acids are the most important components in the seeds. The seeds also contain avenasterol and α- and β-amyrins. The leaves contain free amino acids and organic acids.

Used in: Ayurveda.

Parts used: Fruits, leaves and seeds.

Uses: The fruits are useful in curing asthma, cough, diabetes, hemorrhages from internal organs, epilepsy, fever, heart diseases, diabetes mellitus and urinary diseases. They are traditionally used as a laxative, diuretic, tonic, aphrodisiac, cardiotonic and to cure urinary calculi and blood diseases. The juice from the fruits is used for treating epilepsy, schizophrenia, insanity and other nervous diseases. In Ayurveda, the fresh juice of the fruit is administered to treat haemoptysis and other hemorrhages from internal organs as well as burning sensations, diabetes, piles and dyspepsia. The seeds are useful in dry cough, fever, urethrorrhea, syphilis, hyperdipsia and vitiated conditions of *pitta.* They are also useful in expelling tapeworms and curing difficult urination and bladder stones. The rejuvenative drug made from this plant is capable of improving intellect and physical strength. It is a good antidote for many kinds of vegetable, mercury and alcohol poisoning. It is also administered in cough, asthma, respiratory and heart diseases. The leaf juice is rubbed on bruises. The ash of the fruit rind is applied on painful swellings. The rind is used to treat urinary dysfunctions and the seeds for vaginal discharge. The fruit is used to treat summer fevers.

Methods of propagation: Trough seeds.

Photo: Prof Srinath Rao

BERBERIS ARISTATA

Synonyms: *Berberis coccinea*; *Berberis macrophylla.*

Classification:

Kingdom:	Plantae
Subkingdom:	Phanerogamea
Class:	Dicotyledons
Subclass:	Polypetalae
Series:	Thalamiflorae
Order:	Ranales
Family:	*Berberidaceae*
Genus:	*Berberis*
Species:	*Berberis aristata*

Common names: English: Indian barberry, Tree turmeric; **Hindi:** Chitra, Chotra, Dar-chob, Dar-hald; **Kannada:** Bagisutra; **Malayalam:** Kasturimanjal, Maradarisina; **Marathi:** Daruhalad; **Sanskrit:** Daruharidra, Darvi, Darurajani, Darhald; **Tamil:** Kasturimanja; **Telugu:** Daruharidra, Kasthoori pushpa, Kasturipasp.

Distribution: Found in Assam, Uttranchal, Gharwaland, Himachal Pradesh and Bhutan up to an altitude of 1800–3000 m.

Description: It is an erect spiny shrub, 2–3 m in height. It is a woody plant, with bark that appears yellow to brown from the outside and deep yellow from the inside. The bark is covered with 3-branched thorns, which are modified leaves. The leaves are simple with pinnate venation; they are arranged in tufts of 5–8 and are approximately 4.9 cm long and 1.8 cm broad, deep green on the dorsal surface and light green on the ventral surface. The leaves are leathery in texture and toothed, with several to many small indentations along the margin of the leaf. The flowers are yellow, 1.2 cm, in a racemose inflorescence, with 11–16 flowers per raceme, arranged along a central stem.

Flowering: March–April.

Fruiting: May–June.

Chemical constituents: The root bark contains berberine, berbamine, bervulcine, columbamine, oxyberberine, oxycanthine, palmatine, jatrorrhizine, ixotetrandrine. It also contains flavonoids such as quercentin and rutin as well as chlorogenic acid, anthocyanin, glucose, fructose, malic acid, pectin and vitamin C. The root bark is a rich source of the alkaloid berberine (about 6 %). It also contains alkaloids, 1-Omethylpakistanine, pakistanine, pseudopalmatine chloride and pseudoberberine chloride.

Used in: Ayurveda and Siddha medicine.

Parts used: Roots, stem bark and fruit.

Uses: It is used in the treatment of cancer, diarrhea, diabetes, eye and ear infections, gynaecological problems, haemorrhoids, hepatic diseases, HIV-AIDS, jaundice, malarial fever, skin diseases and osteoporosis. The roots of this plant are used for their antiperiodic, antipyretic and diaphoretic properties. The stem bark is used for its antiperiodic, astringent, febrifuge, tonic and stomachic properties as well as to treat menorrhagia and diabetes.

Methods of Propagation: Through seeds.

Photo: Prof Srinath Rao

BERBERIS LYCEUM (ROYAL)

Synonyms: *Berberis angustifolia* Roxb.; *Berberis aurahuacensis* Hook. f.; *Berberis vulgaris* f. *lyceum* (Royale).

Classification:

Kingdom:	Plantae
Subkingdom:	Phanerogamea
Class:	Dicotyledons
Subclass:	Polypetalae
Series:	Thalamiflorae
Order:	Ranales
Family:	Berberidaceae
Genus:	*Berberis*
Species:	*Berberis lyceum*

Common names: Bengali: Darhaldi; **English:** Barberry; **Hindi:** Chitra, Kashmal; **Malayalam:** Maramanjal; **Marathi:** Kattukasturi; **Tamil:** Mullukala.

Distribution: *Berberis lyceum* is native to Nepal, but globally distributed in various parts of the world. It occurs abundantly in the Himalayan region of India and Pakistan. Within India, it is found in Jammu and Kashmir, Himachal Pradesh, Uttar Pradesh, Sikkim, Madhya Pradesh and Tamil Nadu, at altitude ranges of 850–3500 m.

Description: The shrub of *Berberis lyceum* is attractive and easily grown. It grows up to 9–10 feet in height, and is erect or suberect and semideciduous with dimorphic shoot (the long shoot forming the structure of the plant and the short shoot 1–2 mm long). The stem and branches are pale whitish to grayish and contain spines arranged alternately on the stem. The leaves of this plant are brightly coloured, thick, entire or toothed with oblanceolate to oblong-obovate shape. The inflorescence is corymbose racemes with 11–16 flowers per cluster; the flowers are hermaphrodite, which means they have both male and female organs. Usually, the flowers are pale yellow in colour, born in axillary clusters and longer than the leaves. The fruits of the plant are called berries and are ovoid or obovoid-subglobose. They acquire a bright red colour or a purplish colour on ripening. On an average they are 7 mm long and 4 mm in diameter. The colour of the pulp or juice is plum purple. On average, the fruit contains 2–5 seeds. Their colour varies from yellow to pink. The root is hard, 3–8 cm in diameter, branched and gradually tapering, occasionally split longitudinally; however, its wood is smooth and bright yellow in colour. The root bark can be up to 3 mm thick, externally fissured and internally smooth.

Flowering: April–June.

Fruiting: June–August.

Chemical constituents: The various chemical constituents of this plant are berberine, berbamine, chinabine, karakoramine, palmatine balauchistanamine, gilgitine, jatrorrhizne, jhelumine, punjabine, sindamine, umbellatine, oxyacanthine, acetic acid, maleic acid and ascorbic acid. The fruits contain malic, tartaric and citric acids, as well as tannins. Berberine and palmitine are found in the roots (Gupta et al., 2015; Khan et al., 2016).

Used in: Ayurveda and traditional system of medicine.

Parts used: root, bark, stem, leaves.

Uses: The plant is known to prevent liver and abdominal disorders, skin diseases, cough and ophthalmic issues. It has hypoglycemic, antihyperlipidemic, hepatoprotective, anticarcinogenic and antipyretic properties. Traditionally, it has been used against diarrhea, intestinal colic, jaundice, internal wounds, rheumatism, diabetes, gingivitis, throat pain, backache, scabies, bone fractures, sun blindness, pustules and menorrhagia and as diuretic, expectorant and diaphoretic. It is known to possess antidiabetic, antihyperlipidemic, antibacterial, antifungal, anticoccidial, pesticidal, antimutagenic and wound-healing properties. Furthermore, it possesses antihistaminic, stomachic, astringent and diaphoretic properties. Berberine and palmitine alkaloids are effective against eye diseases, fever and piles. The plant has anti-ageing effects and keeps the skin fresh. In the Unani system of medicine, it is used for the treatment of leprosy (Ghosh et al., 1990; Sing and Raat, 2000; Gupta et al., 2015).

Methods of propagation: The seeds of the overripe fruit will take a longer time to germinate. The seedling can be raised in a nursery and then transplanted when small plants have grown and are easy to handle. However, proper ventilation is required to prevent seeds from damping off.

BETULA UTILIS D. DON

Synonym: *Boerhavia repens* Linn.

Classification:

Kingdom:	Plantae
Subkingdom:	Phanerogamea
Class:	Mangoliopsida
Order:	Falages
Family:	Betulaceae
Genus:	*Betula*
Species:	*Betula utilis*

Common names: Hindi: Bhoj patra; **Tamil:** Bhurjjamaram, Purchcham; **Malayalam:** bhujapatram, bhurjjamaram; **Telugu:** Bhujapatri; **Kannada:** bhuyapathra; **Sanskrit:** bahulavalkalah, bahupata, bhurjapatraka.

Distribution: East Asia, from the Himalayas to Southwest China. Widespread throughout the Himalayan region. Moist hillsides at elevations of 3000–4000.

Description: It is a moderate-sized tree that grows up to 20 m in height. The bark is shining, reddish white or white, with white horizontals, smooth, lenticels. The most important characteristic is that the outer bark consists of layers, exfoliating in broad horizontal rolls. The leaves are alternate, petiolate, simple, 3.0–7.5 cm, ovate-acuminate, elliptic, and irregularly serrate. Flowers in pendulous spikes, monoecious (individual flowers are either male or female, but both sexes can be found on the same plant). Seeds are thin and winged.

Flowering: May–June.

Fruiting: September–October.

Chemical constitutions: The bark contains betulin, lupeol, oleanolic acid, acetyloheanolic acid and betulinic acid, lupenone, sitosterol, methyle betulonate, methyl betultriterpenoid and karachic acid. *Betula utilis* contains betulin up to 12 % of its weight.

Used in: Ayurveda and traditional medicine.

Uses: Antiseptic, aromatic, carminative and contraceptive effects. *Betula utilis* contains betulin, which can be easily converted into betulinic acid. Studies revealed that betulinic acid inhibits the growth of malignant melanoma and cancers of the liver and the lung. Based on their chemical structure, betulinic acid derivatives have been reported as inhibitors of HIV-1 and HIV protease. They also have antimicrobial and antioxidant properties. The plant is useful in the treatment of inflammation. Traditionally, the bark has been widely used to heal wounds and disinfect the skin as well as in the treatment of various ailments and diseases such as bronchitis, convulsions, leprosy and diseases of the blood and the ear. The leaves of the plant show efficacy in the treatment of urinary tract infections as well as kidney and bladder stones. The white, paper-like bark of the tree was used in ancient times as a support on which to write Sanskrit scriptures, texts and sacred mantras. It was then placed in an amulet and worn for protection.

An infusion of the bark is antiseptic and carminative. It has been used in the treatment of hysteria and jaundice. It is applied as drops to the ears to relieve earache. A paste made from the bark is used as a poultice on cuts, wounds and burns. The German Commission E Monographs, a therapeutic guide to herbal medicine, approved the *Betula* species for the treatment of infections of the urinary tract, kidney and bladder stones and rheumatism.

Parts used: Bark.

Methods of propagation: Through seeds. The seeds succeeds in a well-drained loamy soil in a sheltered position. They grow well in heavy clay soils. They dislike wet soils and are sown in an outdoor seedbed, either as soon as it they are ripe or in the early spring – do not cover the spring-sown seeds. Grow the plants in a seedbed for 2 years before planting them out into their permanent positions in the winter.

Photo: https://upload.wikimedia.org/wikipedia/commons/7/73/Betula_utilis02_by_Line1.jpg

BOERHAVIA DIFFUSA L.

Synonym: *Boerhavia repens* Linn.

Classification:

Kingdom:	Plantae
Subkingdom:	Phanerogamea
Order:	Caryophyllales
Class:	Dicotyledonae
Subclass:	Polypetalae
Series:	Curvembryeae
Family:	Nyctaginaceae
Genus:	*Boerhavia*
Species:	*Boerhavia diffusa*

Common names: Bengali: Punarnava; **English:** Spreading hog weed; **Hindi:** Sant, Bishkhapra, Gadhapuruna; **Kannada:** Sanadika; **Malayalam:** Tavilama; **Sanskrit:** Punarnava, Sophaghni; **Tamil:** Mukkurattai; **Telugu:** Attamamidi.

Distribution: This species is globally distributed in the Pantropics. Within India, it is found occurring as a weed throughout, ascending to an altitude of 2000 m in the Himalayas. It is also cultivated to some extent in West Bengal.

Description: A perennial herb with stout root stock, divaricately branched. The branches are many and procumbent. The leaves are simple, opposite or subopposite, short-petioled in unequal pairs, ovate-oblong, glabrous above and whitish beneath. The larger leaves are 2.5–3.5 cm long, the smaller ones 1.2–1.8 cm long. The flowers are regular, bisexual, very small, bright-purplish pink, short-stalked, pendunculate, with glomerulate clusters arranged in slender, long-stalked, axillary or terminal corymbs. The fruit are obovoid or sub-ellipsoid, rounded above, slightly cuneate below, broadly and bluntly 5-ribbed, very glandular throughout, highly viscid, easily detachable, 1-seeded, indehiscent with a thin pericarp.

Flowering: Throughout the year in Indian conditions.

Fruiting: Throughout the year in Indian conditions.

Chemical constituents: It contains alkaloids, triacontanol, hentriacontane, β- sitosterol, 8-dimethyl flavones, hypoxanthine-9-L-arabinoside, 4-dimethoxy-6 and ursolic acid. The entire plant is used to make the drug Punarnava, containing 2 quinolizidine alkaloids: punarnavine and punarnavoside. The leaves are found to contain kaempferol and phenolic acids, as well as alkaloids such as punarnavine-1 and punarnavine-2. They also contain phytosterols such as β-sitosterol, β-sitosterol-D-glucoside, sitosteryl oleate and sitosteryl palmitate. The flavonoids present are C-methyl flavone and kaempferol. Also present are quercetin rotenoids such as boeravinone A, boeravinone B, boeravinone C, boeravinone D, boeravinone E and boeravinone F. There are also lignans such as liriodendrin, stringarsionol, mono-D-glucoside and β-D-glucoside. Rotenoids like punarnavoside and salt potassium nitrate are contained in the plant (Nayak and Thirunvoukkrsu, 2016).

Used in: Ayurvedic, Unani, Siddha and traditional medicine.

Parts used: The whole plant.

Uses: The plant is bitter, astringent, anthelmintic, diuretic, cardiac stimulant, aphrodisiac, expectorant, anti-inflammatory, febrifuge and laxative. The root and leaves are used in the form of juice and decoction to treat anaemia, internal abscess, calculi, bronchitis, eye diseases, oedema during pregnancy and haemoptysis as well as to induce sleep. They are useful in case of fever, rheumatic ailments, difficult labor and vaginal pain. They also have rejuvenating effects. According to the Unani system of medicine, the leaves are appetizer, alexiteric and useful in ophthalmia and joint pains (Nayak and Thirunvoukkrsu, 2016).

Methods of propagation: The plant is propagated through seeds. Fresh seeds are collected and sown directly in raised beds or polybags. The best time for sowing is May–June. The seeds germinate within 7–8 days. They show 70–90 % germination.

Photo: Prof Srinath Rao

BOSWELLIA SERRATA ROXB. EX COLEBR.

Synonyms: *Boswellia balsamifera* Spreng.; *Boswellia glabra* Roxb.; *Boswellia thurifera* Roxb.; *Chloroxylon dupada* Buch. Ham.

Classification:

Kingdom:	Plantae
Subkingdom:	Phanerogamea
Class:	Dicotyledonae
Subclass:	Polypetalae
Series:	Disciflorae
Order:	Sapindales
Family:	Burseraceae
Genus:	*Boswellia*
Species:	*Boswellia serrata*

Common names: English: Indian frankincense, **Gujarati:** Salai guggul, Saaledi; **Hindi:** Shallaki; **Kannada:** Guggula mara; **Malayalam:** Kungilyam; **Marathi:** Dhupali, Dhupasali, Kurunda;

Sanskrit: Bhishan, Guggulu, Hastinashana, Shallaki; **Tamil:** Kumancam, Kunkiliyam, Vellai kirai; **Telugu:** Guggilamu, Parangi sambrani chettu.

Distribution: Native to India. It grows in the dry mountainous regions of India. It is also distributed in Northern Africa and in the Middle East.

Description: It is a moderate–large-sized tree, much branched.

Flowering: February–March.

Fruiting: April onwards.

Chemical constituents: The resin contains monoterpenes, diterpenes, triterpenes, tetracyclic and triterpenic acids and 4 major pentacyclic triterpenes, namely acetyl- β-boswelic acid, acetyl-11-keto-β-boswellic acid, β-boswellic acid and 11-keto-β-boswellic acid (https://en.wikipedia.org/wiki/Boswellia_serrata).

Used in: Ayurveda and Unani traditional medicine.

Parts used: Gum resin.

Uses: Gum resin extracted from the tree is used to treat chronic inflammatory illness, osteoarthritis, rheumatoid arthritis, inflammatory bowel disease, asthma, leukemia and breast cancer. It also has anti-depressant effects and is effective in ulcerative colitis. It is also a remedy for boils, bronchitis, cardiovascular diseases, cough, diarrhea, dysentery, fever, hemorrhoids, irregular menses, jaundice and syphilitic diseases. Moreover, it stimulates the liver.

Methods of propagation: Through seeds.

Photo: Milind Girdhari

BRYONOPSIS LACINIOSA LINN.

Synonyms: *Diplocyclos palmatus* (L.) Jeffrey; *Bryonia palmata* Linn.; *Cucumis verrucosus*; *Coccinia palmata.*

Classification:

Kingdom:	Plantae
Subkingdom:	Phanerogamea
Class:	Dicotyledonae
Subclass:	Polypetalae
Series:	Calyciflorae
Order:	Passiflorales
Family:	Cucurbitaceae
Genus:	*Bryonopsis*
Species:	*Bryonopsis laciniosa*

Common names: Bengali: Mala; **English:** Indian bryony, Lollipop climber; **Hindi:** Gargumaru, Ishwaralingi, Shivalingi; **Kannada:** Lingatondeballi, Shivalinga; **Malayalam:** Iyiviralikkova, Namakai, Sivalingakkaya; **Sanskrit:** Apastambhinii, Bahupatra, Chitraphala; **Tamil:** Iyvirali; **Telugu:** Lingadonda.

Distribution: All over India.

Description: It is a perennial climber with tendrils. The plant has a large, tuberous root and is very slender, glabrous, often spotted, with dark-green stems and long internodes. The tendrils are bifid. The leaves are simple, alternate, membranous, 7.5–12.5 cm long and about as broad, ovate-rotundate in outline. They are very deeply cordate at the base, cut near the base into 5 lanceolate or linear, acute, coarsely serrate segments. The petioles are 2.5–3.5 cm long, striate and slender. The flowers are regular and green or pale yellow in colour; both male and female flowers are present. Male flowers are in small fascicles of 3–6, stamens 3, quite distinct. The female flowers are solitary or few and occur in clusters in the same axils. The sepals are 5, fused into a cup-shaped calyx; segments linear, filiform and glabrous. The petals are 5, fused to about ¼ of the way up; segments oval-oblong, acute and pubescent. The ovary is inferior, globose, glabrous; style slender.

The fruit is globose, sub-sessile, 1.8–2.5 cm in diameter and smooth. The immature fruits are green with white blotched stripes and later turn a bright-red or orange colour when ripe.

The seeds are 5–6 mm long, gibbous at the sides with a prominent raised band running round the edge.

Flowering: August–October.

Fruiting: October–December.

Chemical constituents: Bryonin is the bitter principle of the herb. The seeds have been reported to contain 12 % oil, protein, bryonin, goniothalamin, punicic acid and lipids. The seed oil contains punicic acid in large amounts. The leaves contain p-coumaric, caffeic, ferulic, p-hydroxybenzoic and syringic acids. They also contain flavonoids such as quercetin, kaempferol, n-hentriacontane, n-tritriacontane and α- and β-amyrin as well as sitosterol.

Used in: Ayurveda, folk medicine and homeopathy.

Parts used: Fruits, seeds, roots; the whole plant.

Uses: The whole plant is cathartic, galactagogue and laxative. It is also used to treat venereal diseases. It is a bitter plant and possesses tonic properties. The plant is also valued as a remedy for diarrhea, dysentery and convulsion. The leaves exhibit styptic, antiseptic and astringent properties and are also used in diabetes. The paste of the leaves is applied to wounds and abscesses. The leaves are applied topically on inflammations. The powdered seeds and roots are given to help conception in women and the seed powder is used to increase sperm count. The seeds of this plant are available in the market as 'Shivlingi ke beej'. The ripe fruits are also used in pregnancy. A tincture prepared from the roots is used for treating painful menstruation, dysmenorrhea, pain in the ovaries during menstrual period, inflammation of the uterus and burning pain in the fundus of the uterus. The tincture is also prescribed for threatened abortion.

Methods of propagation: Through seeds.

Photo: Prof Srinath Rao

BRYOPHYLLUM PINNATUM (LAM.) OKEN

Synonyms: *Bryophyllum calycinum* Salisb.; *Bryophyllum germinans* Blanco; *Crassula pinnata* (Lam.) L. f.; *Kalanchoe pinnata* (Lam.) Pers.; *Kalanchoe brevicalyx* (Raym-Hamet & H. Perrier) Boiteau.

Classification:

Kingdom:	Plantae
Subkingdom:	Phanerogamea
Class:	Dicotyledonae
Subclass:	Polypetalae
Series:	Calyciflorae
Order:	Saxifragales
Family:	Crassulaceae
Genus:	*Bryophyllum*
Species:	*Bryophyllum pinnatum*

Common names: English: Life plant, Miracle leaf; **Hindi:** Patharchur; **Kannada:** Gandu kalinga; **Marathi:** Panfuti; **Sanskrit:** Param beej, Rakta kusum; **Tamil:** Malaikkalli, Runakkalli;

Distribution: The plant is found in Africa, Arabia, Brazil, Egypt, India, Myanmar, Nepal, Pakistan, Thailand. In India it is cultivated throughout.

Description: It is a succulent perennial plant, about 3–4 feet tall. The stem is fleshy, cylindrical and has a reddish colour. The leaves are about 10–20 cm long and 4–10 cm broad. They are opposite, decussate, thick, fleshy, elliptical, with serrate margins, often reddish. The leaf–stem combinations are called phylloclades. The phylloclades produce bulblets at the margins. The inflorescence is a panicle with many pendant orange-red flowers. The calyx is inflated, cylindric, 4-fid, green with red streaks, papery, calyx tube 2–4 cm. The corolla is tubular, about 4–5 cm long, with pronounced constriction, 4-lobed, yellowish with red-purple streaks. The stamens are 8 in 2 whorls about 4 cm long, welded on the corolla. The ovary has 4 carples, slightly fused together in the center; styles slender.

Flowering: Winter–Spring.

Fruiting: January–February.

Chemical constituents: The compounds present are bryophillin-A,

bersaldegenin-3-acetate and bryophillin-C. The plant contains caffeic, syringic, malic oxalic and ferulic acids, triterpenoids,flavonoids, flavons and steroids.

Used in: Ayurveda and traditional medicine.

Parts used: Leaves and roots.

Uses: The plant is used to treat asthma, cold, cough, cuts, diabetes, dysentery, fever, sore throat and wounds. The plant root is used to treat high blood pressure and constipation. It also protects the liver and is useful for the treatment of hepatitis. The juice of the plant is used to treat stomachache, piles, gray hairs, inflammation, diarrhea and jaundice.

Methods of cultivation: Vegetative, bulblets formed on the margins of the leaves.

Photo: Prof Srinath Rao

BUCHANANIA LANZAN SPRENG.

Synonyms: *Buchanania latifolia* Roxb.; *Spondias elliptica* Rottl.; *Spondias simplicifolia* Rottb.

Classification:

Kingdom:	Plantae
Subkingdom:	Phanerogamea
Class:	Dicotyledonae
Subclass:	Polypetalae
Series:	Disciflorae
Order:	Sapindales
Family:	Anacardiaceae
Genus:	*Buchanania*
Species:	*Buchanania lanzan*

Common names: Bengali: Piyala; **English:** Cuddapah-almond; **Hindi:** Chironji; **Kannada:** Nurkala; **Malayalam:** Mungapera; **Marathi:** Chiraunji; **Tamil:** Katimangu-marum; **Telugu:** Chinnamoral.

Distribution: Throughout the drier parts of the country, in deciduous forests, ascending to 1300 m in the sub-Himalayan tract, Central India.

Description: Deciduous trees, 12–15 m tall, with straight cylindrical trunk and tomentose branches. Bark is rough, dark grey or black, fissured into prominent squares. Leaves alternate, oblong to elliptic-oblong with a rounded base, 4–19 × 2–10 cm, apex obtuse, margin entire, coriaceous, tomentose beneath, pubescent above; petiole swollen, 1.5–2.0 cm long, pubescent. Inflorescence: terminal and axillary pyramidal ferrugineo-pillose panicles. Flower: small, sessile, greenish white, hermaphrodite, bracts small, caducous. Calyx: lobes short, broadly ovate, ciliate, persistent, imbricate. Corolla: petals 5, ovate-oblong, imbricate, subacute. Androecium: stamens 10, inserted at the base of the disk, a little shorter than the petals, filaments flattened, anther about as long as filaments. Gynoecium: carpels 5–6, ovaries 1, perfect, conical, villous, the other 4 reduced to cylindrical filaments. Style: short, stigma truncate. Fruit: drupes, obliquely lentiform, black, stone hard, 2-valved. Seed: gibbous, acute at one end.

Flowering: January to March.

Fruiting: April to May.

Chemical constituents: Isolation of kaempferol and its 7-O-glucoside, gallic acid, quercetin, 3-O-rhamnoglucoside and 7-O-rhamnoside from leaves (Rastogi and Mehrotra, 1985–1989).

Used in: Ayurveda.

Parts used: Seed oil, stem bark.

Uses: Fruits are edible. They taste somewhat like pistachio nuts and are used in sweetmeats. Plant is used in bronchitis, cholera, consumption, cough, dysuria, fever, inflammation of gums and scorpion bite. Stem and bark are also useful in burns, cuts, skin diseases, snake bite and wounds. Roots are useful in venereal complaints.

Methods of propagation: Through seeds.

Photo: Dr R. S. Modi

CALOPHYLLUM INOPHYLLUM L.

Synonyms: *Calophyllum bintagor* Roxb.; *Calophyllu blumei* Wight.

Classification:

Kingdom:	Plantae
Subkingdom:	Phanerogamea
Class:	Dicotyledonae
Subclass:	Polypetalae
Series:	Thalamiflorae
Order:	Guttiferales
Family:	Clusiaceae (Calophyllaceae) Guttiferae
Genus:	*Callophyllum*
Species:	*Callophyllum inophyllum*

Common names: Bengali: Sultanchampa; **English:** Dielo oil tree; **Gujarati:** Udi; **Hindi:** Sultana champa; **Kannada:** Honne; **Malayalam:** Pinna; **Marathi:** Undi; **Sanskrit:** Nagchampa; **Tamil**: Punnai; **Telugu:** Poona.

Distribution: A tree, common on sandy coasts and laterite soils just above the high tide line, sometimes up to 200 m, often leaning out over the sea. Native of East Africa. Distributed from tropical East Africa to Taiwan, the Ryukyu and Line Islands and New Caledonia. In India it is available in Orissa, Maharashtra, Karnataka, Tamil Nadu, Kerala, Lakshadweep and Andaman and Nicobar Islands.

Description: Trees: up to 20 m tall with spreading crown; trunk 1.5 m in girth, wood reddish white to brown, bark brown to pale grey (blackish), smooth, often mottled with wide boat-shaped fissures; exudates milky or yellow sap, clear, very sticky. Leaves: Opposite, dark green, shiny with broad elliptical blades, size 10–15 × 3–10.5 cm, midrib prominent below, venation distinct, running parallel to each other, and perpendicular to the midrib. Petioles 1–1.5 cm long, glabrous, stout, flat. Flowers born on axillary racemes 5–13 cm long, 5–15 marble-white fragrant flowers; polygamous, bisexual, 2–2.5 cm in diameter. Bracts: 3–4 mm long, ovate, soon deciduous. Pedicels: 1.5–4.5 cm long, glabrous. Sepals: 4, reflexed, 2 outer ones, 5.5–10 × 4.5–8 mm, ovate or suborbicular, concave, inner petaloid, 9–15 × 7.5–10 mm, sub-elliptic. Petals: usually 4 (or 3–5), reflexed, 9–16 × 5–10 mm, obovate, elliptic or oblong, rarely glabrous. Stamens (175–) 210–360 (–440); filaments creamy, 5–7 mm long, connate up to 2 mm into 4–6 bundles; anthers 0.7–2 mm long, rounded or retuse at apex, yellow when young, brownish at maturity. Ovary 1.5–3.5 mm long, depressed globose, stipitate, pink or light purple; styles much longer than ovary, 4.5–9 mm long, twisted; stigmas peltate, 3–5-radiate, 0.7–2 mm across. Fruit: drupes about 2.5–5 × 2.5–4 cm, globose or spherical to obovoid, shortly apiculate, smooth, yellowish, brownish green or pale brownish, ochraceous, pulpy; pericarp thick; stone about 2 cm in diameter, subspherical, rounded at apex, smooth, basal plug 4–10 mm long.

Flowering: December–October; probably flowers throughout the year with several flushes.

Fruiting: December–October; probably flowers throughout the year with several flushes.

Chemical Constituents: Neoflavone-calophyllolide isolated from seeds (Garazd, et al., 2003).

Pyranocoumarins, calophyllolide, calophyllin, xanthones and jacareubin are other compounds reported from various parts of the plant.

Used in: Ayurveda, Siddha and Unani medicine.

Parts used: Leaves, fruit oil and bark.

Uses: The oil is wonderfully healing to the skin, reducing the swelling of rashes, insect bites, and sunburns. It also shows antimicrobial, antibacterial, and antifungal qualities, which makes it effective on scabies, psoriasis, ringworm, athlete's foot and jock itch. It is used in curing joint pains caused by arthritis and gout and for relieving nerve pain from sciatica. Various parts of this plant have been reported to have traditional medical uses such as the treatment of venereal disease, high blood pressure, rheumatism, inflammation, eye diseases, varicose veins, haemorrhoids, chronic ulcers, skin infections and wounds. Inophyllum, pyranocoumarins and calophyllolide have anti-viral properties.

Methods of propagation: Through seeds.

Photo: Prof Srinath Rao

CALOTROPIS GIGANTEA (L.) R. BR.

Synonym: *Asclepias gigantea* L.

Classification:

Kingdom:	Plantae
Subkingdom:	Phanerogamea
Class:	Dicotyledonae
Subclass:	Gamopetalae
Series:	Bicarpellatae
Order:	Gentianales
Family:	Asclepiadaceae
Genus:	*Calotropis*
Species:	*Calotropis gigantea*

Common names: Bengali: Akanda; **English:** Gigantic swollow-wart, Madar; **Kannada:** Arkagida; **Malayalam:** Eirkku; **Marathi:** Rui; **Sanskrit:** Arka, Mandara; **Tamil:** Arkkam; **Telugu:** Jilledu.

Distribution: The plant grows throughout India.

Description: Erect shrubs or small trees, 1–3 m. Stems woody, nodes slightly swollen, up to 8 mm thick; internodes terete, sometimes squarish, 1.2–8 cm long and 4–10 mm in diameter, fugaciously cottony pubescent, grooved in the center. Latex milky. Leaves opposite, decussate, subsessile or shortly petiolate, lamina 2.6–20 × 1.2–10.5 cm, broadly obovate or elliptic-oblong, subacute or obtuse at apex, cordate with auricled base, margin entire, fugaciously pubescent on both sides, thick, lateral nerves 4–7 pairs, petioles 3–8 mm long and 2–4 mm in diameter.

Flowers in lateral or axillary umbellate or sub-corymbose cymes (flower bud cylindrical), many-flowered; peduncles squarish, 1.5–13 cm long and 2–4 mm in diameter, fleshy, cottony pubescent; bracts 5–8 × 2–3 mm, ovate or lanceolate, apex acute, caducous, margin ciliate, pubescent without; pedicels in pairs, terete, 2.5–4 cm long and about 2.5 mm in diameter, greenish white, fulvous pubescent. Calyx 5-lobed, divided up to base, lobes 4–5 × 2.5–3 mm, broadly ovate, saucer-shaped, apex acute, margin ciliate, glandular within, cottony pubescent without, glabrous within. Corolla rotate, lobes 5, united up to about 7 mm, 1.2–2.1 × 0.6–1 cm, ovate, apex acute, margin revolute, spreading, glabrous, petals long, recurved. Corona curved, staminal, fleshy, completely adnate, shorter than gynostegium; lobes about 1.3 cm long and 7 mm wide, laterally compressed, margin ciliate, glabrous, with a recurved spur at the base, about 2.5 mm wide and 2 obtuse auricles just below the rounded apex. Pollinia 5, pollen masses solitary in each anther cell, yellow, waxy, oblong, attached by light-brown caudicles, dark-brown corpuscle, winged. Gynostegium 1.2–2.3 cm long. Follicles paired, horizontal, 3.8–10 × 1.8–4 cm, oblong or boat-shaped, with hooked tip, fugaciously cottony pubescent; seeds many, 7–7.5 × 4 mm, oblong, brown coloured, with light-brown margin; coma dull-white, silky, 3–3.5 cm long.

Flowering: Throughout the year.

Fruiting: Throughout the year.

Chemical constituents: The major phytochemicals viz. alkaloids, carbohydrates, glycosides, phenolic compounds/tannins, proteins and amino acids, flavonoids, saponins, sterols, acid compounds, resins in flower, bud, and root of Calotropis. The milky latex of the plant is rich in lupeol, calotropin, calotoxin, and uscharidin, the latex protein.

Used in: Ayurveda and Unani medicine.

Methods used: The seeds freely float in the air and natural regeneration is very common. Vegetative propagation through stem and root cuttings is very useful in large scale multiplication.

Parts used: Leaves, root bark, latex of the plant.

Uses: The plant is purgative, anthelmintic and cures diseases of the spleen, leprosy, leucoderma, liver issues, mumps, painful joints, piles, rheumatism, skin blemishes, tumors, ulcers and vitiligo.

Photo: Prof Srinath

CALYCOPTERIS FLORIBUNDA (ROXB.) LAM.

Classification:

Kingdom:	Plantae
Subkingdom:	Phanerogamea
Class:	Dicotyledonae
Subclass:	Polypetalae
Series:	Calcyciflorae
Order:	Myrtales
Family:	Combretaceae
Genus:	*Calycopteris*
Species:	*Calycopteris floribunda*

Common names: Bengali: Gaichhalata; **English:** Paper flower climber; **Hindi:** Kokkare; **Kannada:** Enjarigekubsa; **Malayalam:** Varavalli; **Marathi:** Ukshi; **Sanskrit:** Sushavi; **Tamil:** Pullanji valli, **Telugu:** Murugudutige.

Distribution: Found extensively in the tropical evergreen forests of the Western Ghats; also found in the sacred groves of Kerala.

Description: Straggling shrubs or large lianes, about 15–20 feet in height. Leaves: simple, opposite, elliptic-oblong or ovate-elliptic, acute or acuminate at apex, rounded at base, silky pubescent above when young, fulvous tomentose beneath with numerous pellucid glands; inflorescence of axillary or subterminal, simple or branched racemes; flowers white, showy; calyx tube elongate, 5-angled, produced above the ovary; petals absent. Stamens 10, arranged in two whorls; fruits narrow, ovoid, 5-angled, crowned by persistent calyx.

Flowering: January–April.

Fruiting: January–April.

Chemical constituents: Presence of flavonoids, alkaloids, tannins and saponins. Calycopterones, calyflorenones, novel biflavonoids, were found in the dried leaves of *Calycopteris floribunda* (Mayer, 1999). 6″-demethoxyneocalycopterone, calyflorenone C, 6″-epi-calyflorenone B, 6″-epi-calyflorenone C, calyflorenone D (Mayer, 2004), neocalycopterone, methyl-neocaly- copterone, penduletin (Mayer, 1999) calycopterone, isocalycopterone, 4-demethylcalycopterone and calycopterin (Rodriguez et al. 1972) were isolated from organic extract of *Calycopteris floribunda* leaves. Calycopterone, isocalycopterone and 4-demethyl-calycopterone showed a wide range of activities against solid tumor cell lines. The main flavonoid, calycopterin, has anthelmintic, anti-viral (inhibition of polio virus in vitro) and anticancer properties (anti-proliferative and anti-aromatase).

Used in: Ayurveda, folk and Unani medicine.

Parts used: Leaves and fruits.

Uses: The stems are full of water; villagers use them to quench their thirst. The leaves are bitter, astringent, laxative, anthelmintic, depurative, diaphoretic and febrifuge. They are also used against intestinal worms, colic, leprosy, malarial fever, dysentery, ulcers and vomiting. The fruits are used in curing jaundice, ulcers, pruritus and skin diseases (Mayer, 1999).

Methods of propagation: Through seeds.

Photo: Dr Kotresh

CANARIUM STRICTUM ROXB.

Synonyms: *Canarium resiniferum* Brace ex. King; *Canarium sikkimense* King.

Classification:

Kingdom:	Plantae
Subkingdom:	Phanerogamea
Class:	Dicotyledonae
Subclass:	Polypetalae
Series:	Thalamiflorae
Order:	Sapindales/ Rutales
Family:	Burseraceae
Genus:	*Canarium*
Species:	*Canarium strictum*

Common names: Bengali: Kala dammer; **English:** Black dammar tree; **Gujarati:** Kala dammer; **Hindi:** Kala dammer; **Kannada:** Halemaddu, Karedupa, Sambrani; **Malayalam:** Karuthakungiliyamtelli, Pantham, Kunthirikkam; **Marathi:** Dhup, Raldhup, Sambran; **Sanskrit:** Mandadhupa, Raladhupa; **Tamil:** Karang, Kunthrikam, Karupa, Kangiliam, Sambrani, Karunkungilium; **Telugu:** Nallarojanamu.

Distribution: Globally, this species is recorded in India, Sri Lanka, upper Myanmar and Bangladesh. The tree is found scattered in the evergreen forest of the Western Ghats in India. It is also found in Assam and Meghalaya, in the Tirunelveli District, Andhra Pradesh, Arunachal Pradesh, Maharashtra, Goa, Rajasthan, Tripura, Karnataka, Kerala and Tamil Nadu. In Karnataka, the districts of Hassan, Chikmagalur, Coorg, Mysore, Shimoga, North Kanara and Kodagu have reported its presence. It is found in the Shevaroys (Endemic) and all over the hills, up to 1200 m.

Status: The Red List status of this species has been assessed as 'vulnerable' for the three states of Southern India, i.e. Karnataka, Kerala and Tamil Nadu and 'moderately threatened' for the Western Ghats, since people burn trees to obtain large amounts of resin in a short time (Desha et al., 2012).

Description: A tall, handsome, evergreen buttressed tree, up to 50 m tall with a straight, white, cylindrical stem; bark hard, light greenish-gray, thin, rough, somewhat aromatic, exfoliating, yielding a black resin when cut. The undersurfaces of both the twigs and leaflets are covered with reddish hairs. The young foliage is bright yellow, later turning crimson. Leaves alternate, large, 0.6–1.2 m long, imparipinnate; leaflets 7–15, opposite or alternate, ovate-oblong, ovate-lanceolate or unequal-sided, 10–20 cm long and 4–10 cm wide, apex acuminate, margin finely serrate, lateral nerves 10–15 pairs, very prominent beneath, petioles 2.5–8.0 cm long. Flowers small, white or bright yellow, faintly fragrant in terminal or axillary panicles shorter than the leaves, male and female flowers separate, pedicel very short, calyx usually 4-lobed, cupular, rusty tomentose outside, tube 3 mm long, lobes triangular, 1 mm long; petals 6cm long and 2cm broad, oblong-obovate, apex rounded. Fruit (drupes) ellipsoid or obovoid, 3.8–5.0cm long, resinous with a hard stone.

Flowering: January–March.

Fruiting: November–January (almost all the trees bears fruits once in every 2–3 years).

Chemical constituents: The plant contains α-amyrin, β-amyrin and its acetate, canarone, epi-khusinol (Suruse et al., 2008 and 2010; Muthuswamy and Senthamarai, 2014).

Used in: Ayurveda, homeopathy and Siddha medicine.

Parts used: Resin.

Uses: Decoction or powder of the resin is given orally as a remedy for rheumatism, cough, fever, epilepsy, asthma, syphilis, blood impurities, various poisons, hernia, chronic skin diseases, haemorrhage. It also improves the complexion.

Methods of propagation: Through seeds.

Photo: http://www.biotik.org/india/species/c/canastri/canastri_17_en.html

CANSCORA DECUSSATA (ROXB.) ROEM & SCHULT

Synonyms: *Canscora alata* (Roth) Wallich; *Pladera decussata* Roxb., *Exacum alatum* Roth.

Classification:

Kingdom:	Plantae
Subkingdom:	Phanerogamea
Class:	Dicotyledonae
Subclass:	Gamopetalae
Series:	Bicarpellatae
Order:	Gentinales
Family:	Gentianaceae
Genus:	*Canscora*
Species:	*Canscora decussata*

Common names: Bengali: Dankuni; **English:** Canscora; **Hindi:** Kalameg, Sankhaphuli; **Kannada:** Sankinisoppu; **Malayalam:** Sankhapushpi, Kanjankora; **Marathi:** Tilavi; **Sanskrit:** Akshapida, Sankhapushpi; **Telugu**: Kanapa, Kanigi.

Distribution: Found throughout India, in damp localities ascending to 2344 m.

Description: Annual, erect, somewhat flaccid, decussate-branched, glabrous herb, 45 cm tall, 2–3 mm in diameter at base; stems 4-angled, angles sharply winged. Leaves: lanceolate, ovate or ovate-lanceolate, sub-sessile, 0.7–3.5 × 0.4–0.9 cm, acute, 3-nerved. Inflorescence: terminal trichotomous cymes; peduncles winged, obsolete, 0.7–4.2 cm long; bracts narrowly ovate to ovate-lanceolate, acute. Flowers: tubular, white pinkish, 0.6–1.6 cm long, pedicellate; pedicels quadrialate, 0.6–2.0 cm long. Calyx: tubular, inflated, 4-winged; tube 0.6–1.2 cm long, contracted at mouth; lobes 4, ovate-lanceolate, pointed, 2–3 × 1–2 mm. Corolla tubular, funnel-shaped, nearly bilabiate; tube nearly equal to calyx, 0.6–1.0 cm long, cylindrical; lobes 4, 4–6 × 1–2 mm, obovate, obtuse, upper 2 equal, separate, lower 2 combined nearly up to tip. Stamens 4, only 1 fertile, inserted higher up; filaments filiform, 3–4 mm long, Ovary oblong, obtuse, sessile, 6–7 × 2–3 mm; style slender, linear, about 5 mm long; stigma bilamellate, rotundate, revolute, minutely papillose. Fruits: capsules oblong-ellipsoid, as long as the calyx, 1.2–7.0 × 0.3–0.4 cm. Seeds numerous, minute, blackish brown, compressed; testa nearly reticulate.

Chemical constituents: The whole plant/aerial parts contain xanthones, glycoflavones such as vitexin and isovitexin, terpenoids such as gluanone and canscoradione. Sitosterol, stigmasterol, campesterol, friedlin and β-amyrin. The alkaloids present are gentianine and another glycoalkaloids. Apart from these, the plant also contains phenolic acids, carotenoids and sugar alcohols.

Flowering: November–December.

Fruiting: November–December.

Used in: Ayurveda, Siddha, Unani, folk and traditional medicine; homeopathy.

Parts used: The whole plant.

Uses: The drug made from the plant is bitter, pungent, oleaginous, heating, laxative, alexiteric, anthelmintic and febrifuge It is also a nervine tonic. It improves taste and appetite. It also cures tridosha, leucoderma, leprosy, tuberculosis and fevers. It is reported to promote memory power. It is also used in inflammation and abdominal troubles. The plant is used to cure leprosy, oedema and urinary disorders, and is hypertensive. The fresh juice of the herb is given in all cases of insanity, in doses of about 1 ounce. The plant is pounded together with *Cyperus rotundus* and *Withania somnifera* and used to cure delirium.

Methods of propagation: Through seeds.

Photo: Prof Srinath Rao

CARALLUMA STALAGMIFERA C.E.C. FISCH.

Synonyms: *Caralluma stalagmifera* var. Intermedia Karupp. & Pull.; *Caralluma stalagmifera* var. longipetala Karupp. & Pull.

Classification:

Kingdom:	Plantae
Subkingdom:	Phanerogamea
Class:	Dicotyledonae
Subclass:	Gamopetalae
Series:	Bicarpellatae
Order:	Gentianales
Family:	Asclepiadaceae
Genus:	*Caralluma*
Species:	*Caralluma stalagmifera*

Common names: English: Dark purple caralluma; **Kannada**: Mangana kodu gida; **Telugu:** Kundeti kommulu.

Distribution: Africa, Egypt, East Europe, India, Pakistan, Sri Lanka, UAE. In India found in all states growing wild or cultivated.

Description: It is a perennial succulent plant, slightly branched to about 60 cm tall stem; axis tapers sharply toward the tip, diameter of 5–8 mm. It is greenish-coloured, the edges being overrun with reddish. The leaves are narrow and small, triangular, with a length of 2–3 mm. The inflorescence bears usually only 1, at most 2 flowers. The small bracts are subulate-triangular. The pedicel is about 5 mm long. Flowers: hermaphrodite, radially symmetrical. The 5 sepals are 1.5–2 mm long. The 5 petals are fused basally, the central part of a short funnel-shaped. The fleshy crown is wheel-shaped, with a diameter of about 1 cm. The colour of the crown is greenish outside, inside uniformly dark cream to dark purple and the corolla tube is cream-coloured. Petals: oval-lanceolate, hairy at the top, the edges are folded outwards and the inner surfaces, a silky shine. The mobile hairs the top of the petal are up to 2 mm long, white to dark purple and hang down. The corona is dark purple. The interstaminal secondary crown tip are filled at the base, cup-shaped and with nectar. Towards the peak, they are deeply notched. If you stand upright and face towards it, the general shape is lanceolate-linear. The secondary crown tip are triangular-lanceolate, emarginate at the upper end or fine tridentate. The lateral edges may overlap somewhat. They are often hairy, fine and fluffy. The pollinia measure 0.25 × 0.3 mm. The follicles are stained green with purple stripes.

Flowering: Throughout the year.

Fruiting: Throughout the year.

Chemical constituents: Steroidal glycosides such as indicoside I–II and stalagmoside I–V along with well-known steroidal glycosides such as lasianthosides A–B and carumbelloside III.

Used in: Ayurveda, folklore medicine.

Parts used: Aerial parts.

Uses: The plant is used in traditional Indian medicine and Ayurveda. The juice mixed with black pepper is orally taken against migraine. A decoction of the fresh stems is used orally for treating diabetes. It also shows anti-inflammatory activity and is considered stomachic, carminative, and tonic. The extracts of *Caralluma* are known to stimulate weight loss by suppressing the appetite. It is reported to have antiarthritic properties.

Methods of cultivation: Through stem cuttings.

Photo: Prof Srinath Rao

CARDIOSPERMUM HALICACABUM L.

Synonyms: *Cardiospermum corycodes* Kunze; *Cardiospermum glabrum* Schum. & Thonn; *Cardiospermum luridum* Bl.

Classification:

Kingdom:	Plantae
Subkingdom:	Phanerogamea
Class:	Dicotyledonae
Subclass:	Polypetalae
Series:	Disciflorae
Order:	Sapindales
Family:	Sapindaceae
Genus:	*Cardiospermum*
Species:	*Cardiospermum halicacabum*

Common names: Bengali: Lataphatkari; **English:** Ballon vine; **Hindi:** Kapalaphore-bel; **Kannada:** Batte kaayi balli; **Malayalam:** Jyotishmati; **Marathi:** Kapalaphodi; **Sanskrit:** Indravalli; **Tamil:** Alappukkrran; **Telugu:** Budda gudichi.

Distribution: A native of tropical America. Globally, the species has been recorded throughout the tropical regions. Within India, it is found in the tropical and subtropical regions throughout the plains, generally on fences, ascending up to an altitude of 1300 m. in the Himalayas.

Description: It is an annual scrambling climber, with a wiry stem. Leaves: deltoid, bipinnate, leaflets deeply lobed or dentate. Flowers born in umbellate cymes, small in size, white, polygamous. The lower pair of peduncles of the inflorescence is modified to tendrillar hooks. Fruit: a capsule, 3 in number, 3-valved, pyriform and inflated. Seeds: globose, smooth, with a heart-shaped aril.

Flowering: July–March.

Fruiting: July–March.

Chemical constituents: Leaves contain flavones such as apigenin and acacetin and phenolic acids such as vanillic, syringic, mellitic, ρ-coumaric and ferulic acids. Also present are β-sitosterol, saponins and alkaloids. The seeds yield a fatty oil (28 %), which is unique in having 11-eicosenoic acid as the major component (48 %) and oleic, arachidic and linoleic acids as minor components.

Used in: Ayurveda, Siddha, Unani, folk and traditional medicine; homeopathy.

Parts used: Roots, leaves and seeds.

Uses: The roots and leaves are diuretic, laxative, stomachic and are recommended to improve hair growth, rheumatism, nervous disorders, piles and fever. The plant extract exhibits anti-inflammatory and sedative effects on the central nervous system. The root is used in the form of powder to treat udara roga, rat poisoning, snake poisoning, cough with fever, scrotal enlargement, alopecia, anemia and jaundice.

Methods of propagation: Through seeds.

Photo: Prof Srinath Rao

CARISSA CARANDAS (L.)

Synonyms: *Carissa congesta* Wight; *Carissa salicina* Lam.; *Ardunia carandas* (L.) Baill.; *Capparis casandas* (L.) Burm. f.

Classification:

Kingdom:	Plantae
Subkingdom:	Phanerogamea
Class:	Dicotyledonae
Subclass:	Gamopetalae
Series:	Bicarpellatae
Order:	Gentianales
Family:	Apocynaceae
Genus:	*Carissa*
Species:	*Carissa carandas*

Common names: Bengali Koromcha; **English:** Bengal currant, Christ's thorn; **Hindi:** Karonda; **Kannada:** Kaulihannu, Karandekai, Garchinikai; **Malayalam:** Kalavu, Karekkai; **Marathi:** Karanda, Boranda; **Sanskrit:** Krishnapaka phala; **Tamil:** Kilakkai, Karventa; **Telugu:** Vakkai.

Distribution: Found in Afghanistan, Burma, Nepal, Sri Lanka and all over India in tropical climate.

Description: It is an evergreen shrub growing up to 10–15 feet, rich in white, gummy latex. Highly branched with sharp thorns simple or forked in the axils of the leaves. Leaves are evergreen, opposite, elliptic or oval, 3–6 cm long, leathery and dark green, glossy on the upper surface and dull at the lower surface. Flowers are borne in terminal clusters of 2–12, tubular, 5-lobed, twisted to the left. They are fragrant, white in colour, often tinged with pink colour. Fruits are green when unripe and turn to reddish black when ripe. They exude latex when cut from the branch. Seeds 2–8, small, flat and brown in colour.

Flowering: May–July.

Fruiting: July–September.

Chemical constituents: The roots contain salicylic acid and cardiac glycosides. Other compounds are also reported, such as carissone, the D-glycoside of β-sitosterol; glucosides of odoroside H; carindone, a terpenoid; lupeol; ursolic acid and its methyl ester; also carinol, a phenolic lignan. Fresh fruits are rich in ascorbic acid and the seeds are rich in potassium.

Used in: Ayurveda.

Parts used: Leaves, fruits, dried stem bark and roots.

Uses: The plant is highly effective in treating anemia, anorexia and anxiety. It an effective analgesic, antidiarrheal, anthelmintic and antipyretic. It is used in treating biliousness, cancer, diabetes, heart diseases, inflammation within the body, scurvy and stomachache. The juice of the stem bark is used to treat skin diseases and wounds. The fruit juice is used to improve digestion and to strengthen the cardiac muscles. The root paste is applied to treat diabetic ulcers. A decoction made from the leaves is used in the treatment of diarrhea, earache, fever and oral inflammation.

Methods of propagation: Through seeds and young stem cuttings.

Photo: Prof Srinath Rao

CARUM CARVI (L.)

Synonyms: *Carum aromaticum* Salisb.; *Carum decussatum* Gilib.; *Carum velenovskyi* Rohlene; *Foeniculum carvi* (L.) Link.; *Bunium carvi* (L.) M. Bieb.

Classification:

Kingdom:	Plantae
Subkingdom:	Phanerogamea
Class:	Dicotyledonae
Subclass:	Polypetalae
Series:	Calyciflorae
Order:	Umbellals
Family:	Umbelliferae (Apiaceae)
Genus:	*Carum*
Species:	*Carum carvi*

Common names: Bengali So-jeera; **English:** Black caraway, Persian cumin; **Gujrati:** Shaahjeera; **Hindi:** Syahjeera, Kalijeera; **Kannada:** Shahjirige, Karijirege; **Malayalam:** Karunjeeraka, Karinijeerakam; **Marathi:** Shahaajire; **Sanskrit:** Karavi, Krishna jeerka; **Tamil:** Karamjeeragam, Shimayishiraga; **Telugu:** Sheemajilkara.

Distribution: Native to Western Asia, Europe and Northern Africa. In India it is found in the Northern Himalayas. In the plains it is cultivated during the cold season, while in the hills of Chamba, Garhwal and Kashmir it is cultivated as summer crop at an altitude of 3000–4000 m.

Description: A biennial plant, 50–60 cm long. Stem: erect, cylindrical, smooth, furrowed. Leaves: alternate, finely divided into pinnate or bipinnate and feathery, fern-like. Flowers: small white or pink in umbels, each umbel may contain 6–14 unequal radiuses. The fruits are achenes (erroneously called seeds), crescent-shaped, laterally compressed, somewhat horny, around 2 mm long, with 5 pale ridges, light to dark brown and with a pungent, anise-like flavour and aroma. The root is tap root, thick and tapering, yellow in colour, about 8–9 in long.

Flowering: May–June.

Fruiting: July–August.

Chemical constituents: The seeds contain 40 % proteins, 10 % resin and volatile oil. The chemicals present in the essential oil are acetaldehyde, cumunic aldehyde, R-carvone (79.72 %), D-limonene (19.71 %), β-myrcene (0.07 %), Z-β-ocimene (0.33 %) and α-pinene (0.01%). The whole plant yields a volatile oil consisting of cadinene, carvone and dihydrocarveol. The roots contain umbelliferone and scopoletin. The leaves contain glycosides of kaempferol and quercetin.

Used in: Ayurveda.

Parts used: Fruit and seeds.

Uses: It is used for the treatment of digestive problems such as bloating, gas, loss of appetite, and mild spasms of the stomach and intestines. It helps improve the symptoms of irritable bowel syndrome and also to treat heartburn. Caraway oil is also taken by mouth to help cough up phlegm, improve the control of urination, kill bacteria in the body and relieve constipation. Caraway oil helps start menstruation and relieve menstrual cramps. It is also used to increase the flow of breast milk. Moreover, it is commonly employed as a fragrance in cosmetics, soap and toothpaste. In foods, caraway is used as a spice. The essential oil is an effective tissue regenerator and useful for fighting oily skin while alleviating bruises, reducing boils and cleaning infected wounds. Furthermore, it relieves itching skin and helps to clear acne as well as scalp problems. The fruits are used to improve eyesight. The essential oils extracted from the dried seeds help to relieve mental strain and emotional fatigue. The roots are edible.

Methods of propagation: Through seeds/bulbs. The optimum time for seed sowing or transplanting bulbs is mid-October to mid-November. The seeds are sown in lines 20–30 cm apart; 5.0–7.5 kg of seeds are required for a hectare. The seeds germinate in March–April and develop small bulbs in the first year and remain dormant till the following February. Later, when the plants are 5–7 cm in height, they are thinned to keep 20 cm distances in the lines.

CASSIA AURICULATA (L.)

Synonyms: *Senna auriculata* (L.) Roxb.; *Cassia densistipula* Taub.

Classification:

Kingdom:	Plantae
Subkingdom:	Phanerogamea
Class:	Dicotyledonae
Subclass:	Polypetalae
Series:	Calyciflorae
Order:	Rosales
Family:	Leguminosae (Fabaceae)
Subfamily:	Caesalpinoidae
Genus:	*Cassia*
Species:	*Cassia auriculata*

Common names: English: Avaram senna, Tanner's cassia, Mature tea tree; **Gujarati:** Aawal; **Hindi:** Tarva; **Kannada:** Avarika; **Malayalam:** Avara; **Sanskrit:** Charmaranga, Avaraki; **Tamil:** Avarai; **Telugu:** Tamgedu.

Distribution: It is distributed throughout the hot deciduous forests of India and Sri Lanka.

Description: It is a much-branched shrub with smooth brown bark and pubescent branchlets. The leaves are alternate, stipulate, paripinnate compound, very numerous, closely placed; rachis 8.5–12.5 cm long, narrowly furrowed, slender, pubescent, with an erect linear gland between the leaflets on the rachis of each pair, leaflets 8–12 pairs, very shortly stalked, 2–2.5 cm long, 1–1.3 cm broad, slightly overlapping, oval-oblong, obtuse at both ends, mucronate, glabrous, dull green, paler beneath; stipules very large, reniform-rotund, produced at base on side of next petiole into a filiform point and persistent. The flowers are irregular, bisexual, bright yellow and large (nearly 5 cm across); the pedicels glabrous and 2.5 cm long. The racemes are few-flowered, short, erect, crowded in axils of upper leaves so as to form a large terminal inflorescence, stamens barren; the ovary is superior, unilocular, with marginal ovules. The fruit is a short legume, 7–10 cm long, 1.5 cm broad, oblong, obtuse, tipped with long style base, flat, thin, papery, undulately crimpled, pilose, pale brown. 12–20 seeds per fruit are carried each in its separate cavity.

Flowering: Throughout the year.

Fruiting: Throughout the year.

Chemical constituents: This plant is said to contain a cardiac glucoside (sennapicrin) and sap, leaves and bark yield anthraquinones, while the bark is one of the most priceless of Indian tans, containing nearly 18 % tannin; hence, the plant is also known as 'Tanner's Cassia'.

Used in: Ayurveda and Siddha medicine.

Parts used: Leaves, flowers, roots and bark.

Uses: holds a very prestigious position in the Ayurveda and Siddha systems of medicine. The plant is used in the traditional system of medicine for urinary disorders, female antifertility, leprosy, worm infestation and diarrhea; the leaves have laxative properties; the bark is used in skin conditions and as astringent; the leaves, flowers and fruits as anthelmintic; the seeds for eye troubles and diabetes. The dried flowers are used as tea to treat urinary discharges, nocturnal emissions, diabetes and throat irritation; they are also believed to improve the complexion in women. A powder of the seeds is applied to the eyes to cure conjunctivitis. It is one of the constituents of the polyherbal formulation 'Diasulin'. It is also used in rheumatism, eye diseases, gonorrhea and gout.

Methods of propagation: Through seeds.

Photo: Prof Srinath Rao

CASSIA FISTULA LINN.

Synonyms: *Cassia bonplandiana* DC.; *Cassia excels* Kunth.; *Cassia rhombifolia* Roxb.; *Bactyrilobium fistula* Willd.

Classification:

Kingdom:	Plantae
Subkingdom:	Phanerogamea
Class:	Dicotyledonae
Subclass:	Polypetalae
Series:	Calyciflorae
Order:	Rosales
Family:	Caesalpinaceae
Genus:	*Cassia*
Species:	*Cassia fistula*

Common names: Bengali: Bundaralati, Sondala; **English:** Golden shower, Purgingcassia; **Gujurati:** Garamala, Garamalo: **Hindi:** Amaltas; **Kannada:** Aragvadha; **Malayalam:** Konna, Kritamalam; **Marathi:** Amaltas bahava, Garamala; **Sanskrit:** Aragvadha, Krtamala, Nripadruma; **Tamil:** Irjviruttam, Konai, Sarakonrai, Sarak kondi; **Telugu:** Aragvadhamu, Kondrakayi, Rela chettu.

Distribution: Native of tropical Asia, widely cultivated and naturalized in the tropics, including India.

Description: Tree about 30–35 feet tall. Leaves alternate, pinnate, 30–40 cm long, with 4–8 pairs of ovate leaflets, 7.5–15 cm long, 2–5 cm broad, entire, petiole 2–6 mm long. Inflorescence a raceme, long, drooping. Flowers yellow, petals 5, yellow; sepals 5, green, the flower stalks 3–6 cm long. Stamens 10, 3 with longer stalks. Fruits pendulous, cylindrical, brown, septate, 25–50 cm long, 1.5–3 cm in diameter, resembling somewhat drumstick fruits. Seeds lenticular, light brown and lustrous.

Flowering: April–December.

Fruiting: April–December.

Chemical constituents: The plant is rich in alkaloids, anthraquinones, flavonoids, flavan- 3-ol derivatives, saponins, steroids, tannins and terpenoids. It also contains barbaloin, chrysophanic acid, emodin, fistulic acid, fistuacacidin, galactomannan, hexacosano, lupeol, oxyanthraquinone substances, phlobaphenes, rhein, rhein glucoside, sennosides A and B, β-sitosterol and tannin.

Used in: Ayurveda.

Parts used: Leaves, flowers, root, fruit pulp.

Uses: Antifeedant, anti-inflammatory, anti-microbial, antipyretic, antioxidant, antiulcer cardio-protective, hypoglycemic, liver protective, larvicidal and wound healing activities; employed as a remedy for tumors of the abdomen, glands, liver, stomach and throat, as well as for burns, cancer, constipation, convulsions, delirium, diarrhea, dysuria, epilepsy, gravel, hematuria and pimples. The seeds have anti-bilious, aperitif, carminative and laxative properties. The leaves are employed to treat erysipelas, malaria, rheumatism and ulcers. The flower buds are used for biliousness, constipation, fever, leprosy and skin disease. The root is used for adenopathy, burning sensations, leprosy, skin diseases, syphilis and tubercular glands. The leaves are employed for erysipelas, malaria, rheumatism and ulcers. The fruit is used for abdominal pain, constipation, fever, heart disease and leprosy.

Methods of propagation: Through seeds. Germination is poor – soaking the seeds in sulphuric acid for 15–20 minutes will enhance germination.

Photo: Prof Srinath Rao

CASSIA OCCIDENTALIS L.

Synnonym: *Senna occidentalis* (L.) Roxb.

Classification:

Kingdom:	Plantae
Subkingdom:	Phanerogamea
Class:	Dicotyledonae
Subclass:	Polypetalae
Series:	Calyciflorae
Order:	Rosales
Family:	Leguminosae (Fabaceae)
Subfamily:	Caesalpinioidae
Genus:	*Cassia*
Species:	*Cassia occidentalis*

Common names: Bengali: Kalkashunda; **English:** Coffee senna; **Hindi:** Badikasondi; **Kannada:** Kolthagache, Anecogate, Doddatagase; **Malayalam:** Natramatakara, ponnaviram; **Marathi:** Kasoda; **Sanskrit:** Kasamarda, Arimarda, Kasamardaka; **Tamil:** Ponthagarai, Nattandagarai; **Telugu:** Kasinda peddakasinda.

Distribution: Found throughout India, up to an altitude of 1500 m.

Description: An erect, fetid, annual herb or under shrub, 1–1.5 m, with reddish stems without hairs. Leaves: compound, pinnate, 15–20 cm long; each leaf has a distinct spherical-shaped gland, which is located about 0.3–0.5 cm from the base of the petiole. Each leaf has 3–5 pairs of lanceolate or ovate-lanceolate leaflets which are glaucous, membranous, with a long, fine pointed tip arranged oppositely. Inflorescence: a terminal or axillary raceme. Flowers: yellow and about 2 cm long and 3–4 cm wide.

Fruits: A pod, recurved, glabrous, compressed, 10–13 cm long and about 1 cm broad, curved slightly upwards. Each pod contains 20–30 dark seeds, ovoid, compressed, hard, smooth, shining and dark olive-green.

Flowering: March, June.

Fruiting: August–January.

Chemical constituents: The leaves contain anthraquinones such as chrysophanol, amodin and physcion. Also present are tetrahydroanthracene derivatives, germichrysone and occidentalins A and B. The flowers contain β-sitosterol, emodin, physcion and its 1-beta-D-glucoside. The seeds contain a gum –galactomannan, a fixed oil and anthraquinones, physcion, rhein, aloe emodin, chrysophanol and steroidal glycosides. The roots yield pinselin (cassiollin), rhein, aloe emodin and their glycosides, chrysophanol, physcion, emodin, islandicin, helminthosporin, xanthorin, sitosterol, campesterol, stigmosterol and quercetin.

Used in: Ayurveda, Siddha and traditional medicines.

Parts used: All parts of the plant.

Uses: Almost all parts of the plant are purgative, tonic, febrifugal, expectorant and diuretic. The plant is used to cure sore eyes, haematuria, rheumatism, typhoid, asthma and disorders of the haemoglobin; it is also reported to cure leprosy. A decoction of the plant is used in hysteria, dysentery and other stomach troubles and also as an application to sores, itch and inflammation of the rectum. The leaves have purgative, febrifugal, tonic, sudorific, diuretic and stomachic properties. They are used internally and externally in skin diseases, such as itches, yaws, scabies and ringworm They are also used in inflammatory swellings, rheumatism, wounds, sprains and wrenches, and are given in jaundice, pleurisy, headache and toothache. The plant is employed in dropsy and as a vermifuge. The leaves are anti-cancerous in nature due to the presence of germichrysone and cassia occidentalin A, B. The seeds are considered to be a blood tonic and excellent diuretic, and are useful against cough and whooping cough, convulsions and heart diseases. Their powder is externally applied in cutaneous diseases and eruptions. An infusion of the bark is given in diabetes. The roots are considered bitter, tonic, purgative, anthelmintic and diuretic, and are also used to treat dysentery and diarrhea associated with malaria. The pounded root mixed with black pepper is used in case of headache; it is given to nursing women to purify the milk. The root bark is used as a quinine substitute to cure fever and its decoction is given for chest complaints, as a stimulant and as a specific medicine for gonorrhea and hepatic troubles. The herb forms an ingredient of the patented indigenous herbal drug 'Liv-52', which shows a marked effect in early cases of hepatic cirrhosis with steatorrhea. It is also an ingredient of another patented drug: 'Geriforte'.

Methods of propagation: Through seeds from mature pods.

Photo: Prof Srinath Rao

CASSIA SENNA (L.)

Synonyms: *Cassia acutifolia* Deli; *Cassia angustifolia* Vahl; *Cassia alexandrina* (Garsault) Thell.; *Senna alexandrina* Garsault; *Senna angustifolia* (Vahl) Batka.

Classification:

Kingdom:	Plantae
Subkingdom:	Phanerogamea
Class:	Dicotyledonae
Subclass:	Polypetalae
Series:	Calyciflorae
Order:	Fabales (Rosales)
Family:	Fabaceae (Leguminosae)
Genus:	*Cassia*
Species:	*Cassia senna*

Common names: English: Alexandrian senna, Sen; **Hindi:** Bhuikhakhasa, Hindisana; **Kannada:** Avarike; **Malayalam:** Sunnamukki; **Sanskrit:** Pitapushpi, Swarnamukhi, Swarnapatrika; **Tamil:** Avarai; **Telugu:** Tamgadu.

Distribution: The plant is cultivated throughout South India and few parts of Rajasthan and Gujrat.

Description: This is an erect branching shrub about 1.5 m in height. Leaves: pinnate, with 3–5 pairs of lanceolate acute leaflets of about 6 mm. Flowers: brilliant yellow borne in terminal racemes. Fruit: A pod over 5 cm long, flat, containing 5–7 obovate dark-brown and smooth seeds.

Flowering: August–September.

Fruiting: August–September.

Chemical constituents: The leaves and the shells of the fruits contain sennosides and related anthraquinones. The plant contains sennoside A–D (Srivastava et al., 2010). Also present are free anthraquinones such as rhein, chrysophanol, emodin and aloe emodin. The flavonoids present are isorhamnetin and kaempferol. The roots contain physcion, physcionin, chrysophanin and sennidin C.

Used in: Ayurvedic and Unani medicine.

Parts used: Dried leaves and fruits.

Uses: The leaves and shells are used as tonics in Ayurvedic and Unani medicine. This is an efficient purgative, but due to its tendency to cause gripe, it is usually administered with carminatives. This plant is also used as a febrifuge, in splenic enlargement, anaemia, jaundice, rheumatism, tumours and bronchitis as well as against skin diseases. The plant pacifies vitiated *pitta*, *vata*, constipation, jaundice, dyspepsia, helminthiasis, cough and bronchitis (https://www.stylecraze.com/articles/benefits-and-uses-of-senna-plant/).

Methods of propagation:

It is usually propagated by seeds. As a pure crop, it is raised on red loams including coarse gravelly soils or alluvial loams.

Steps to follow:

The seeds have a hard seed coat, which may be softened by rubbing the seeds against a rough surface such as sandpaper.

The seeds are then soaked in cold water for 8 hours.

- Dibble the seeds on seed beds at a distance of 1 foot and 2 cm deep. Irrigate the beds lightly.
- Sow the seeds either during the months of January–February or during June–July.
- Germination starts after 13–15 days of sowing and is over in 1 week.

Photo: Prof Srinath Rao

CATHARANTUS ROSEUS (L.) G. DON.

Synonym: *Vinca rosea* L.; *Locnera rosea* (L.) Reich.

Classification:

Kingdom:	Plantae
Subkingdom:	Phanerogamea
Class:	Dicotyledonae
Subclass:	Gamopetalae
Series:	Bicarpellatae
Order:	Gentianales
Family:	Apocynaceae
Genus:	*Catharantus*
Species:	*Catharanthus roseus*

Common names: English: Periwinkle; **Hindi:** Sadabahar; **Kannada:** Sadapushpi; **Malayalam:** Ushamalari; **Sanskrit-:** Sadafuli, Nityakalyani; **Tamil:** Sudukadu mallikai; **Telugu:** Billaganneru.

Distribution: The plant is a native of Madagascar, but has become naturalized throughout the tropics of both hemispheres.

Description: An erect herb to about 1 m high. Leaves: ovate, opposite; flowers in axillary clusters of 2 or 3 petals, white or rose-pink, in one variety petals white with pink or reddish tinge at base. Calyx: 5 in number and fused. Corolla: 5 in number, gamopetalous and pink in colour. Ovary: Ovary superior, carples 2 free at base, ovules many. Fruit: A follicle, many-seeded.

Flowering: Throughout the year.

Fruiting: Throughout the year.

Chemical constituents: More than 400 indole alkaloids have been reported from this plant (Sain and Sharma, 2013), but the most important are indole dimeric alkaloids such as vinblastine and vincristine, which are present in the aerial parts. The roots contain ajmalicine, serpentine and reserpine. Vincristine and vinblastine are oncolytic alkaloids effective against leukaemia cells. They are also used against Hodgkin's disease. The leaves contain essential oils such as linolenic acid ethyl ester, stearic acid, phytol and hexadecanoic acid. The flowers contain limonene, phytol and linolenic acid (Lawal, et al., 2015).

Used in: Ayurvedic, Siddha and traditional medicine; homeopathy.

Parts used: Roots and leaves.

Uses: The roots of the plant constitute the drug. They are known to possess toxic and stomachic properties. The alkaloids possess hypotensive, sedative and tranquillizing properties; they are also diuretic, hypoglycemic and antimicrobial.

Other uses: This is a handsome garden plant which flowers throughout the year. This gives the plant the name 'Sadabahar'. The plant is quite hardy and popular for ornamental planting.

Methods of propagation: Through seeds.

Steps to follow:

- Freshly collected seeds are sown in the nursery beds (seed viability is 6–8 months).
- Soaking the seeds in water for 12–24 hours, before sowing, increases the germination percentage.
- Young seedlings are later transplanted in the field.

Photo: Prof Srinath Rao

CAYRATIA PEDATA (WALL.) GAGNEP.

Synonyms: *Cissus pedata, Cissus serratifolia, Vitis pedata.*

Classification:

Kingdom:	Plantae
Subkingdom:	Phanerogamea
Class:	Dicotyledonae
Subclass:	Polypetalae
Series:	Disciflorae
Order:	Celastrales
Family:	Vitaceae
Genus:	*Cayratia*
Species:	*Cayratia pedata*

Common names: English: Birdfoot grape vine; **Kannada**: Kaama pattige; **Malayalam:** Velutta sori valli; **Marathi:** Gorpadvel; **Sanskrit:** Suvaha; **Tamil:** Kattu pirandi, Pannikkodi, Tiripatakam; **Telugu:** Gummadi tega.

Distribution: Distributed in the tropical evergreen and semi-evergreen forests of India, Sri Lanks and Malaysia. It is included in the IUCN list of vulnerable plants.

Description: It is a large, weak climbing shrub. Tendrils are opposite to leaf, branched and coiled. Leaves alternate, pedately 5–7 foliate, leaf stalk long about 2–7 cm long, leaflets 5–12 × 3–6 cm across, elliptic to egg-shaped, base rounded, margin toothed. Inflorescence a corymb, present on the leaf axil. Flowers bisexual, very small, about 3mm, Yellowish green.

Flowering: March–December.

Fruiting: March–December.

Chemical constituents: Rich in vitamins. The plant contains alkaloids, coumarins, flavonoids, phenols, phytosterols, phlobatannins, quinines, saponins, steroids and terpenoids (Aswathy et al., 2019).

Used in: Folk and traditional medicine.

Parts used: The whole aerial plant and the roots.

Uses: The aerial parts of the plant are used to treat asthma, bronchitis, cough, joint pain and uterine reflexes. Leaves astringent, diuretic, spasmolytic and refrigerant, used to treat diarrhea and ulcers. Young shoots powdered and fried in coconut oil are given to children to remove intestinal worms and to cure diarrhea. Roots astringent. The dried powder of tuberous roots is mixed with sugar and given to treat fever and swellings.

Methods of cultivation: Through seeds.

Photo: Prof Srinath Rao

CELASTRUS PANICULATUS WILD.

Synonym: *Celastrus dependens* Wall.

Classification:

Kingdom:	Plantae
Subkingdom:	Phanerogamea
Class:	Dicotyledonae
Subclass:	Polypetalae
Series:	Disciflorae
Order:	Celastrales
Family:	Celastraceae
Genus:	*Celastrus*
Species:	*Celastrus paniculatus*

Common names: Bengali: Malkanjri; **English:** Intellect tree; **Gujarati:** Malkangana; **Hindi:** Malkangani; **Kannada:** Kariganne; **Malayalam:** Polulavam; **Marathi:** Malkangoni; **Sanskrit:** Jyotishmati; **Tamil:** Valuluvai; **Telugu:** Malaria teega.

Distribution: In sub-Himalayan tracts up to 2000 m in Central India, Western and Eastern Ghats extending up to the Rajmahal Hills in Bihar and Orissa (Wealth of India, 1992).

Description: It is a deciduous woody vine with stems up to 10 cm in diameter and 6 m long with rough, pale-brown exfoliating bark covered densely with small, elongated lenticles. Leaves: simple, alternate, broad and oval, obovate or elliptic in shape, with toothed margins, 6 × 11 cm. Flowers: borne in terminal large panicles. Unisexual, small, yellowish green. Fruit: a capsule, 3-celled, yellow to orange, globose. Seeds: reddish brown in colour, ellipsoid or ovoid, enclosed in scarlet.

Flowering: December–May.

Fruiting: December–May.

Chemical constituents: The seeds constitute the drug; they possess aphrodisiac, emetic, diaphoretic, sodorific, febrifugal and nervine properties and are also used for curing sores, ulcers, rheumatism and gout. The oil is stomachic, tonic, good for cough and asthma; used in leprosy; cures headache and leucoderma. The seeds contain fatty acids and alkaloids, and have sedative and antidepressant actions. They have dose-dependent cholinergic activity, thereby improving memory performance. The leaves are emmenagogue. The seeds contain the alkaloids celastrine and paniculatin.

Used in: Ayurveda, Siddha, Unani and tradional medicine; homeopathy.

Parts used: Bark, leaves and seeds.

Uses: According to Ayurveda, the leaves are emmenagogue whereas the seeds are acrid, bitter, hot, appetizer, laxative, emetic, aphrodisiac and a powerful brain and liver tonic. They also cause a burning sensation. The seed oil possesses sedative and anticonvulsant properties. It also enriches the blood and cures abdominal complaints. According to the Unani system of medicine, the seeds are expectorant and cure joint pains, paralysis and weakness. The oil is stomachic, tonic, good for cough and asthma; used in leprosy, cures headache and leucoderma. The oil has been reported to increase the intelligence quotient (IQ) of children with learning disabilities (Gaitonde et al., 1957).

Methods of Propagation: Through seeds.

Photo: Prof Srinath Rao

CENTELLA ASIATICA (L.) URBAN

Synonym: *Hydrocotyle asiatica* L.

Classification:

Kingdom:	Plantae
Subkingdom:	Phanerogamea
Class:	Dicotyledonae
Subclass:	Polypetalae
Series:	Calyciflorae
Order:	Umbellales
Family:	Apiaceae (Umbelliferae)
Genus:	*Centella*
Species:	*Centella asiatica*

Common names: Bengali: Thankuni, Tholkuri; **English:** Indian pennywort; **Gujarati:** Moti brahmi; **Hindi:** Brahmamanduki, Brahmi, Bemgsag; **Kannada:** Brahmisoppu, Urage; **Malayalum:** Kutannai, Muttil; **Marathi:** Brahmi; **Sanskrit:** Brahmamanduki; **Tamil:** Akantapuntu, nirpirami; **Telugu:** Mandook brahmi, Sarswataku.

Distribution: It is distributed in temperate to tropical regions. It is generally found near streams. It spreads on the ground on moist soil, especially along bunds and canals. This species is globally distributed. Within India, it is commonly found as a weed in crop fields and other waste places throughout, up to an altitude of 600 m.

Description: It is a creeping stoloniferous perennial herb with a prostate stem rooting at the nodes. Stem: glabrous, pink and striated. Leaves: simple reniform, fleshy, crenate-dentate, base cordate and often lobed, long-petioled, smooth on the upper surface and sparsely hairy on the lower side. Flowers: irregular, reddish, minute, in umbels of 3. Calyx: 5 sepals. Corolla: 5 petals. Ovary inferior or sunk in the hypanthium or disc. Fruit: laterally compressed, mericarp, oblong, dull brown.

Flowering: Throughout the year.

Fruiting: Throughout the year.

Chemical constituents: A number of triterpenoid glycosides are present in the plant, such as asiaticoside, indocentelloside, brahmoside and brahminiside (Zheng and Qin, 2007). Other constituents reported include flavonoids such as quercetin and kaempferol (Hunter, 2007; Zheng and Qin, 2007; Longmore, 2005) and some phytosterols such as campesterol, sitosterol and stigmasterol (Longmore 2005).

Used in: Ayurveda, folk, Tibetan, Siddha and modern medicine; homeopathy.

Part used: The whole plant.

Uses: The whole plant is used in the form of juice to treat cough, insanity, boils, chronic coryza and jaundice. For an intellect-promoting rasayana, the juice is mixed with milk. This is lifegiving, destroys diseases and promotes strength, digestive power, complexion and voice. A local application of the juice cures boils. The juice is also mixed with honey, haridra (*Curcuma longa*), amalaki (*Phyllanthus emblica*) or milk and taken in the morning to cure jaundice.

Methods of propagation: It is propagated through seeds as well as by stem divisions.

Steps to follow:

- The fresh seeds are collected during February–March and sown on nursery beds or directly in the fields. 60 % germination is obtained within 10 days.
- Stem pieces with 3–4 nodes are used as the planting material; 100 % sprouting will be seen in 7–10 days.

Photo: Dr Kotresh

CENTRATHERUM ANTHELMINTICUM (L.) KUNTZE

Synonyms: *Vernonia anthelmintica* (L.); *Conyza anthelmintica* L.

Classification:

Kingdom:	Plantae
Subkingdom:	Phanerogamea
Class:	Dicotyledonae
Subclass:	Gamopetalae
Series:	Inferae
Order:	Asterales
Family:	Compositae Asteraceae
Genus:	*Centratherum*
Species:	*Centratherum anthelminticum*

Common names: Bengali: Somraj, Kalijiri; **English:** Iron weed, Purple fleabane; **Gujarati:** Kali jiri; **Hindi:** Kaliziri, Bukshi; **Kannada:** Kadujirige; **Malayalam:** Kalajirakum; **Marathi:** Kalenjiri, Kadujera; **Sanskrit:** Aranyaajarika, Vanya jirika; **Tamil:** Kattuchiragam; **Telugu:** Adavajilakara.

Distribution: Throughout India, Afghanistan, Pakistan and Sri Lanka.

Description: An erect, pubescent, annual herb up to 60–90 cm tall. Leaves alternate, elliptic-lanceolate, 5–8 cm long and 2.5–3 cm wide, obovate to lance-shaped, base tapering into petiole, margins serrate, pubescent on both surfaces. Florets violet or purple, tubular, 5-lobed in many-flowered, homogamous, solitary, axillary or terminal heads, bract linear; involucral bracts linear, hairy. Flower's head 1.5–2.5 cm in diameter; each head with 30–40 minute purplish flowers. Fruits are small, 4.5–6.5 mm long, cylindrical and hairy with 10 narrow ridges. The plant has an unpleasant odour due to presence of volatile oils.

Flowering: Rainy season.

Fruiting: Spring.

Chemical constituents: The seeds are rich in flavonoids containing the alkaloid veronine, brassicasterol, butin, and stigmasterol as well well as linoleic, oleic, monohydroxy-oleic, myristic, palmitic, vernolic and stearic acids (Akihisa et al., 1992).

Used in: Ayurveda, folk medicine.

Parts used: Seeds, leaves.

Uses: It is used for the treatment of skin diseases; it helps to reduce itching and skin irritation. It is a blood purifier; it removes toxins from the blood. It is used in removing roundworm, tapeworm and threadworm. It increases appetite.

This plant also promotes hair growth. It helps to control the blood glucose level and is beneficial for people with postprandial hyperglycemia (rise in blood glucose level following a meal). The tender leaves are used in the treatment of cough and diarrhea. They are anthelmintic, stomachic and anti-phlegmatic (Mehta and Mehta, 2004; Nadkarni, 1927). The seeds are used as anthelmintic, antihyperglycemic, antipyretic, anti-inflammatory and to cure eczema, skin ulcers, leucoderma and psoriasis (Arya et al., 2012). Kali jeeri (kalijiri or *Centratherum anthelminticum*) is often confused with black caraway seed or black cumin seeds (kala jeera).

Methods of propagation: Through seeds.

Photo: Ayurvedaherb.blogspot.com. Seeds of Kalijiri

CHLOROPHYTUM BORIVILIANUM SANTAPAU & FERNANDES

Synonyms: No synonyms are recorded for this.

Classification:

Kingdom:	Plantae
Subkingdom:	Phanerogamea
Class:	Dicotyledonae
Subclass:	Polypetalae
Series:	Coronarieae
Family:	Liliaceae
Genus:	*Chlorophytum*
Species:	*Chlorophytum borivilianum*

Common names: English: Land-calotrops; **Hindi:** Safed musli; **Kannada:** Dravanti; **Malayalam:** Shedheveli; Sanskrit; Swetha Musli; **Tamil:** Taniravi, Thang, Tiravanticam; **Telugu:** Swetha musli.

Distribution: This species is endemic to India.

Description: It is a small perennial herb with 6–12 sessile linear radical spreading leaves, solitary terminal scape containing white flowers in clusters of 3. Fruit: a capsule 3-angled, dehiscent, loculicidal. Seed: many, small, black and angled.

Flowering: June–September.

Fruiting: June–September.

Chemical constituents: The tubers contain saponins (2–3 %), proteins (8 %), carbohydrates (40 %), mucilage fiber (25–35 %), alkaloids (15–25 %), saponins (2–20 %) and proteins (5–10 %). This plant is a rich source of over 25 alkaloids, vitamins, proteins, carbohydrates, steroids, saponins, potassium, calcium, magnesium, phenol, resins, mucilage and polysaccharides, and also contains a high quantity of simple sugars, mainly sucrose, glucose, fructose, galactose, mannose and xylose.

Used in: Folk, Ayurvedic, Unani and allopathic medicine; homeopathy.

Parts used: Tuberous roots.

Uses: The tubers of this plant have aphrodisiac and galactagogue properties (Bhandary et al, 1995). The powder of the tubers is used to treat freckles, deafness, physical illness, weakness, diabetes, arthritis, natal and postnatal problems, rheumatism and joint pains. It is also used as a revitalizer and a general sex tonic. It has antimicrobial, anti-inflammatory and antitumor properties. It fights diarrhea, dysentery, gonorrhea and leucorrhea, and increases immunity. Moreover, it has spermatogenic properties and is found useful in curing impotency. The leaves are eaten by the tribes of the Western Ghats as an expectorant (Sebastian and Bhandari, 1984). The root is used as aphrodisiac, astringent and diuretic, useful in dysentery, as an antidiabetic and as an appetizing agent (Chetty and Rao, 1989).

Methods of propagation: The plant is usually propagated through seeds as well as by root divisions. It prefers well-drained, loamy to sandy soil rich in organic matter. The plant exhibits poor flowering and seed setting.

- Since the fruits dry and split open to disperse the seeds, it is not easy to collect them. Cover the inflorescence with a bag after the fruits are developed. This will ensure the collection of the seeds. Otherwise they will be dispersed.
- The seeds are to be sown in seedbeds, which need to be heavily manured with farmyard manure or well-rotten leaf manure. The first or second week of June is ideal for sowing.
- The germination extends up to 16 days with 11–24 % germination.
- More vigorous plants are produced when the root cuttings are used. Root cuttings are to be taken from the tough root system before the plant begins new growth. Collect the fleshy root bunches from the ground during May and store them in polythene bags. Even a small (1 cm), fleshy root has the capacity to reproduce a new plant.
- The seedlings emerge out of the ground within 4–6 days after rains.

CHLOROXYLON SWIETENIA DC.

Synonym: *Swietenia chloroxylon* Roxb.

Classification:

Kingdom:	Plantae
Subkingdom:	Phanerogamea
Class:	Dicotyledonae
Subclass:	Gamopetalae
Series:	Disciflorae
Order:	Geraniales
Family:	Rutaceae
Genus:	*Chloroxylon*
Species:	*Chloroxylon swietenia*

Common names: English: Ceylon satinwood; **Hindi**: Bhirra; **Kannada:** Bittulla; **Malayalam:** Varimaram; **Marathi:** Halda; **Sanskrit:** Bherul, Bhillotaka; **Tamil:** Vaaimaram; **Telugu:** Billudu.

Distribution: It is a native of Sri Lanka and South India, extending in the north to the Satpuras, Chota Nagpur and Chhattisgarh.

Description: A medium-sized deciduous tree, up to 15 m tall, with a short, straight, clear bole and spreading crown. Bark thick, corky rough, furrowed, pale yellow or light gray, aromatic; leaves alternate, pinnate, 15–22 cm long, aromatic; leaflets 10–20 pairs, oblong, unequal-sided, gland-dotted,

1.5–3.0 cm long and 0.6–1.5 cm wide. Flowers: white or cream, born in terminal or axillary, pyramidal, panicles up to 15 cm long. Fruits: capsules, 3-segmented, oblong, glabrous, 2.5 cm long, dark brown when ripe, dehiscent; seeds oblong, brown, winged. The tree is usually leafless during February–April.

Flowering: March–April.

Fruiting: May–August.

Chemical constituents: The bark contains 17 % tannins. The alkaloid skimmianine is contained in the heartwood. The leaves, bark and roots of the plant include alkaloids (Ravi Kiran et al., 2006), cumarin (Govindachari et al., 1967) fluoroquinolones (Vorko and Sedmera, 1977) and lignans. Some active constituents, such as coumarins xanthyletin, xanthoxyletin and 7- demethylsuberosin (King, et al., 1954) are present. The flowers have been found to contain gossypetin and its glycoside gossypetin-8-O-β-D glucopyranoside 3-sulphate.

Used in: Ayurveda, folk and Siddha medicine.

Parts used: Leaves, bark and seeds.

Uses: The bark of the tree is used as an astringent. The paste made from the leaves is used to treat wounds, snake bites and rheumatism. The plant extract also has anti-fertility activity. The extract from the bark is used to treat chest pain and asthma (Nikitha Devi and Krishna Mohan, 2018.

Methods of propagation: Through seeds.

Photo: Dr Kotresh

CHONEMORPHA FRAGRANS (MOON) ALSTON

Synonyms*:* *Chonemorpha macrophylla*, *Chonemorpha grandiflora* (Roth).

Classification:

Kingdom:	Plantae
Subkingdom:	Phanerogamea
Class:	Dicotyledonae
Subclass:	Gamopetalae
Series:	Bicarpellatae
Order:	Gentianales
Family:	Apocynaceae
Genus:	*Chonemorpha*
Species:	*Chonemorpha fragrans*

Common names: Bengali: English: Frangipani vine; **Hindi:** Moorva; **Kannada:** Manijinaru; **Malayalam:** Perumkurumba; **Marathi: Sanskrit:** Murva; **Tamil:** Perumkurumpa; **Telugu:** Chaga.

Distribution: Indian and Malayan forests. Found in dense mountain forests, often clinging to trees.

Description: A latex-containing woody climber. It climbs to the tops of the tallest trees, about 20 m in height, branchlets tomentose. Leaves large, shiny green, 20–25 cm long and 15–20 cm broad, elliptic or ovate-orbicular, apex shortly acuminate, base cordate, pubescent above and tomentose beneath, veins prominent. Flowers pure white with a yellow center, waxy texture, highly fragrant in terminal tomentose cymes; calyx 1 cm long, glabrous, glandular within; corolla fairly large, salver-shaped, lobes longer than tube; stamens included, anthers lanceolate, sagittate; carpels distinct, stigma bifid, fruit a follicle, spindle-shaped, 30 × 2 cm long, coma 4–5 cm long, white.

Flowering: April–December.

Fruiting: April–December.

Chemical constituents: Steroidal alkaloids, such as chonemorphine and funtumafrine camptothecin.

Parts used: Roots, leaves.

Uses: The root extract is used in the treatment of diabetes, bronchitis, cough, constipation, dipsia, dysphasia, cancer, jaundice, leprosy and scabies. It is anthelmintic, antipyretic, antispasmodic, febrifuge, digestive, laxative and a blood purifier.

Methods of propagation: Through seeds. The seeds are sown in a well-drainied soil mixture, and should be kept damp in bright light.

CINNAMOMUM ZEYLANICUM BLUME

Synonym: *Cinnamomum verum* Presl.

Classification:

Kingdom:	Plantae
Subkingdom:	Phanerogamea
Class:	Magnoliopsida
Subclass:	Dicotyledonae
Order:	Daphnales
Family:	Lauraceae
Genus:	*Cinnamomum*
Species:	*Cinnamomum zeylanica*

Common names: English: Cinnamon, Cylon; **Hindi:** Dalchini; **Kannada:** Dalchini, Lavangpatti, Dalcini; **Malayalam:** Elavangam; **Marathi:** Dalchini; **Sankrit:** Darusita; **Tamil:** Kannalavangpatti; **Telugu:** Chekka dasini.

Distribution: This tree occurs in South India up to an altitude of about 1500 m, but it is more common at lower altitudes, even below 200 m. It is also cultivated in certain parts of India.

Description: It is an evergreen tree, i.e. it does not become totally leafless at any time of the year; about 6–8 m high. Leaves: large, ovate thick, leathery, pointed at tip, shining green, lighter coloured beneath; main nerves on the leaves, 3 or 5, running from base of leaf to middle or tip. Flowers: hermaphrodite or polygamous, minute, in large hairy clusters. Calyx 6, cleft, with the limb deciduous. Stamina 12, in 4 rows; the 9 external ones fertile, the 3 inner ones capitate, abortive; the 3 most internal of the fertile stamina having 2 sessile glands at the base; anthers 4-celled, the 3 inner turned outwards. Ovary 1-celled, with 1 ovule. Fruit: oblong or ovate, about 1.5–2 cm long, dark purple, seed 1.

Flowering: March–April.

Fruiting: March–April.

Chemical constituents: The most important constituents of cinnamon are cinnamaldehyde and trans-cinnamaldehyde (Cin), which are present in the essential oil, thus contributing to the fragrance and to the various biological activities observed in cinnamon (Yeh et al., 2013). The bark contains procyanidins and catechins.

Used in: Ayurveda, Siddha and Unani medicine.

Parts used: Bark and oil.

Uses: The branches of the trees are lopped and their bark is removed; the dried inner bark constitutes the drug Cinnamon. This drug is used in diarrhea, nausea and vomiting. It is commonly used as a condiment.

The oil obtained from the bark, called cinnamon bark oil, has similar properties, and has even certain advantages over the bark. It is used as a stomachic and carminative, it cures gastric debility and flatulence and also has the property of destroying certain germs and fungi.

The oil obtained from the leaves is used as a flavouring agent and preservative for sweets, soaps, etc., and for local application on certain rheumatic pains. Cinnamon extracts lower blood sugar and cholesterol levels (Blevins et al., 2007). The aqueous extract has the potential to cure Alzheimer's disease (Peterson et al., 2009).

Other species: The genus *Cinnamomum* has several other useful species. *C. camphora* (L.) Presl (Camphor tree, *Kapur*) is cultivated in India in Nilgiris and also in some botanical gardens of Northern India. Camphor is obtained through the distillation of its wood and leaves. This helps treat inflammation and rheumatic pains. It is also given internally in certain types of diarrhea or as cardiac stimulant. Camphor is used in many other ways and more than 5000 quintals of it are imported into India each year. Camphor is now also being obtained from certain species of *Ocimum*, the genus of the well-known *Tulsi* plant.)

C. tamala Nees & Eberm. (Hindi – *Tejpat*, Sanskrit – *Tamalpatra*, Tamil – *Talishpattiri*) or *Indian cassia* occurs in the outer and middle Himalayas, and in parts of Assam and Bengal. The leaves are mainly used as a spice, but also as a carminative and in colic and diarrhea.

Methods of propagation: Through seeds.

CISSUS QUADRANGULARIS (L.)

Synnonyms: *Vitis quadrangularis* (L.) Wall. ex Wight & Arn.; *Vitis succulent* Galpin; *Cissus bifida* Schumach. & Thonn.; *Cissus edulis* Dalzell.

Classification:

Kingdom:	Plantae
Subkingdom:	Phanerogamea
Class:	Dicotyledonae
Subclass:	Polypetalae
Series:	Disciflorae
Order:	Celastrales
Family:	Vitaceae
Genus:	*Cissus*
Species:	*Cissus quadrangularis*

Common names: Bengali Harajora; **English:** Adamant creeper, Devil's backbone; **Gujarati:** Asthisrnkhala; **Hindi:** Hadjod; **Kannada:** Mangarvalli, Vajravalli; **Malayalam:** Changalamparanda, Patahtulang; **Marathi:** Ghanasavel, Ghonuskande; **Sanskrit:** Asthisamdhani, Asthisamharaka; **Tamil:** Pirandai, Arukani, Kirutti; **Telugu:** Gudametige, Kirutti.

Distribution: It is probably native to India or Sri Lanka, but is also found in Africa, Arabia, and Southeast Asia. It has been imported to Brazil and the Southern United States.

Description: The plant reaches a height of 1.5 m and has quadrangular-sectioned branches with internodes 8–10 cm long and 1.2–1.5 cm wide. Along each angle is a leathery edge. Leaves 2–5 cm wide, toothed, trilobed, appear at the nodes. Each has a tendril emerging from the opposite side of the node. Racemes of small, white, yellowish or greenish flowers; globular berries are red when ripe.

Flowering: June–July.

Fruiting: August–September.

Chemical constituents: The aerial parts have been found to contain carotenoids, triterpenoids and ascorbic acid. The plant also produces resveratrol, ketosteroids and quadrangularin A., as well as high amounts of vitamins C and E.

Used in: Ayurveda, Siddha and traditional medicine.

Parts used: Leaves and stem.

Uses: *Cissus* has been used in various ayurvedic classical medicines to heal broken bones and injured ligaments and tendons, to treat obesity, diabetes, metabolic syndrome, high cholesterol, osteoporosis, scurvy, cancer, upset stomach, hemorrhoids, stomach ulcer, menstrual discomfort, asthma, malaria and pain. Gargling with the juice of this plant helps to treat pyorrhea and reduces the inflammation of the gums. It clears plaque and tartar. It also has an antimicrobial action that prevents and stops microbial growth. In traditional medicine, fresh stems of C. *quadrangularis* are warmed keeping them in burning coals for a few minutes. Then the juice is extracted from them. 20 ml of extracted juice is added to 20 ml of ghee and taken after meals twice daily for 7 days. This combination helps to treat syphilis and reduces its symptoms (Mallika et al., 2005).

Photo: Prof Srinath Rao

CITRULLUS COLOCYNTHIS (L.) SCHRAD.

Synonyms: *Colocynthis officinalis* Schrad.; *Colocynthis vulgaris* Schrad.

Classification:

Kingdom:	Plantae
Subkingdom:	Phanerogamea
Class:	Dicotyledonae
Subclass:	Polypetalae
Series:	Calyciflorae
Order:	Passiflorales
Family:	Cucurbitaceae
Genus:	*Citrullus*
Species:	*Citrullus colocynthis*

Common names: English: Desert gourd, Wild gourd

Distribution: It is native to the Mediterranean Basin and Asia. It is distributed in North America, Egypt, India, Pakistan, Sahara and Southern European countries. It grows from sea level up to 1500 m above sea level on sandy loam, subdesert soils.

Description: It is an annual or perennial plant. The tap roots are long, large, fleshy and perennial. The leaves are palmate and angular with 3–7 divided lobes (similar to watermelon). The flowers are yellow and solitary in the axes of leaves and are borne on greenish yellow peduncles, monoecious, i.e. male and female flowers are on the same plant. Calyx 5-partied. Corolla campanulate, 5-lobed. Stamens 5, 4 grouped and 1 single with monadelphous anther. The female flowers have 3 staminoids. Ovary inferior, hairy, 3-carpellate. Fruits smooth, 5–10 cm in diameter, green with undulate yellow stripes. Mesocarp filled with spongy white pulp. Seeds embedded in the fruit pulp, 3 per carpel, grey, 5 mm long and 3 mm wide.

Flowering: October–November.

Fruiting: December.

Chemical constituents: The plants contain alkaloids, cucurbitacins, flavonoids, glycosides, saponins and tannins. The seeds contain linoleic, oleic, palmitic and stearic acids, and also good amounts of arginine, tryptophan and sulphur containing amino acids. The fruits contain constituents such as glycosides, flavonoids, alkaloids and terpenoids as well as curcurbitacins A– L, and colocynthosides A and B (Hussain et al., 2014).

Used in: Traditional medicine.

Parts used: Fruits and seeds.

Uses: Traditionally, this plant is used to cure anemia, asthma, bronchitis, cancer, constipation, diabetes, dyspepsia, enlargement of spleen, jaundice, leprosy, leucoderma, rheumatism, tuberculosis, ulcers and urinary discharge (Hussain et al., 2014; Kumar et al., 2017).

Methods of cultivation: Through seeds. The seeds have a very poor germination percentage. It is also propagated vegetatively.

Photo: Prof Srinath Rao

CLERODENDRUM INFORTUNATUM (L.)

Synonyms: *Clerodendrum viscosum* Vent.; *Clerodendrum calycinum* Turcz.

Classification:

Kingdom:	Plantae
Subkingdom:	Phanerogamea
Class:	Dicotyledonae
Subclass:	Gamopetalae
Series:	Bicarpellate
Order:	Lamiales
Family:	Verbenaceae
Genus:	*Clerodendrum*
Species:	*Clerodendrum infortunatum*

Common names: Bengali: Bhant, Ghentu; **English:** Hill glory bower; **Gujarathi:** Bhanua; **Hindi:** Bhant, Ghato; **Kannada:** Basavana Pada, Ibbane; **Malayalam:** Peruku, Peravellam; **Marathi:** Bhandira, Kari, Bhat; **Sanskrit:** Bhandir, Bhandi, Bhantaka; **Tamil:** Perugilai, Karukanni; **Telugu:** Gurrapu.

Distribution: It is found in the Malayan region and Sri Lanka, as well as throughout the temperate region of India, commonly in forest and on the roadside, as a weed.

Description: A shrub or an under shrub with a fetid smell. Stem grayish, lenticellate; branchlets slender, obtusely quadrangular; nodes swollen, gregarious and forming dense thickets. Leaves simple, opposite, decussate, ex stipulate, broadly cordate, orbicular or ovate-cordate, 7–28 cm long, 6–25 cm broad, base sub-truncate, rounded or cordate, margin entire or subentire when young, apex acuminate or acute, chartaceous, dark green above, slightly pubescent, lateral veins 6–9 on either side of the midvein, impressed and prominent above, distinct beneath, ascending, margins arcuate, petiole slender, cylindric, densely pubescent, about 3–15 cm long. Inflorescence in terminal panicles of pedunculate of 6–14 cymes, obtusely quadrangular, about 5 cm long, bracts elliptic, foliaceous, apex acuminate, densely pubescent, bractlets caducous. Flowers slightly fragrant, bisexual, zygomorphic, pedicel slender, about 1–2 cm long, calyx 5-toothed, divided, pubescent, lanceolate; corolla 5-lobed, 1 upper lobe slightly bigger than the other 4 lobes which are opposite, lobes elliptic, apex acute, white with pinkish throat; corolla tube narrow, cylindric, pubescent, about 1.5–2 cm long, stamens 4, didynamous to subequal, exserted; filaments filiform, creamish white; anthers bilobed, elliptic; ovary oblong, 4-lobed, 4-loculed; style filiform, stigma shortly 2-fid; fruit a drupe, subglobose, about 1–1.5 cm in diameter, purplish black when ripe, calyx shallow, cupular, pinkish. Fruit usually with 4 dry nutlets.

Flowering: Almost throughout the year.

Fruiting: Almost throughout the year.

Chemical constituents: The major compounds of the roots are sterols, sugars, flavonoids and saponins. Novel crystalline compounds such as clerodolone, clerodone, clerodol and clerosterol are also present (Bhattacharjee et al., 2011). 7 sugars, namely raffinose, lactose, maltose, sucrose, galactose, glucose and fructose were also identified from the roots. The flowers contain acacetin, apigenin, caffeic acid esters, β-sitosterol, β-sitosterol glucoside and fumaric acid. The leaves contain saponins, scutellarin and hispidulin-7-O-glucuronide, as well as 24-β-ethylsterols, clerosterol and 22-dehydroclerosterol. 24-methyl-sterols and 24 β-ethyl-22-dehydrocholestanol are found in the seeds.

Used in: Ayurveda and Unani medicine.

Parts used: Leaves, roots and flowers.

Uses: This herb is used for the treatment of diarrhea, wounds, skin diseases, liver disorders, intestinal worms, fever, joint pain and venereal and scrofulous complaints (Bhattacharjee et al., 2011). The fresh juice of the leaves is applied over the wounds for quick healing and employed to treat constipation, fever, liver disorders, as well as to remove intestinal worms. The leaves are slightly bitter and cure inflammation and skin diseases. They are useful against smallpox (Sannigrahi et al., 2009). The paste of the leaves is applied over the area of the swelling to reduce it. It is an ingredient in several Ayurvedic formulations.

Methods of propagation: Through seeds.

Photo: Prof Srinath Rao

CLERODENDRUM INERME (L.) GAERTN.

Synonym: *Volkameria inermis* L.

Kingdom:	Plantae
Subkingdom:	Phanerogamea
Class:	Dicotyledonae
Subclass:	Gamopetalae
Series:	Bicarpellate
Order:	Lamiales
Family:	Verbenaceae
Genus:	*Clerodendrum*
Species:	*Clerodendrum inerme*

Common names: Hindi: Sangkupi; **Tamil:** Peechangu.

Distribution: Widely distributed in the tropical and subtropical regions of the world.

Description: This is an evergreen plant, a hardy, straggling shrub that reaches a height of 3–4 m with closely arranged, almost round, shiny, deep-green leaves. Leaves opposite, simple, ovate to elliptical about 5–10 cm long, acute to acuminate tip, margin smooth, entire, slightly shiny upper surface, pinnate venation. The plant is always in flower. The flowers are white with 5 corolla lobes, 3 cm long, white tubes and long purple stamens. Cyme or umbel usually comprised of 3 flowers joined at a common base point; corolla white, fused, with 5 lobes; stamens 4, reddish to purple and upwardly curved. Fruit green turning black, 1–1.5 cm long, obovoid.

Flowering: Throughout the year.

Fruiting: Throughout the year.

Chemical constituents: The plant contains carbohydrates, flavonoids, steroids, volatile constituents, terpenes, salidroside, jinoside-D, acetoside, and steroids such as octacosanol, β-sitosterol, γ-sitosterol, bungein A, acteoside, betulinic acid, campesterol, clerosterol, colebrin A-E, 4-α-methylsterol and cholestanol. Terpenes such as α-amyrin, β-amyrin, caryoptin, clerodendrin A, B and C, clerodin, clerodermic acid and iridoids are reported. The major flavonoids present are apigenin, cynaroside, kaempferol and luteolin. Other constituents include ribosome-inactivating proteins (Avani et al., 2006).

Used in: Ayurveda, Siddha, Folk, and Chineese medicine.

Parts used: Leaves.

Uses: Different parts of this plant are traditionally used for the treatment of beriberi, boils, cough, fever, rheumatism, skin rashes, skin diseases, tumors and venereal infections. They are also used to treat umbilical cord infection and for cleaning the uterus in local medicine. The leaves and roots of the plant are used in rheumatism and skin diseases.

Methods of cultivation: Suckers and seeds.

Photo: Prof Srinath Rao

CLITORIA TERNATEA L.

Synonyms: *Clitoria albiflora* Mattei; *Clitoria bracteata* Poir.; *Clitoria mearnsii* De Wild.

Classification:

Kingdom:	Plantae
Subkingdom:	Phanerogamea
Class:	Dicotyledonae
Subclass:	Polypetalae
Series:	Calyciflorae
Order:	Rosales
Family:	Leguminosae (Fabaceae)
Genus:	*Clitoria*
Species:	*Clitoria ternatea*

Common names: Bengali: Aparajita; **English:** Butterfly-pea, Mezereon; **Guajarati:** Garnee; **Hindi:** Aparajita, Koyala; **Kannada:** Girikarniballi; **Malayalam:** Sangu pushpam; **Marathi:** Gokarnee; **Sanskrit:** Girikarnikaa, Kokilaa, Yonipushpaa; **Tamil:** Kakkanam; **Telugu:** Dintena, Sankupushpam.

Distribution: Almost throughout India, wild or cultivated. Western–North India (Marwad), Egypt, Syria, Mesopotamia, Iraq, Persia, Arabia, Afghanistan, etc. Dates and Dhamasa grow in sandy regions.

Description: It is a perennial herbaceous plant. Its leaves are elliptic and obtuse. It grows as a vine or creeper, doing well in moist neutral soil. The most striking feature about this plant is its vivid deep-blue flowers. They are solitary, with light yellow markings. They are about 4 cm long by 3 cm wide. There are some varieties that yield white flowers. The fruits are 5–7 cm long, flat pods with 6–10 seeds in each pod. They are edible when tender.

Flowering: March–May.

Fruiting: Throughout the year.

Chemical constituents: The active constituents include tannins, resins, starch, taraxerol and taraxerone. Several biologically active peptides called cliotides have been isolated from the heat-stable fraction. The plant contains compounds such as kaempferol-3-rhamnosyl, 1–6 glycoside, 1–6 galactoside, kaempferol, 3-glucoside, kamempferol-3-0 rutinoside, kaempferol-3-neohesperidoside and clitorin (Rastogi and Mehrotra, 1980).

Used in: Ayurveda.

Parts used: Flowers, roots.

Uses: In traditional Ayurvedic medicine, it is considered as 'Medhya drug' and has been used for centuries as a memory enhancer, nootropic, anti-stress, anxiolytic, anti-depressant, anticonvulsant, tranquilizing and sedative agent. The root juice in milk is taken orally to cure chronic bronchitis; the juice is also taken mixed with honey for improving brain function and memory. Moreover, it is used as nasal drops for migraine, chronic headache and leucoderma. Leaf juice is taken with water to get relief in case of heavy or frequent periods and over-bleeding. The leaves are boiled in water and after adding some alum the water is used to wash the affected area to cure vaginal infection, gonorrhea and hydrocele. The herb has been found to be effective in treating female sexual disorders such as infertility and menstrual issues. The flowers are aphrodisiac. The herb is also effective in treating male infertility and reported to increase sperm mobility. It has a tranquillizing effect on the brain, hence it is used in symptoms like syncope, vertigo and brain weakness. It is an antiemetic, anti-dyspeptic, mild-laxative and cholagogue, used in emesis, constipation, jaundice and piles. It is also used in healing ulcers of the pylorus, duodenum, etc. Being a hemostatic and blood purifier, it is useful in hemorrhagic disorders and vatarakta. A hot infusion of dhamasa is given to prevent smallpox. It is used in common cold, cough and asthma as it acts as an expectorant and reduces the irritation in respiratory organs. Besides this, the whole plant is used for smoking. A decoction is used for gargling in throat issues. The extracts from the roots of this herb are used as an antidote in treating poisonous bites of snakes and insects. The root and seed powder are mixed with water and taken orally to cure filariasis.

Photo: Prof Srinath Rao

COCCINIA INDICA L.

Synonyms: *Cephalandra indica*; *Coccinia grandis* (L.) J. Voigt.

Classification:

Kingdom:	Plantae
Subkingdom:	Phanerogamea
Class:	Dicotyledonae
Subclass:	Polypetalae
Series:	Calyciflorae
Order:	Passiflorales
Family:	Cucurbitaceae
Genus:	*Coccinia*
Species:	*Coccinia indica*

Common names: Bengali: Bimbu; **English:** Ivy gourd; **Gujarati:** Galedu; **Hindi:** Tinda; **Kannada:** Tondaekai; **Marathi:** Bimbi; **Sanskrit:** Bimbaka; **Tamil:** Kavoi.

Distribution: *Coccinia grandis* is distributed from Africa to Asia, including India, the Philippines, China, Indonesia, Malaysia, Thailand, Vietnam, eastern Papua New Guinea and the Northern Territories, It is cultivated abundantly in India as a vegetable.

Description: A twiner. Leaves: 5–10 cm, long and broad, bright green above, paler beneath, studded and sometimes rough with papillae, palmately 5-nerved from a cordate base, often with circular glands between the nerves, obtusely 5-angled or sometimes deeply 5-lobed, the lobes broad, obtuse or acute, apiculate, more or less sinuate-toothed, petioles 2–3.2 cm long. Flowers: male flowers: peduncles are 2–3.8 cm long and sub-filiform. Calyx-tube: glabrous, broadly campanulate and 4–5 mm long. Corolla: 2.5 cm long, veined, pubescent inside and glabrous outside. Female flowers: peduncles are 1.3–2.5 cm long. Ovary is fusiform, glabrous and slightly ribbed. Fruits: fusiform-ellipsoid, slightly beaked, 2.5–5 ×1.3–2.5 cm, marked when immature with white streaks, bright scarlet when fully ripe. Seeds: obovoid and rounded at the apex, slightly papillose, much compressed and yellowish grey. Roots: the fresh root is thick, tuberous, long, tapering, more or less tortuous with a few fibrous rootlets attached to it.

Flowering: July–September.

Fruiting: July–September.

Chemical constituents: Roots contain triterpenoid, saponin coccinioside, lupeol, β-amyrin and β-sitosterol. Fruits contain taraxerone, β-carotene, lycopene, cryptoxanthin, apo-6'-lycopenal, β-sitosterol and taraxerol. Leaves contain heptacosane, cephalandrol, β-sitosterol alkaloids cephalandrine A and cephalandrine B. Several amino acids such as aspartic acid and glutamic acids, asparagine, tyrosine, histidine, phenylalanine, threonine, valine and arginine are reported (Basu et al., 1972; Kundu et al., 1987; Vaishnav et al., 2001).

Used in: Ayurvedic, Unani and Siddha medicine.

Parts used: Leaves, fruits and roots.

Uses: In traditional medicine, the fruits have been used to treat leprosy, fever, asthma, bronchitis and jaundice. It is famous for its hypoglycemic and antidiabetic properties in the Ayurvedic system of medicine. Other applications include the therapy of various conditions such as skin diseases and gonorrhea. It regulates and reduces blood sugar levels naturally, improves blood lipid levels and reduces insulin resistance as well as the reduces complication of diabetes mellitus. The roots are used to treat osteoarthritis and joint pain. A paste made from the leaves is applied to the skin to treat scabies_(Hossain et al., 1992).

Methods of propagation: Through seeds and stem cuttings. Seeds are sown in situ or in containers. When using containers, place 2–3 seeds in each pot. The seeds usually germinate within 2–4 weeks. They can be planted when 12–15 cm tall. Stem cuttings, 12–15 cm long, can be planted in situ as long as they are not allowed to become dry whilst they are rooting.

Photo: Dr R. S. Modi

COLEUS AROMATICUS BENTH.

Synonyms: *Coleus amboinicus Lour.*; *Plectranthus amboinicus* (Lour.) Spreng.

Classification:

Kingdom:	Plantae
Subkingdom:	Phanerogamea
Class:	Dicotyledonae
Subclass:	Gamopetalae
Series:	Bicarpellate
Order:	Lamiales
Family:	Labiatae
Genus:	*Coleus*
Species:	*Coleus aromaticus*

Common names: English: Country borage; **Hindi:** Patta ajavayin; **Kannada:** Doddapatre soppu; **Malayalam:** Panikoorka; **Sanskrit:** Parnayavani, Pashanabhedi; **Tamil:** Karpooravalli; **Telugu:** Sugandhavalkam.

Distribution: This grassy plant is found or cultivated throughout India.

Description: It is a large succulent herb, fleshy and highly aromatic, much-branched, possessing short soft erect hairs. The stem is fleshy, about 30–90 cm, either with long rigid hairs (hispidly villous) or tomentose (densely covered with soft, short and erect hairs, pubescent). Leaves are undivided (simple), broad, egg/oval-shaped with a tapering tip (ovate) and very thick; they are pubescent (thickly studded with hairs), with the lower surface possessing the most numerous glandular hairs, giving a frosted appearance. The taste of the leaf is pleasantly aromatic with an agreeable and refreshing odour. Flowers are on a short stem (shortly pedicelled), pale purplish in dense whorls at distant intervals in a long slender raceme. Corollas are pale purplish, tube short, throat inflated, lips short. Stamens are shortly exserted.

Flowering: August–November.

Fruiting: August–November.

Chemical constituents: Butylaniside, -caryophyllene, carvacrol, 1-8-cineole, p-cymene, ethyl salicylate, eugenol, limonene, myrcene, and -pinenes, -selenene, -terpinene, terpinen-4-ol, thymol, verbenone (essential oil), apigenin, chrysoeriol, 5,4-dihydroxy-6,7-dimethoxy-flavone (cirsimaritin), eriodictyol, 6-methoxygenkawanin, luteolin, quercetin, salvigenin, taxifolin; oxaloacetic, crategolic, euscaphic, 2-3-dihydro-olean-12-en-28-oic, pomolic, oleanolic, tormentic, 2, 3, 19, 23-tetrahydroxyurs-12-en-28-oic and ursolic acids, as well as sitosterol-D-glucoside were isolated from the leaves (Chatterjee and Pakrashi, 2001).

Used in: Traditional and Ayurvedic medicine.

Parts used: Leaves and roots.

Uses: The leaves have had many traditional medicinal uses, especially for the treatment of coughs, sore throats and nasal congestion, but also for a range of other problems such as infections, rheumatism and flatulence. They are used in traditional medicine to treat malarial fever, hepatopathy, renal and vesical calculi, chronic asthma, hiccough, bronchitis, helminthiasis, colic, convulsions and epilepsy. Moreover, they are hepatoprotective. The juice of the leaves is used in India for the treatment of skin allergies and skin ulcerations, scorpion bite, wounds, diarrhea.

Methods of propagation: The Indian Borage is very commonly grown as a potted plant. It is a fast-growing plant. Propagation is via stem cuttings. To encourage a bushy plant, cut the tip of the top, insert into the soil and, instantly, you have another plant, as the cutting will grow within days. Ideally it should be grown in a semi-shaded and moist location as the leaves will remain a beautiful jade-green colour. If it getsgetting too much sun, the leaves turn yellow, start curling and look unsightly; if it does not get enough sun, the leaves turn a dark shade of green and become spaced out. The herb grows easily in a well-drained, semi-shaded position.

Photo: Prof Srinath Rao

COLEUS FORSKOHLII BRIQ.

Synonyms: *Plectranthus barbatus* (Andr.) Benth.; *Plectranthus forskohlii.*

Classification:

Kingdom:	Plantae
Subkingdom:	Phanerogamea
Class:	Dicotyledonae
Subclass:	Gamopetalae
Series:	Bicarpellatae
Order:	Lamials
Family:	Labiatae
Genus:	*Coleus*
Species:	*Coleus forskohlii*

Common names: English: Coleus; **Gujarati:** Maimool; **Hindi:** Patharchur; **Kannada:** Makandiberu; **Marathi:** Maimnul; **Sanskrit:** Makandi, Pashan bhedi; **Tamil:** Marundhu koorkan.

Distribution: It grows perennially over tropical and subtropical regions of India, Pakistan, Sri Lanka, East Africa and Brazil at 600–1800 m elevation. In India, the crop is cultivated in parts of Gujarat, Maharashtra, Rajasthan, Karnataka and Tamil Nadu.

Description: It is a perennial plant that grows to 45–60 cm tall and is aromatic in nature. It has 4-angled stems that are branched and its nodes are often hairy. Leaves are 7.5–12.5 cm in length and 3–5 cm in width, usually pubescent, narrowed into petioles. The inflorescence is a raceme, 15–30 cm in length; flowers are stout, 2–2.5 cm in size, usually perfect; calyx hairy inside. Upper lip of calyx is broadly ovate. The blue or lilac corolla is bilabiate. Lower lobes are elongated and concave so that they enclose the essential organs. The ovary is 4-parted and stigma is 2-lobed; stamens 4 didynamous, dedinate; the filaments unite at their base to form a sheath around the style. The roots are tuberous, thick, fibrous, brown in colour, orange-red within and strongly aromatic.

Flowering: August–October.

Fruiting: November–December.

Chemical constituents: The tuberous root extracts of *P. Forskohlii* contain minor diterpenoids, such as deactylforskolin, 9-deoxyforskolin, 1,9-deoxyforskolin, 1,9-dideoxy-7deacetylforskolin in addition to forskolin (7-acetoxy-8, 13-epoxy-1,6,9- trihydroxylabd-14-en-11-one).

Used in: Ayurveda.

Parts used: Roots.

Uses: The whole plant is used to treat dyspepsia, indigestion, dysentery, vomiting, thirst fever, dermatitis, ulcer and in paediatric disorders such as diarrhea, thirst, vomiting and fever. Water processed with *hribera* and *sunthi* (*Zingiber officinale*) should be given to drink. According to ancient Sanskrit texts, *Coelus* has been used as a medical herb to treat heart and lung diseases, intestinal spasms, insomnia and convulsions. It also lowers blood pressure, is an antispasmodic and dilates the blood vessels. It is found to be effective in skin conditions such as eczema and psoriasis. *Coelus* is primarily indicated for cardiovascular diseases including hypertension, congestive heart failure and angina. It is also used for asthma, intestinal colic and uterine cramps. In Ayurvedic medicine, the *Coelus* species have been used to treat convulsions, spasmodic pain and painful urination. Moreover, the plant is used in glaucoma, cardiac problems, eczema, asthma, and hypertension as well as in the treatment of certain types of cancers. It is useful in the treatment of congestive heart failure.

Methods of propagation: It is propagated through seeds as well as through stem cuttings.

Steps to follow: (through seeds)

- Seeds are collected in spring.
- The fresh seeds are sown in the nursery beds for germination.
- Seeds germinate in 15–20 days.
- The seedlings have to be maintained in the nursery for 6–8 weeks.

Vegetative propagation:

Semi-hardwood stem cuttings are used.

- These are planted in summer.
- Cuttings root quickly and easily even in water. Cuttings 4–5 in long comprising 3–4 pairs of leaves are taken from the tip portion of rapidly growing shoots and planted in well-prepared nursery beds under shade. Regular watering should be ensured.

- Cuttings produce sufficient roots in a month's time and they are ready for transplanting. When the seedlings are 45 days old and 8–10 cm tall, they can be transplanted.

Photo: Prof Srinath Rao

COMMIPHORA WIGHTII (ARNOTT)

Synonyms: *Balsamodendron mukul* (Hook. ex Stocks); *Commiphora mulul* (Hook. ex Stocks); *Commiphora roxburghii* (Santapau).

Classification:

Kingdom:	Plantae
Subkingdom:	Phanerogamea
Class:	Dicotyledonae
Subclass:	Polypetalae
Series:	Disciflorae
Order:	Geraniales
Family:	Burseraceae
Genus:	*Commiphora*
Species:	*Commiphora mukul*

Common names: Bengali: Guggul; **English:** Indian bdellium; **Hindi:** Gugal; **Kannada:** Guggul; **Malayalam:** Gulgulu; **Marathi:**, Kattukasturi; **Sanskrit:** Gugulu, Devadhupa; **Tamil:** Gukkulu; **Telugu:** Gukkulu.

Distribution: The center of origin of *Commiphora* is believed to be Africa and Asia. It is a

widely adapted plant, well distributed in the arid regions of Africa. In India it is distributed in the Rajasthan, Gujarat, Maharashtra and Karnataka states. It has been listed as a threatened plant by the Botanical Survey of India and is included in the Red Data Book (IUCN) and in the list of over-exploited species in the country.

Description: It is a small tree, 3–4 m in height with spinescent branching. Stem is brownish or pale yellow with ash-coloured bark peeling off in flakes. Leaves are alternate, 1–3 foliate, obovate, leathery and serrate. Lateral leaflets when present are only less than ½ the size of the terminal ones. Flowers small, brownish red, with short pedicel, seen in fascicles of 2–3. Calyx campanulate, glandular, hairy and 4–5-lobed. Corolla with brownish red, broadly linear petals reflexed at apex. Stamens 8–10, alternatively long and short. Ovary oblong, ovoid and stigma bifid. Fruit is a drupe and red when ripe, ovate in shape with 2–3-celled stones.

Flowering: Throughout the year.

Fruiting: Throughout the year.

Chemical constituents: The gum resin contains guggul sterols Z and E, guggul sterols I-V, two diterpenoids – a terpene hydrocarbon named cembrene A, and a diterpene, alcohol-mukulol; α-camphrone and cembrene. Major components from the essential oil of gum resin are myrcene and dimyrcene. The plant without leaves, flowers and fruits contains myricyl alcohol, β-sitosterol and 15 amino acids. Flowers contain quercetin and its glycosides as major flavonoid components, other constituents being ellagic acid and pelargonidin glucoside. The gum resin is bitter, acrid, astringent, thermogenic, aromatic, expectorant, digestive, anthelmintic, anti-inflammatory, anodyne, antiseptic, demulcent, carminative, emmenagogue, haematinic, diuretic, lithontriptic, rejuvenating and a general tonic. Guggulipid is hypocholesterolemic.

Used in: Ayurveda.

Parts used: Gum resin.

Uses: Highly effective in the treatment of Alzheimer's disease, brain stroke, rheumatism, obesity, neurological and urinary disorders, tonsillitis, arthritis and a few other diseases. The fumes from burning guggul are recommended in hayfever, chronic bronchitis and phthisis. It is alexiteric, anti-aging, hypocholesterolemic and alterative.

Methods of propagation: Plants are propagated vegetatively from stem cuttings from the semi-woody (old) branch. For this purpose, a 1 m-long woody stem of 10 mm thickness is selected and the cut end is treated with IBA or NAA and planted in a nursery bed during June–July; the beds should be given light irrigation periodically. The cuttings initiate sprouting in 10–15 days and grow into a good green sprout in 10–12 months. The plants are also propagated by seeds.

CORDIA WALLICHII G. DON.

Synonyms: *Cordia obliqua* Willd. var. wallichii; *Cordia myxa* Linn.; *Cordia latifolia* Roxb.

Classification:

Kingdom:	Plantae
Subkingdom:	Phanerogamea
Class:	Dicotyledonae
Subclass:	Gamopetalae
Series:	Bicarpellatae
Order:	Polemoniales
Family:	Boraginaceae
Genus:	*Cordia*
Species:	*Cordia wallichi*

Common names: Bengali: Bahubara, Bohari; **English:** Sebestan; **Gujarati:** Gundomoto, Lepistan; **Hindi:** Lasura, Bhirala, Rasalla; **Malayalam:** Celu, Cheruviri; **Marathi:** Bargund, Bhokar; **Sanskrit:** Bahuvaraka, Laghupichhila, Uddaalaka, Bahu-vaaraka; **Tamil:** Perunaruvili. Naruvili, Virisu; **Telugu:** Bankanakkera, Chinnabotuku.

Distribution: It is a medium-sized deciduous tree, found scattered throughout the mid-Himalayas up to elevations of 1470 m.

Description: It is a medium-sized deciduous tree. Leaves are alternate, entire to slightly dentate and glabrous, shape from elliptic-lanceolate to broad ovate, often with a rounded or cordate base, basal nerves 3, rarely 5, petioles 3–5 cm long. Flowers are actinomorphic, bisexual, complete, short-stalked, white and glabrous, 6–8 mm in diameter. The inflorescence is terminal or an axillary cyme. It has 14 flowers per cluster. The calyx is cup-shaped, sepals united, 4–6 mm in length, slightly dentate, light green in color. The corolla has 4 creamish white color petals which are 6–8 mm in length and polypetalous. The androecium contains 2 stamens, epipetalous, each with a very small filament. The gynoecium is bifurcated, 4 mm in length and with a globose-shaped ovary at the base. Fruit is a drupe, 1.5–2.5 cm long, when ripe yellowish brown, pink or nearly black, shiny but minutely rugose. Epicarp is thick while mesocarp is mucilaginous and endocarp is hard and stony and contains 2 seeds, which are separated by a stony septum. The seeds are mildly sweet in taste.

Flowering: April–June.

Fruiting: July–August.

Chemical constituents: Phytochemical investigation shows the presence of alkaloids, flavonoids, phenolics, phenylpropanoid derivatives, pyrrolizidine tannins and triterpenes. Seeds contain various constituents such as α-amyrin, betulin, β-sitosterol, β-sitosterol-3-glucoside, hentriacontanol, hentricontane, lupeol-3-rhamnoside, octacosanol, taxifolin-3,5-dirhamnoside and hesperitin-7-rhamnoside (Thirupathi et al., 2008).

Used in: Ayurveda, Unani and traditional medicine.

Parts used: Leaves, fruits, bark.

Uses: It possesses anthelmintic, purgative, diuretic, expectorant, antipyretic, hepatoprotective and analgesic properties. The leaves are useful as an external application to treat ulcers and headaches. The fruits are edible and used as pickle; they are sweet and possess astringent, anthelmintic, demulcent, diuretic, expectorant, maturant and purgative properties. They are also useful against dry cough, biliousness, chronic fever and diseases of the chest and urethra. They are also beneficial against pains in the joints. The juice of the bark is given in gripes, along with coconut oil. The bark and unripe fruit are used as a mild tonic. The kernels are a good remedy in the treatment of ringworm. The Santals use a powder of the bark for external application in prurigo.

Methods of propagation: Through seeds.

Photo: Dr R.S Modi

CORIANDRUM SATIVUM (L.)

Synonyms: *Coriandropsis syriaca* H. Wolff.; *Coriandrum diversifolium* Gilib.; *Coriandrum majus* Gouan.

Classification:

Kingdom:	Plantae
Subkingdom:	Phanerogamea
Class:	Dicotyledonae
Subclass:	Polypetalae
Series:	Calyciflorae
Order:	Umbellale (Apiales)
Family:	Umbelliferae (Apiaceae)
Genus:	*Coriandrum*
Species:	*Coriandrum sativum*

Common names: Bengali: Dhaanyaka; **English:** Coriander; **Gujarati:** Dhaanaa, Kothmir; **Hindi:** Dhaniya, Kothmir; **Kannada:** Kottmbari; **Malayalam:** Koththampaa;, **Marathi:** Kotimir; **Sanskrit:** Dhanayaka, Kusthumbari; **Tamil:** Cukkuma-pattiram; **Telugu:** Daniyalu, Kotimiri.

Distribution: Cultivated throughout India and many other countries.

Description: It is an annual plant, with an erect branching stem rising about 2 feet; the leaves are alternate, and the first ones are often gathered in a rosette. The blade shape of the basal leaves is usually either undivided with 3 lobes, or 3-pinnatifid. The higher the leaves are inserted, the more pinnate they are. The upper leaves are deeply incised with narrow lanceolate or even filiform-shaped blades. The lower leaves are stalked, while the petiole of the upper leaves is reduced to a small, nearly amplexicaul leaf sheath. The inflorescence is a compound umbel. The flowers are protandrous, white or whitish pink. The central flowers of the umbellets are staminiferous or sometimes sterile. Calyx with 5 sepals. The calyx teeth are of different length, and the ones surrounding the stylopodium are still visible in the ripe fruit. The flowers have 5 petals. The peripheral flowers of every umbellet are asymmetric, as the petals toward the outside of the umbellets are lengthened. The central flowers are circular, with small inflexed petals. The colour of the petals is pale pink or sometimes white. The ovary is inferior; the fruit is globular, and composed of 2 concavo-convex mericarps. Mericarps usually coherent; cremocarp nearly globular, 3–5 mm in diameter; externally light brown or rose-coloured; summit with 5 calyx teeth and a short stylopodium; each mericarp has 5 prominent, straight, longitudinal, primary ribs and 4 indistinct, undulate secondary ribs; mericarps can be easily separated; deeply concave on the inner surface.

Flowering: June–July.

Fruiting: August–September.

Chemical constituents: The main compounds present are ascorbic acid (vitamin C), linoleic, oleic, palmitic and stearic acid. Other components such as borneol, cineole, citronellol, limonene, α-pinene and β-phellandrene are present (Chithra and Leelamma, 1999a).

Used in: Ayurveda and folk medicine.

Parts used: Leaves, seeds.

Uses: Its use is recommended in allergies, amoebic dysentery, burns, cough, cystitis, sore throat, urethritis, urinary tract infection, urticaria, vomiting, indigestion, nosebleed, hay fever and dizziness. It has anti-carcinogenic, anticonvulsant, antihistaminic, antihypertensive, antioxidant, antimicrobial, antimutagenic, carminative, diaphoretic, diuretic, hypoglycemic, hypolipidemic, hypnotic and postcoital stimulant activity. Coriander helps cure ulcer, inflammation, spasm and acts as an expectorant, protects and soothes liver. It is and. The seeds are chewed as a remedy for halitosis (Gray and Flatt, 1999; Ramadan et al., 2002).

Methods of propagation: Through seeds.

Photo: Prof Srinath Rao

COSTUS PICTUS D. DON.

Synonyms: *Costus congestus* Rowlee; *Costus mexicanus* Liebm.; *Costus igneus* Nak.; *Pyxa picta* (D. Don.) S. R. Dutta & S. Yadav.

Classification:

Kingdom:	Plantae
Subkingdom:	Phanerogamea
Class:	Monocotyledonae
Subclass:	Commelinids
Series:	Coronarieae
Order:	Alismatales
Family:	Zingiberaceae (Costaceae)
Genus:	*Costus*
Species:	*Costus pictus*

Common names: English: Spiral ginger; Insulin plant; **Hindi:** Keu; **Kannada:** Nirvala; **Malayalam:** Nirmathalam; **Sankskrit:** Kemuka; **Tamil:** Nandiyavattai; **Telugu:** Nandivardhana.

Distribution: Ornamental plant. A native of Mexico, now newly introduced to India, and being grown in gardens especially in the states of Kerala and Gujarat.

Description: It is an erect herb growing up to 3 m tall, with a stem horizontally striped at the base; leaves narrowly lanceolate, dark green above, lighter green below; small leaves are present on the basal part; bracts green, with outer margin maroon. Flowers: yellow; lip with maroon striations, darker yellow stripe down the middle region; anther cream-coloured.

Chemical constituents: Methyl tetracosanoate, a bio-active molecule isolated from *Costus pictus* extract, has anti-diabetic properties (Shilpa et al., 2009). The leaves are found to contain flavonoids such as kaempferol, 3', 4'-di O-Me-quercetin and 4'-OMe-Kaempferol and phenolic acids such as gentisic, 2, 5-dihydroxy benzoic, o-coumaric, melilotic, α-resorcyclic, 3,5-dihydroxy benzoic acid, p-hydroxy benzoic and cis and trans-p-coumaric acids. Stem is rich in hexadecanoic acid (28.3 %), 9,12-octadecadienoic acid (18.33 %), linalyl propionate (6.03 %), dodecanoic acid (5.62 %), tetradecanoic acid (4.82 %), α-eudesmol (3.55 %), γ-eudesmol (3.21 %) and 4-ethoxy phenol (3.06 %). The essential oil from the leaf contains hexadecanoic acid (24.51 %), 2-pentanol (22.48 %), β-ionone (8.69 %), α-ionone (8.01 %), farnesyl acetone (7.04 %) and dodecanoic acid (3.96 %). Hexadecanoic acid (25.26 %), dodecanoic acid (16.56 %), tetradecanoic acid (10.20 %), linalool (8.48 %), 9, 12-octadecadienoicacid (7.74 %), and α-terpineol (4.44 %) are the main constituents of the rhizome essential oil (Jose and Reddy, 2009a).

Used in: Traditional and Ayurveda medicine; homeopathy.

Parts used: Rhizomes.

Uses: Properties such as anti-diabetic, anti-inflammatory, anticancer, antihepatotoxic and anti-depressant have been reported from the rhizomes. It is also used in asthma, eye complaints and snake bite.

Methods of propagation: Through rhizomes.

Photo: Prof Srinath Rao

COSTUS SPECIOSUS (KOEN. EX RETZ.) SMITH

Classification:

Kingdom:	Plantae
Subkingdom:	Phanerogamea
Class:	Monocotyledonae
Subclass:	Zingiberidae
Series:	Coronarieae
Order:	Alismatales
Family:	Zingiberaceae (Costaceae)
Genus:	Costus
Species:	*C. speciosus*

Common names: English: Wild zinzer, Spiral; **Hindi:** Keokanda; **Kannada:** Cangalvakostu, Kangalakoshta; **Malayalam:** Ankua; **Sanskrit:** Kembuk; **Tamil:** Kostam; **Telugu:** Cengalva kostu.

Distribution: This species is widely distributed in India, Bangladesh, Myanmar (earlier Burma), Malaysia, New Guinea, Australia, China, Sri Lanka and Taiwan. It enjoys a wide distribution throughout India up to an altitude of 1200 m, mostly occurring in moist, partially shaded localities. Within India, its geographical distribution is in the sub-Himalayan tract from the Kangra District of Himachal Pradesh eastwards to Arunachal Pradesh; and in the Western Ghats in Maharashtra, Goa, Karnataka, Kerala and Tamil Nadu.

Description: An erect plant, up to 2.7 m high; succulent with tuberous rhizomes, spirally twisted stem. Leaves: Simple, oblong, acuminate, 15–20 × 6–7 cm, sub-sessile, oblong, spirally arranged, silky-pubescent beneath; sheaths coriaceous. Flowers: in dense terminal spikes, many, white,; bracts ovate, mucronate, bright red; corolla tube short, lobes ovate-oblong subequal; lip white with yellow center, crisped, concave. Fruit: a capsule, ovoid, 3-valved. Seeds: black with a white aerial.

Flowering: August–October.

Fruiting: November–December.

Chemical constituents: Rhizomes contain saponins; dioscin and gracilin based on diosgenin and also tigogenin, β-sitosterol and a volatile oil. Roots yield aliphatic hydroxyl ketones. Seeds contain diosgenin.

Used in: Traditional and Ayurvedic medicine; homeopathy.

Part used: Stem and rhizomes.

Uses: The rhizome is bitter, astringent and used for dyspepsia, fever, cough and other respiratory diseases, diabetes, oedema, blood and skin diseases. It also improves the complexion. Saponins and prosapogenins cause spasmodic uterine contraction and are estrogenic. They are anti-inflammatory, hypotensive and bradycardic in nature.

Methods of propagation: Through the vegetative method.

Photo: Prof Srinath Rao

CROTALARIA PALLIDA AITON

Synonyms: *Crotalaria brownei* DC.; *Crotalaria fertilis* Delile; *Crotalaria hookkeri* Arn.; *Crotalaria mucronata* Desv.; *Crotalaria striata* Schumach. & Thonn.; *Crotalaria javanica* Jungh.

Classification:

Kingdom:	Plantae
Subkingdom:	Phanerogamea
Class:	Dicotyledonae
Subclass:	Polypetalae
Series:	Calyciflorae
Order:	Rosales (Fabales)
Family:	Leguminosae (Fabaceae)
Genus:	*Crotalaria*
Species:	*Crotalaria pallida*

Common names: English: Smooth rattlepod, Striped crotalaris; **Hindi:** San; **Malayalam:** Kilukkampettiched; **Marathi:** Jungali tag; **Tamil:** Kilukkillkki; **Telugu:** Killukkachedi.

Distribution: Native to tropical Africa, China, Indian subcontinent, Indochina; widely naturalized elsewhere in the tropics.

Description: It is a terrestrial, annual, erect herb reaching a height of 50–60 cm. Stem is grooved, solid and glabrous. Leaves: trifoliate, stipulate, alternately arranged and stalked, about 10 cm long; leaflets elliptic or obovate; apex rounded to acute, more than 2 cm long/wide, hairy on upper surface, margin entire, apex obtuse or rounded, base acute, pinnately veined. Flowers are bisexual, grouped together in a terminal raceme. Calyx: 5–7 mm long, upper 2 lobes united, adpressed, tomentose. Petals 5, yellow, with a prominent reddish vein, standard about 1.–1.5 cm broad; elliptical, keel curved and beaked. Stamens 10 monadelphous. Style abruptly incurved at the base; stigma oblique. Fruit a rounded pod, 3–4 cm long, 0.5–0.8 cm broad, exserted, glabrescent; seeds brown, 20–30 per pod. Root: a taproot white, or brown in colour.

Flowering: September–January.

Fruiting: December–February.

Chemical constituents: The presence of saponins, flavonoids, terpenoids, steroids, phenol, alkaloids and coumarins is reported. Seeds of many *Crotalaria* spp., including this species, contain a number of pyrrolizidine alkaloids, such as mucronatine and monocrotaline. The flavonoids apigenin and vitexin have been isolated from the bark and the leaves.

Used in: Ayurveda and folk medicine.

Parts used: Leaves, flower, and roots.

Uses: It is used as blood purifier, abortifacient, astringent, demulcent, emetic, purgative and in the treatment of anemia, impetigo, menorrhagia, urinary problems and psoriasis. It possesses anticancer properties. The plant is said to have anti-HIV properties. The flowers are used as a vegetable. A poultice made from the roots is applied to painful swelling of joints. An extract of the leaves is taken as a vermifuge. Alkaloids from the seed have shown antitumor activity (Govindappa et al., 2011).

Methods of propagation: Through seeds. Pre-soaking the seed for 12 hours in warm water can help to reduce the germination time. If you have sufficient seeds, then they can be sown in situ. Otherwise, sow in trays in a nursery and, when the seedlings are large enough to handle, pot them up into individual pots. Plant out when 15 cm or taller.

Photo: Prof Srinath Rao

CROTON TIGLIUM L.

Synonyms: *Croton acutus* Thurnb.; *Croton arboreus* Shecut. Nom. Rej.; Croton officinalis (Klotzsch) Alston., *Tiglium officinale* Klotzsch.

Classification:

Kingdom:	Plantae
Subkingdom:	Phanerogamea
Class:	Dicotyledonae
Subclass:	Monochlamydeae
Order:	Unisexulaes
Family:	Euphorbeaceae
Genus:	*Croton*
Species:	*Croton tiglium*

Common names: English: Croton; **Gujarati:** Jamalagota, Nepalo, Nepala;

Hindi: Jamalgota; **Kannada:** Berada, Jaapalada beeja; **Malayalam:** Cadalavanacu; **Marathi:** Arabi-erand, Jamalgota; **Sanskrit:** Dantibija, Dravanti, Mukula, Tintiphala; **Tamil:** Chiduram; **Telugu:** Kayapalum.

Distribution: This species is native to China, globally distributed in Indo-Malesia. Within India, it has been recorded in Arunachal Pradesh, Meghalaya, Bengal, Assam, Kerala and Tamil Nadu, either in a naturalized or cultivated state.

Description: A small tree or shrub with a few spreading branches bearing alternate petiolate leaves which are ovate, serrate, smooth, dark green on the upper surface and pale beneath. 2 glands are present at the base. Flowers born in erect terminal racemes, small, lower ones female, upper ones male; male flowers stellately hairy with narrow oblong straw-coloured petals. Stamens 15–20, female flowers apetalous. Fruit a capsule scabrid, with stellate hairs, triangular, 15–20 mm long, 10–15 mm broad, oblong or ellipsoid, 3-lobed; seeds 3 per fruit.

Flowering: Feburary–June.

Fruiting: Feburary–June.

Chemical constituents: The oil contains 3.4 % toxic resin. Of the acids, 37.0 % is oleic, 19.0 % linoleic, 1.5 % arachidic, 0.3 % stearic, 0.9 % palmitic, 7.5 % myristic and 0.6 % acetic, 0.8 % formic, with traces of lauric, tiglic, valeric, and butyric acids.

Used in: Ayurveda, folk, Tibetan, Unani and Siddha medicine; homeopathy.

Parts used: Seeds.

Uses: The seeds and oil of this plant are used in the form of paste and oil to treat udara roga, wounds and snake poison. It is a powerful drastic purgative, evacuating the bowels in less than 1 hour, and has been used as a counter-irritant in rheumatism, gout, neuralgia and bronchitis. It is typically used to treat skin conditions, such as itching, scabies, carbuncles, eczema and rashes. It is also reported to treat cancerous lesions and tumors and is a common homeopathic remedy for digestive problems, such as dysentery, diarrhea and nausea. The medicinal properties of *Croton tiglium* are found in the oils of the seeds.

Methods of propagation: Propagated through seeds. The seed are sown directly in seedbeds and the young plants are planted in the desired places.

CRYPTOLEPIS BUCHANANI ROEM. & SCHULT.

Synonyms: *Cryptolepis dubia* (Burm.f.) M. R. Almedia; *Cryptolepis reticulate* (Roth) Wall.; *Periploca dubia* Burm.f.; *Nerium reticulatum* Roxb.

Classification:

Kingdom:	Plantae
Subkingdom:	Phanerogamea
Class:	Dicotyledonae
Subclass:	Gamopetlae
Series:	Bicarpellatae
Order:	Gentianales
Family:	Asclepiadaceae
Genus:	*Cryptolepis*
Species:	*Cryptolepis buchanani*

Common names: Bengali: Kalasariba; **English:** Wax-leaved climber; **Hindi:** Karanta; **Kannada:** Kareballi; **Sanskrit:** Jamba sariba; **Tamil:** Paalkodi, Maattaankod.

Distribution: The plant is distributed in Myanmar (Burma), Sri Lanka and China. It is also found in Assam, Meghalaya, Arunachal Pradesh and Nagaland in Northeast India, Kashmir in Northern India and Travancore in Southern India. Found growing in moist and swampy areas as a climber on some trees up to an altitude of 200 m.

Description: It is a large evergreen twiner and grows to about 10 m. The plant is shade-loving and found as a climber on plants. Sometimes, it is also found as a creeper on the ground. The mature plant has a stem of about 30 cm in diameter. The bark of the plant is smooth, copper-coloured, peeling off in papery rolls in old stems. Branches lenticellate. Leaves 10.5 × 4.5 cm, opposite decussate, oblong-elliptic, acute, apiculate, coriaceous, smooth and glossy above, glaucous beneath, base acute, petiole 1.5–2.0 cm long. Flowers pale yellow in a lax dichotomous cyme. Follicles 4 × 7 cm, stout, divaricate, tapering and pointed at apex. Seeds compressed, oblong-ovate.

Flowering: May–July.

Fruiting: September–November.

Chemical constituents: Glycosides such as sarverogenin and isosarverogenin; cardenolides such as cryptosin and buchanin; a novel pyridine alkaloid, buchananine, and a serine, protease cryptolepain, are reported in the leaves and roots of this plant.

Used in: Ayurveda, and traditional medicine.

Parts used: Laves and roots.

Uses: The plant possesses anti-diarrheal, anti-ulcerative, anti-inflammatory, anti-cough, antibacterial, demulcent, diaphoretic and diuretic properties. It is a blood purifier, and is also used in the treatment of rickets in children. Traditionally, it is used by tribes for the treatment of bone fractures. It is used in the treatment of inflammatory conditions such as arthritis, and muscle and joint pain (Panthong et al., 1986). In Ayurvedic practice, the root is used as a substitute for *Hemidesmus indicus* to treat gout, polyuria, wounds and leprosy.

Photo: Dr R. S. Modi

CUMINUM CYMINUM L.

Synonyms: *Cumin odorum*

Classification:

Kingdom:	Plantae
Subkingdom:	Phanerogamea
Class:	Dicotyledonae
Subclass:	Polypetalae
Series:	Calyciflorae
Order:	Umbellales
Family:	Umbellifera (Apiaceae)
Genus:	*Cuminum*
Species:	*Cuminum cyminum*

Common names: English: Cumin; **Hindi:** Jira; **Kannada:** Jirigae; **Malayalam:** Jeerakam; **Marathi:** Jeera; **Sanskrit:** Deeraghjeerak, Jeerak, Jaran; **Tamil:** Jeerak; **Telugu:** Jilakara.

Distribution: It is believed to have originated in Turkestan from where it spread to Egypt, Ethiopia, China, India and Indonesia. In India it is cultivated for its seeds.

Description: It is an annual herbaceous plant with slender, glabrous, branched stems growing up to 30–50 cm tall. The stem is grey or dark green in colour. The leaves are alternate, 5–10 cm long, pinnate or bipinnate, and contain thin, thread-shaped leaflets. This is a characteristic feature of this herb. The inflorescence is a terminal umbel; each umbel may contain 3–5 radios. Flowers are very small, hermaphrodite, white or pink in colour, sepals 5, petals 5 and stamens 5. The root is fleshy. The fruit is a diachene, light brown, oval, narrow, 5–6 mm long and 1–2 mm thick, convex, ridged and rough.

Flowering: March–May.

Fruiting: March–May.

Chemical constituents: The plant contains minerals, the most important being selenium, tannins, luteolin, xanthophylls, cuminic aldehyde, anisaldehyde, α- and β-pinene, α- and β-phellandrene, carveol, gammaterpenine, limonene, linalool, p-Cymene, myrcene and sabinene. It also contains bisabolene, caryophyllene, eugenol, farnesol and glycosides such as epiginol and luteolo.

Used in: Ayurveda, Unani and Siddha medicine.

Parts used: Leaves, seeds and roots.

Uses: Cumin possesses many health benefits, such as antiallergic, anti-diabetic, anti-obesity, anti-oxidant, anti-platelet aggregation, carminative, cardiac stimulant, emmenagogue and hypoglycemic properties. The seeds have a cooling effect used to cure catarrh, cough, chronic diarrhea, digestive disorders, dyspepsia, fever, hoarseness of voice, gonorrhea, indigestion, inflammations, lung diseases and uterine infections. A decoction from the leaves of *Cuminum cyminum* and *Phyllanthus fraternus* with sugar is given orally to treat gonorrhea and syphilis. Cumin is used for insomnia, colds and fevers, and, mixed into a paste with onion juice, it has been applied to scorpion stings. The seeds can be taken to improve breast-milk production. Cumin seed is much used for flavouring curries in India and Ceylon.

Methods of propagation: Through seeds. Seeds are sown in beds during April–May and when the plants are 8–10 cm long they are transferred to a field. Cumin grows well on sandy loam to loamy soil with a pH range of 6.8–8.3.

CURCULIGO ORCHIOIDES GAERTIN

Synonyms: *Curculigo brevifolia* Dryand. ex W. T. Aiton; *Curculigo densa* W. Bull.; *Curculigo pauciflora* Zipp. ex Kurz.; *Hypoxis dulcis* Steud. ex Baker; *Hypoxis orchiodes* (Gaertn.) Kurz.

Classification:

Kingdom:	Plantae
Subkingdom:	Phanerogamea
Class:	Monocotyledonae
Series:	Epigynae
Family:	Amaryllidaceae
Genus:	*Curculigo*
Species:	*Curculigo orchioides*

Common names: English: Golden eye grass, Black musli; **Hindi:** Kali Mushali; **Kannada**: Nelatengu, Kadu irulli; **Malayalam:** Nelppana; **Sanskrit:** Mushali, Talamuli; **Tamil:** Nilappanakkilanku; **Telugu:** Nelatadi.

Distribution: This species is globally distributed from Indo-Malesia to Australia. Within India, it has been recorded in the sub-tropical Himalayas, eastwards to Assam and throughout the Western Ghats on marshy sides, especially on lateritic soil.

Description: It is a small perennial herb of about 30 cm, with a long cylindrical rootstock bearing a large number of fleshy lateral roots which are black in colour. Leaves: simple, 15–30 cm long, sessile, linear/lanceolate, sessile or with small petiole, and crowded on the short stem (rhizome) with the tips often bearing bulbils. Flower: small, yellow, distichous, the upper ones are male flowers which are smaller in size, while the lower ones are bigger, maybe hermaphrodite, female. The flowers are borne on a short scape hidden in the leaf sheath. Fruit: a capsule 1–2 cm long with a short slender beak bearing 1–4 seeds.

Flowering: October–January.

Fruiting: October–January.

Chemical constituents: Tubers contain a glycoside, yuccagenin, an alkaloid, lycorine, and flavones.

Used in: Ayurveda, folk, Tibetan, Unani and Siddha medicine.

Parts used: Rhizomes.

Uses: The tuberous root of the plant is used in the form of powder and decoction to treat retention of urine, as rejuvenative and as aphrodisiac. As Rasayana: satavari (*Asparagus racemosus*), mundi (*Sphaeranthus indicus*), guduci (*Tinospora cordifolia*), hastikarnapalasa (*Butea monosperma*) and musali are powdered in equal parts. It is mixed with ghee or honey and taken. It promotes body tissues and strength. Regular use of the powder of krsna musali (talamuli) mixed with ghee acts as an aphrodisiac. It is slightly bitter, viriligenic, diuretic and used in general debility, cough, piles, impotence, jaundice, urinary diseases, leucorrhea and menorrhagia.

Methods of propagation: Propagation of this plant is through rhizomes. It prefers well-drained sandy loam to loamy soil.

Steps to follow:

The rhizomes are obtained from the previous year's growth.

- The rhizomes are cut into pieces and planted horizontally in nursery beds.
- August–September is the most favorable time for planting.
- Root initiation is seen 40 days after planting while sprouting is seen after 60 days of planting.

Photo: Dr Kotresh

CURCUMA LONGA L.

Synonym: *Curcuma domestica* Valet.

Classification:

Kingdom:	Plantae
Subkingdom:	Phanerogamea
Class:	Monocotyledonae
Series:	Coronarieae
Order:	Alismatales
Family:	Zingiberaceae
Genus:	Curcuma
Species:	*Curcuma longa*

Common names: English: Turmeric; **Hindi:** Haldi; **Kannada:** Arishina; **Malayalam:** Manal, Karmmatti; **Marathi:** Halad; **Sanskrit:** Agara, Haridra; **Tamil:** Agdial, Agi, mancal; **Telugu:** Pasupu, Thilla.

Distribution: This is a native of China and the East Indies, now cultivated everywhere for its rhizome.

Description: The plant is a perennial with a short thick rhizome, dark yellowish inside. Leaves: Simple, alternate, long-petioled, ovate-lanceolate, 75–100 cm long, 35–45 cm wide, divided into leaf sheath, petiole and leaf blade. Flowers: hermaphrodite, zygomorphic; 1–1.5 cm long, pale yellow, borne in dense bracteates, strobiliform spikes; sepals 3, fused, with unequal teeth, white, with fluffy hairs; petals 3, bright yellow, fused into a tube, 3 cm long, only 1 stamen fertile, staminoids present; carpels 3, fruit a capsule.

Flowering: September–October.

Fruiting: September–October.

Chemical constituents: The rhizomes contain curcuminoids, volatile oils, sterols, sugars and starch. Curcuminoids are the principal colouring materials of which curcumin forms the major part, with desmethoxy curcumin, *bis*-desmethoxy curcumin and dihydrocurcumin forming the rest. The volatile oil contains high amounts of bisabolene derivatives along with borneol, camphene, linalool and α-phellandrene.

Used in: Folk, traditional, Ayurveda and Siddha medicine.

Parts used: Rhizome.

Uses: *Curcuma* is an excellent anti-inflammatory agent due to curcumin and its derivatives. It is also antihepatotoxic, antiulcer, anti-cancerous and an excellent antioxidant.

Methods of propagation: Through rhizomes.

Photo: Prof Srinath Rao

CYCLEA PELTATA (POIR.) HOOK. F. & THOMS.

Synonyms: *Cyclea barbata* Miers; *Menispermum pelatatum* Poir.; *Cyclea burmanii* (DC.) Wight & Arn.; *Cocculus peltatus* (Poir.) DC; Cyclea *arnottii* Miers.

Classification:

Kingdom:	Plantae
Subkingdom:	Phanerogamea
Class:	Dicotyledonae
Subclass:	Polypetalae
Series:	Thalamiflorae
Order:	Ranales
Family:	Menispermaceae
Genus:	*Cyclea*
Species:	*Cyclea peltata*

Common names: English: Pata root; **Gujarati:** Kanipet, Kalepat; **Hindi:** Raj Patha; **Kannada:** Padubal balli, Hade balli; **Malayalam:** Paataththaali; **Marathi:** Mothi pahadvel, Padathalli; **Sanskrit:** Bhruhat patha, Raj Patha; **Tamil:** Malaitanki; **Telugu:** Patatige, Paatathige.

Distribution: Found throughout India, Sri Lanka and Java, up to 800–900 m elevation. A hedge climber with slender stem and light-green leaves, common on bushes and such low supports in moist and dry deciduous forest areas, especially along roadsides, forest boundaries and waste places.

Description: Shrubby climbers, 1–2.5 m high; branchlets hairy-pubescent, pale green. Leaves simple, alternate, 2.5–6 × 2–5.5 cm, peltate, undulate along the margins, smooth and shiny above, pubescent and pale green beneath, truncate or shallowly subcordate at base, acute at apex; petioles 2–6.5 cm long, slender, pubescent. Flowers greenish white, in small clusters forming axillary panicles with pubescent branches; male flowers with calyx campanulate and 4-segmented, corolla 4-fid with inflexed lobes and short staminal column; female flowers solitary or in twins forming racemes with oblong, hairy sepals, orbicular, short petals, and subulate carpel gibbous at base with short 3–5-fid style; staminodes 6. Drupes white, 0.3–0.5 cm across, subglobose, compressed, hairy or pilose with sub-basal style scars; seeds horseshoe-shaped with transversely ridged endocarp.

Flowering: October–March.

Fruiting: March–May.

Chemical constituents: Fangchinoline, cycleapeltine, cycleadrine, cycleacurine, cycleanorine, cycleahomine chloride, chondocurine, magnoflorine, isotetradrine, perpamine, cycleamine, burmannaline.

Parts used: Roots, leaves.

Used in: Ayurveda and traditional medicine.

Uses: Roots, which contain the alkaloid cycleapeltine are medicinal, mainly used as an antifebrile. The plant is also used against a variety of abdominal ailments such as liver disorders, jaundice, stomachache and dysentery. The drug made from this plant possesses anti-hypertensive, anti-inflammatory and immunomodulatory properties. In Ayurveda, a decoction prepared from the root is recommended for diabetes. The roots are also used in the treatment of asthma and leprosy. The leaf juice of the plant is also used to cure wounds and in the preparation of beverages.

Methods of Propagation: Through seeds.

Photo: Prof Srinath Rao

CYMBOPOGON CITRATUS (DC. EX NEES) STAPF.

Synonyms: *Andropogon citratus* DC; *Andropogon fragrans* C. Cordem; *Andropogon roxburghii* Nees ex Steud.

Classification:

Kingdom:	Plantae
Subkingdom:	Phanerogamea
Class:	Monocotyledon
Order:	Poales
Family:	Poaceae
Genus:	*Cymbopogon*
Species:	*Cymbopogon citratus*

Common names: Bengali: Gandhabena; **English:** Lemongrass; **Gujarati:** Lilicha; **Hindi:** Gandhatrians; **Kannada:** Majjigehullu; **Malayalam:** Vasana pullu; **Marathi:** Olecha; **Sanskrit:** Bhustrina; **Tamil:** Karppurappul; **Telugu:** Nimmagaddi.

Distribution: *Cymbopogon citratus* is native to Indonesia and Malaysia and has been introduced to many tropical regions.

Description: It is a fast-growing, lemon-scented, perennial grass reaching a height of 1 m. It has distinct bluish-green leaves and usually does not produce seeds. Both these grasses produce many bulbous stems that increase the clump diameter as the plants mature.

Flowering: September–October.

Fruiting: September–October.

Chemical constituents: Tannins, sterols, terpenoids, phenols, ketones, flavonoids and sugars. It contains citral, myrcene, citronella, citronellol and geraniol.

Used in: Folk medicine.

Parts used: Leaves.

Uses: Analgesic, anti-depressant, antimicrobial. antipyretic, antiseptic, astringent, bactericidal, carminative, deodorant, diuretic, febrifuge, fungicidal, galactagogue, insecticidal, nervine, nervous system sedative and tonic.

Methods of cultivation: It is recommended to grow lemongrass by slips obtained by dividing well-grown clumps. Tops of clumps should be cut off within 20–25 cm of the root. The latter should be divided into slips and the lower brown sheath should be removed to expose the young roots.

Photo: Dr R.S Modi

DATURA INOXIA

Classification:

Kingdom:	Plantae
Subkingdom:	Phanerogamea
Class:	Dicotyledonae
Subclass:	Gamopatalae
Series:	Bicapellatae
Order:	Poleomoniales
Family:	Solanaceae
Genus:	*Datura*
Species:	*Datura inoxia*

Distribution: It is native to Central and South America, and has been introduced in Africa, Asia, Europe and Australia.

Description: This is an annual shrubby plant that typically reaches a height of 3–4 feet. Its stems and leaves are covered with short and soft grayish hairs, which gives the whole plant a grayish appearance. Leaves are elliptic, entire-edged, with pinnate venation. All parts of the plant emit a foul smell. The flowers are white, trumpet-shaped, 12–19 cm long with 10 lobes, sometimes tinged with rose or violet colour. They first grow upright, and later incline downward. The fruit is an egg-shaped spiny capsule, about 5 cm in diameter. It splits open when ripe, dispersing the seeds. The seeds have hibernation capabilities, and can last for years in the soil.

Datura inoxia differs from *D. stramonium* and *D. metel* in that it has about 7–10 secondary veins on either side of the midrib of the leaf, which anastomose by arches at about 1–3 mm from the margin. No anastomosis of the secondary veins is seen in the other species of *Datura*.

Flowering: July–September.

Fruiting: July–September.

Chemical constituents: The plant contains the highly toxic alkaloids atropine, hyoscine (scopolamine) and hyoscyamine.

Used in: Ayurveda.

Parts used: The whole plant.

Uses: Employed to cure skin diseases as well as anxiety and respiratory disorders. The seeds are chewed to relieve dental pain; they are also used in poultice, plasters and ointments.

Methods of propagation: The plant is usually grown from seeds. They take 3–4 weeks to sprout. Its perennial rhizomes can also be used for propagation.

Photo: Prof Srinath Rao

DATURA METEL L.

Synonym: *Datura fastuosa* sensu Griseb. non L.

Classification:

Kingdom:	Plantae
Subkingdom:	Phanerogamea
Class:	Dicotyledonae
Subclass:	Gamopetalae
Series:	Bicarpellateae
Order:	Pleomoniales
Family:	Solanaceae
Genus:	*Datura*
Species:	*Datura metel*

Common names: English: Devil's trumpet, Thorn apple; **Hindi:** Safed dhatura; **Kannada:** Kedage or Taale Hoovu; **Malayalam:** Ummata; **Sanskrit:** Dhastura; **Tamil:** Madulam, Ummattai; **Telugu:** Tellavummutta, Ummetta.

Distribution: Native of India, wild and naturalized throughout India. Found in the Western Himalayas. In the hilly regions of the western parts of the Deccan Peninsula, Devil's trumpet grows naturally in disturbed areas such as eroded sites, old fields, vacant lots, overgrazed pastures, rangeland, roadsides, abandoned roadbeds and fencerows.

Description: Under shrubs, 8–10 feet in height; the stems are semi-woody and suffruticose.

The plant has smooth, violet or dark-purple branches. Leaves: Simple, alternate, ovate-elliptic, entire or sinuately-toothed edges; 6–15 cm long, matte green colour; petiole of the leaves 3–7 cm long. Flowers: axillary, solitary, white, tubular, shortly pedunculate, peduncle 0.5cm, 2–3cm in long, 5-lobed; lobes about 0.5 in long, lanceolate-acute. Calyx 6–8 cm long, tubular, 5-lobed, minutely pubescent. Corolla large, 12–15 cm long, trumpet- or funnel-shaped, white, violate or tinged with purple, 5 in number, more or less cuspidate. Stamens 5, filaments attached to the corolla tube for ½ their length; anthers 0.3 in long. Ovary glabrous; style as long as the corolla; stigma capitate, bifid. Fruit a globose capsule; it grows upward from the plant and slightly droops when mature, about 3 cm, with small short round thorns that resemble bumps, glabrous, indehiscent, opening irregularly at the top. Seeds flattened, furrowed, tubercled and deep yellow.

Flowering: September–January.

Fruiting: February–March.

Chemical constitution: Leaves contain alkaloids (0.55 %) such as norscopolamine and scopolamine, the latter being the major one. It also contains meteloidine and hydroxyl-6-hyoscyamine. The alkaloid content of the stem is about 0.4 % and that of the seed 0.19 %. Other than alkaloids, fresh leaves contain withanolides such as withametelin, withametelin-B and datumetin (Gupta et al., 1991).

Used in: Ayurveda.

Parts used: The whole plant.

Uses: Employed to cure skin diseases as well as anxiety and respiratory disorders. The seeds are chewed to relieve dental pain. Also used in poultice, plasters and ointments.

DATURA STRAMONIUM L.

Synonyms: *Datura bernhardii* Lundstr., *Datura bertolonii* Parl. ex Guss., Datura cabanesii P.Fourn., *Datura capensis* Bernh., *Datura inermis* Juss. ex Jacq., *Datura laevis* L.f.

Classification:

Kingdom:	Plantae
Subkingdom:	Phanerogamea
Class:	Dicotyledonae
Subclass:	Gamopetalae
Series:	Bicapellatae
Order:	Poleomoniales
Family:	Solanaceae
Genus:	*Datura*
Species:	*Datura stramonium*

Common names: English: Jimson weed, Downy thorn apple; **Hindi:** Safed Dhatura; **Malayalam:** Ummattai; **Sanskrit:** Dhatura; **Tamil:** Ummathai; **Telugu:** Datturamu.

Distribution This plant occurs in the temperate Himalayas up to 2500 m and in the hilly regions of Central and Southern India.

Description: A bushy plant up to about 1 m high. The stem shows dichasil branching. Leaves: alternate, rarely opposite, simple, large, 12–15 cm long, ovate, toothed, dark green above and silvery grey below, foul smelling. Flowers: very large, white, solitary, in the axil of the leaves. Calyx: long and tubular, swollen at the bottom and sharply angled, surrounded by 5 sharp teeth. Corolla: folded and only partly open, white, funnel-shaped with 5 prominent ribs, 6–8 cm long. The flowers open at night emitting a pleasant smell. Stamens 5, free, mostly erect. Fruit: a capsule, ovoid, 4–8 cm in diameter, deeply divided into 4, covered with long or short prickles, at maturity it splits into 4 chambers, seeds black to brown, many in number. *Datura stramonium* differs from *Datura inoxia* in that it has smaller flowers and tooth-edged leaves.

Flowering: July–October.

Fruiting: August–October.

Chemical constituents: Alkaloids such as hyoscyamine and hyoscine. Hyoscine has been found in higher concentrations in young stems. Leaves contain flavonoids such as chrysin, kaempferol, naringenin, quercetin and a withanolide. Roots contain ditigloyl esters of 3,6-dihydroxy tropane.

Used in: Traditional and Ayurveda medicine.

Parts used: Dried leaves, flowering tops and seeds.

Uses: The drug made from the plants consists of dried leaves, flowering tops and seeds. The main active principle in the leaves is hyoscyamine; the drug is useful in bronchitis or asthma, and controls salivation; it is antispasmodic and narcotic. The inhalation of smoke from the burning leaves is also good for relieving asthma. An ointment made from the seed paste is applied to cure boils, pimples and swellings. A poultice made from the leaves checks inflammation of breasts caused by excessive formation of milk. The powdered leaves are applied to hemorrhoids.

Methods of propagation: Mostly through seeds; vegetative methods are also reported.

Photo: Dr R.S. Modi

DESMODIUM GANGETICUM (L.) DC.

Synonyms: *Hedysarum gangeticum* L.; *Hedysarum maculatum* L.; *Meibomia gangetica* (L.) Kuntze.

Classification:

Kingdom:	Plantae
Subkingdom:	Phanerogamea
Class:	Dicotyledonae
Subclass:	Polypetalae
Series:	Calyciflorae
Order:	Rosales
Family:	Leguminosae (Fabaceae)
Genus:	*Desmodium*
Species:	*Desmodium gangeticum*

Common names: Bengali: Salpani**; English:** Ticktrefoil; **Gujarati:** Shaalwan; **Hindi:** Satpura; **Kannada:** Murelchonne; **Malayalam:** Moovila; **Marathi:** Satpura; **Sanskrit:** Shalparnih, Sthira, Somya, Guha; **Tamil:** Pulladi, Sirupulladi, Moovilai, Orilai; **Telugu:** Kolakuponna, Nakkotokaponna.

Distribution: Tropical Africa, Indian subcontinent, China, Japan, Myanmar, Thailand, Cambodia, Laos, Vietnam, Malaysia, Indonesia, Philippines and Australia. It is found in the forests and waste lands of India, from the plains and Western Ghats up to 1500 m in the north up to Sikkim.

Description: It is a 2–4 feet high erect under shrub with woody stem, branches covered with soft hairs. Leaves are unifoliate, ovate, oblong, obtuse, pubescent beneath and up to 15 cm long. Its inflorescence is many-flowered, terminal or axillary, slender and 15–30 cm long. The flowers are small and pink-violet. Pods are thin, pointed, 1.5– 2 cm long clothed with minute hooked hairs.

Flowering: August–October.

Fruiting: October–November.

Chemical constituents: The roots contain several alkaloids, flavonoids, pterocarpanoids, gangetin, gangetinin and desmodin. The aerial parts contain indole-3-alkylamines and their derivatives.

Used in: Ayurveda.

Parts used: Roots, leaves; the whole plant.

Uses: The juice extracted from whole plant shows anti-rheumatic and anti-osteoarthritic properties. Roots are bitter, alterative, tonic with anti-inflammatory, anti-catarrhal, carminative, diuretic, digestive, emmenagogue, fever reducing and laxative properties. The roots are used traditionally for treating cough, diarrhea, dysentery, biliousness, vomiting, piles, bronchitis, asthma, intestinal parasites, arthritis, rheumatism and post-delivery disorders. The decoction of leaves is taken orally to treat diarrhea, emesis, fluid retention and constipation. The leaf paste is applied topically along with *Aloe vera* paste to stop hair fall and to cure eczema.

The juice of roots or the decoction of leaves is given twice/thrice a day to cure diarrhea and dysentery. Root powder mixed with honey is applied to cure ulcers. The roots are chewed to prevent premature ejaculation. ½ cup root decoction is taken orally as an antidote for snake bite (https://www.bimbima.com/herbs/shalaparni-desmodium-gangeticum-information-medicinal-uses-and-caution/61/).

Methods of propagation: This plant can be propagated by seed and stem cuttings. Seed germination is poor. Treating the seeds with sulphuric acid for 5–10 minutes improves seed germination.

Photo: Prof Srinath Rao

DESMODIUM TRIFLORUM (L.) DC.

Synonyms: *Hedysarum triflorum* L.; *Hedysarum granulatum* Schum.; *Desmodium stipulaceum* Burm. f.; *Desmodium albiflorum* Cordem.

Classification:

Kingdom:	Plantae
Subkingdom:	Phanerogamea
Class:	Dicotyledonae
Subclass:	Polypetalae
Series:	Calyciflorae
Order:	Rosales
Family:	Leguminosae (Fabaceae)
Genus:	*Desmodium*
Species:	*Desmodium triflorum*

Common names: Bengali: Kudaliya; **English:** Creeping tick trefoil, Beggarweed; **Hindi:** Kudaliya, Motha; **Kannada** Kaadu menthe; **Malayalam:** Cherupalladi, Nilamparanda; **Marathi:** Jungalimethi, Raan methi; **Sanskrit:** Hamsapaadi, Tripaadi; **Tamil:** Ankuca-Pati; **Telugu:** Muntamandu, Moordoo.

Distribution: This plant occurs in the humid tropics and warmer subtropics with more than 1200 mm rainfall; it behaves as a perennial plant under well-distributed rainfall and as an annual plant under seasonal rainfall.

Description: The herb is highly branched and creeping, rooting at the nodes, forming a mat. The leaves are clover-like, divided in to 3 leaflets, the lower leaves sometimes not divided, slightly hairy below, stipules broad, acuminate, strongly nerved. Leaflets are inverted-egg- or heart-shaped with a notch at the tip, about 1 cm long and up to 8 mm broad, sometimes with 2 white marks. Flowers are reddish violet or pale pink in colour. Calyx hairy with long white hairs, corolla slightly longer than calyx. Flowers arise in fascicles, opposite the leaves on 5–8 mm stalks. Standard petal about 4–5 mm long. Pods falcate, about 1.5 mm long and 2 mm broad, 3–7-jointed.

Flowering: Throughout the year.

Fruiting: Throughout the year.

Chemical constituents: The plant contains ursolic acid, vitexin, genistin, fucosterol and rare diholosylflavane and 2-0 glucosylvitexin. The leaves contain phenethylamine (major alkaloid), tyramine, trigonelline, hypaphorine and choline. The root contains the hypaphorine (major alkaloid), N, N-Dimethyl tryptophan betaine and choline (Ghosal et al., 1971; Adinarayana and Syamsundar 1982).

Used in: Ayurveda.

Parts used: Roots, leaves; the whole plant.

Uses: The leaves are used in diarrhea, convulsions and as a galactagogue. The fresh leaves of the plant are applied to wounds and abscesses. The paste is applied to sores and itches. The paste is applied on the forehead to bring down high fever. The fresh juice of the plant is recommended for use in dysentery and as a laxative. The dried powder of whole plant is given on an empty stomach to cure bone fractures. The infusion of root is used to promote labour and treat vertigo. The roots are reputedly carminative, tonic and diuretic. They are used in bilious complaints. The leaves are ground with cow's milk and given daily in the morning. The main actions include antispasmodic, sympathomimetic, stimulation of the central nervous system, curare-mimetic activity and diuretic (Rao et al 2013).

Methods of propagation: Through seeds.

Photo: Prof Srinath Rao

DIGITALIS PURPUREA L.

Synonyms: *Digitalis alba* Schrank.; *Digitalis carnea* Meing & Weing.; *Digitalis speciosa* Salisb.

Classification:

Kingdom:	Plantae
Subkingdom:	Phanerogamea
Class:	Dicotyledonae
Subclass:	Gamopatalae
Series:	Bicarpellate
Order:	Personales
Family:	Scrophulariaceae
Genus:	*Digitalis*
Species:	*Digitalis purpurea*

Common names: English: Foxoglove; **Hindi:** Tilpushpi; **Malayalam:** Hridpatri; **Sanskrit:** Hritpatri, Tilpushpi.

Distribution: This plant is not indigenous to India; it is cultivated chiefly in the hilly regions of Northern India, such as Kashmir; it was also grown in Darjeeling and in the Nilgiri Hills, but these cultivations have become naturalized in some places.

Description: A biennial or perennial herb, about 1.2 m high; lower basal leaves long-stalked, hairy, ovate, 15–30 cm long, the upper leaves almost without stalks, becoming smaller in size as they grow upwards. Flowers 5–8 cm long, 1-sided bunches. Fruit: ovoid.

Flowering: July–August.

Fruiting: July–August.

Chemical constituents: The dried leaves, which form the drug digitalis, contain more than 40 cardiac glycosides based on 4 genins: digitoxigenin, gitoxigenin, gitaloxigenin and gitaligenin, with their formyl or acetyl derivatives. The 2 primary glycosides – purpurea glycosides A and B – form the chief constituents of the fresh leaves. However, in dried leaves, triglycosides like digitoxin and gitoxin are prominent.

Used in: Folk, Ayurveda and Siddha medicine.

Parts used: Tubers.

Uses: The dried leaves of the plant constitute the drug. The leaves must be dried at about 60 °C and as soon after collection as possible. The main use of this drug is in heart diseases. The drug promotes and stimulates the activity of all muscle tissues. It is used in cases of congested heart failure. All these glycosides improve the rhythm of the heartbeats, making the contraction of the heart more powerful and helping the heart to pump the blood at the time of cardiac failure. It improves the blood supply to the kidneys and thus promotes urination, removing obstructions in the kidneys.

Digitalis is used in some ointments for local application on wounds and burns. This drug, however, has some cumulative toxic effects and causes headache, giddiness, etc. These side effects must be watched.

Other species:

Another species of the genus *Digitalis*, viz. Digitalis fucata Ehrh. (Woolly Foxglove), is being cultivated and also occurs in Kashmir. Its flowers are smaller, hairy, and cream, yellow-, or purple-coloured. This plant has stronger medicinal properties, and its side effects are not as toxic as those of *D. purpurea*.

Methods of propagation: Through seeds.

Photo: https://pixabay.com/images/search/foxglove-digitalis-purpurea-/

DICHROSTACHYS CINEREA (L.) WIGHT & ARN.

Synonyms: *Cailliea glomerata* (Forssk.) J.M. Macbr.; *Dichrostachys glomerata* (Forssk.) Chiov.; *Dichrostachys nutans* (Pers.); *Mimosa cinerea* L.; *Mimosa glomerata* Forssk.

Classification:

Kingdom:	Plantae
Subkingdom:	Phanerogamea
Class:	Dicotyledonae
Subclass:	Polypetalae
Series:	Calyciflorae
Order:	Rosales
Family:	Leguminosae (Fabaceae)
Subfamily:	Mimosoidae

Genus: *Dichrostachys*

Species: *Dichrostachys cinerea*

Common names: Gujarati: Mordundiyun; **English:** Sickle bush, Kalahari Christmas tree; **Hindi:** Kunali; **Kannada:** Vaduvarada gida; **Malayalam:**Vitattal; **Marathi:** Durangi babool; **Sanskrit:** Vellantaru; **Tamil:** Veduttalam; **Telugu:** Nellajammi, Velturu chattu.

Distribution: Distributed in the forests of Africa, Australia, India and parts of Southeast Asia. It is native to Africa but has been introduced to India.

Description: Small, much-branched thorny trees; 6–7 m high, spines axillary, straight, often leaf-bearing; bark on young branches green and hairy but dark grey/brown and longitudinally fissured on older branches and stems; peeling off in thin flakes; branchlets densely pubescent; strong alternate thorns, up to 8 cm long, almost at right angles, slightly recurved, growing out of the branches, may bear leaves at the base. Leaves bipinnate, rachis 4–8 cm, with 5–15 pairs of pinnae with a small erect gland between each pair of pinnae, each pinna bearing 12–22 pairs of leaflets; terminal pair of pinnae shorter, dark green, underside pale. Leaflets about 8 × 2.5 mm wide; leaflets and petioles very tomentose and ciliate. Lamina 0.2–0.3 × 0.1 cm, oblong, base obtuse, apex acute, margin entire, ciliate, tomentose, chartaceous; nerves obscure. Inflorescence axillary or extra-axillary; spikes (like a bottlebrush), with upper half of the spike yellow and lower red. Flowers very characteristic, bicoloured and fragrant. Terminal flowers bisexual, yellow; bracts oblong; calyx tube campanulate, lobes 5; petals 5, lanceolate connate below, pubescent, valvate; exserted, stamens 10, yellow, free anthers ending in stalked glands; ovary sub-sessile; style 3 mm; stigma terminal, truncate. Fruit a pod, linear, flat, in dense, stalked, intertwined clusters; up to 100 × 15 mm, indehiscent or opening from apex; seeds 4–6, ovoid, compressed with a spot at one end.

Flowering: July–December.

Fruiting: July–December.

Chemical constituents: The bark contains friedelin, friedelan-3-β-ol, β-sitosterol and α-amyrin. The heartwood contains octacosanol and sitosterol. The leaves contain hentriacontanol, β-amyrin, stigmasterol and β-sitosterol.

Used in: Folk medicine.

Parts used: Bark, root and leaves.

Uses: The bark is used to treat dysentery, headaches, toothaches and elephantiasis. It also acts as a vermifuge. Root infusions are taken for leprosy, syphilis, coughs and as anthelmintic, purgative and strong diuretic. Pounded roots and leaves are used to treat epilepsy. The roots are chewed and placed on the sites of snake bites and scorpion stings, and the leaves, which are believed to produce local anaesthesia, are used for the same purpose and also as a remedy for sore eyes and toothache. The leaves are taken as a diuretic and laxative, and used for gonorrhea and boils; the powder from the leaves is used in the massage of fractures. The plant is used as a veterinary medicine in India.

Methods of propagation: Seeds as well as root and stem cuttings. The seeds are hard and should be scarified by mechanical or chemical means. Seeds per kg: Rs. 39000.

Photo: Prof Srinath Rao

DIGERA ARVENSIS FORSSK.

Synonyms: *Digera muricata* (L.) Mart.; *Digera alternifolia* L. Aschers.; *Digeria frutescens* Juss.; *Achyranthes alternifolia* L.; *Achyranthes digera Poir.*; *Celosia muricata* Spreng.

Classification:

Kingdom:	Plantae
Subkingdom:	Phanerogamea
Class:	Dicotyledonae
Subclass:	Monochlamydeae
Series:	Curvembryae
Family:	Amaranthaceae
Genus:	*Digera*
Species:	*Digera arvensis*

Common names: Bengali: Lata mouri, Ful, Gun gutiya; **English:** False amaranth; **Hindi:** Chanchali, Lesua; **Kannada:** Chenchali soppu, Goraje palle; **Marathi:** Getan, Kunjar; **Sanskrit:** Aranya, Aranyavastuka, Kunanjara; **Tamil:** Toyya kirai; **Telugu:** Chenchalichettu, Chnchali, Koora,

Distribution: Widespread in Southern Asia from Arabia, Afghanistan, India, Sri Lanka and Yemen. Seen growing wild in waste areas throughout India.

Description: It is an annual herb with erect or spreading branches, up to 50 cm tall. Stem ridged. Leaves simple, alternate, broadly ovate to lanceolate, apex acute, rounded or cordate at base, up to 8 cm long, glabrous or hairy on veins below; petiole slender, glabrous, grooved on upper surface. Flowers bisexual, about 2 mm long, pink or rose-coloured, in 3-flowered clusters, in axillary pedunculate spikes, central flowers fertile with unequal perianth, outer 2 flowers much reduced. Perianth segments 3–5 mm long and 2–3 inner, slightly shorter, hyaline segments; stamens usually 5, free or slightly connate at base; ovary superior, 1-celled, style filiform, up to 4 mm long, stigmas 2, divergent. Fruit compressed, 2–2.5 mm long, yellowish green, notched at the apex. Seeds black.

Flowering: August–September.

Fruiting: September– October.

Chemical constituents: The plant contains α- and β-spinasterol29. The analysis of various fractions of the plant indicated the presence of flavonoids, alkaloids, terpenoids, saponins, coumarins, tannins, cardiac glycosides and anthraquinones. Rutin and hyperoside flavonoids have been identified in the hexane extract of this plant (Sharma and Vijayvergia, 2013).

Used in: Traditional medicine.

Parts used: Leaves, shoots and root.

Uses: Laxative, used internally to cure digestive disorders. Flowers and seeds are given for urinary diseases. All parts of the plant have been used as crude drug for the treatment of kidney stone and urinary tract disorders. This plant is also used as an alternative for secondary infertility. Leaf paste is applied locally to prevent pus formation.

Methods of propagation: Through seeds.

Photo: Prof Srinath Rao

DIOSCOREA DELTOIDEA

Synonyms: *Tamus nepaliensis*; *Diascoria nepalensis*.

Classification:

Kingdom:	Plantae
Subkingdom:	Phanerogamea
Class:	Monocotyledonae
Subclass:	Alismatidae
Series:	Epigynae
Order:	Alismatales
Family:	Dioscoraceae
Genus:	*Dioscorea*
Species:	*Dioscorea deltoidea*

Common names: English: Yam; **Hindi:** Khamalu; **Kannada:** Vanagenasu; **Malayalam:** Kaavathu; **Sanskrit:** Aluka, Varahi; **Tamil:** Perumvalli, Kishongu; **Telugu:** Pendalamu.

Distribution: Yam is found in the Himalayas, from Kashmir to Assam, Indo-China and Western China, Western Pakistan and Afghanistan at altitudes of 450–3100 m.

Description: Nepal yam is a hairless vine, twining clockwise. Tubers are ligneous, irregular. Leaves: simple, alternately arranged, 5–11.5 cm long, 4–10.5 cm broad, ovate or triangular-ovate, often heart-shaped, the basal lobes rounded or sometimes dilated outwards, 7–9-nerved, long-pointed, hairless above, velvety on the nerves beneath. Leaf stalks are 5–10 cm long, slender. Male flower spikes occur solitary in leaf axils, simple or sometimes branched, slender, 7.5–25 cm long. Flowers are in small distant clusters; stamens 6, antheriferous. Female spikes are solitary, slender, up to 15 cm long, few-flowered. Fruit: a capsule, 2 cm long, 3 cm broad, obovate or obcordate. Seeds are winged unequally all round.

Flowering: May–July.

Fruiting: July–August.

Used in: Ayurveda, Siddha, Unani, folk and modern medicine.

Parts used: Roots, leaves and tubers.

Uses: The juice of the root tuber is taken in the evening in the treatment of roundworm. It is also used to alleviate constipation. The roots of most, if not all, members of this genus, contains diosgenin. This is widely used in modern medicine in order to manufacture progesterone and other steroid drugs. These are used as contraceptives and in the treatment of various disorders of the genitourinary organs as well as in a host of other diseases such as asthma and arthritis. The roots of this species contain an average of 4.8 % diosgenin.

ELETTARIA CARDAMOMUM (L.) MATON

Synnonyms: *Amomum cardamum* L.; *Alpina cardamomum* (L.); *Cardamomum officinale* Salisb.; *Ziniber cardamomum* (L.) Stokes.

Classification:

Kingdom:	Plantae
Subkingdom:	Phanerogamea
Class:	Monocotyledonae
Subclass:	Zingiberidae
Series:	Epigynae
Order:	Alismatales
Family:	Scitamineae (Zingeberaceae)
Genus:	*Elettaria*
Species:	*Elettaria cardamomum*

Common names: English: Cardamom; **Hindi:** Chhoti Ilaichi; **Kannada:** Yelakkai; **Malayalam:** Yelam; **Marathi:** Ilaichi; **Sanskrit:** Elo, Yelakkai; **Tamil:** Elatarri; **Telugu:** Elaki.

Distribution: It occurs wild in Southern India, particularly in the moist forests of the hilly regions in Mysore and Kerala. It is also cultivated elsewhere in India.

Description: A herb with thick, fleshy-branched rhizome and several erect stems, going sometimes up to 3 m high. Leaves very large, 30–90 cm long, narrow, with one strong median nerve and numerous faint side nerves. Flowering stock arises near the base of the stem and lies on the ground; flowers about 4 cm long, white or pale green, borne in 30–90 cm-long bunches. Fruits about 1.5 cm long, pale green/yellow, ovoid, 3-called, many-seeded. Seeds triangular, brownish black. The size of the plant and of its parts varies in different localities and with different cultivated races. The fruits and seeds, however, do not vary much. Often, the fruits offered in a market are bleached with sulphur fumes and made white.

Flowering: September–February.

Fruiting: September–February.

Chemical constituents: The dried fruits of the plant constitute the drug. The seeds are actually used, but they are taken out from the fruits just when required. *Cardamom* is used chiefly for relieving flatulence or feeling of overfullness in the stomach, i.e. to promote digestion. It is administered with purgatives, and as a flavouring agent. Powdered with cloves, ginger and caraway, it is good in indigestion, etc.

Used in: Traditional and Ayurveda medicine.

Parts used: Seeds and oil.

Uses: The seeds are aromatic, acrid, sweet, cooling, stimulant, carminative, digestive, stomachic, cardiotonic, expectorant, diuretic, abortifacient and are useful in asthma, renal and vesical calculi, cardiac disorders, hyperdipsia, anorexia and vitiated conditions of *vata*.

Other uses: Cardamom is largely used in cooking and making cakes and confectionery. The oil obtained from the seeds is used for flavouring beverages.

Methods of propagation: Through seeds.

Photo: Prof Srinath Rao

EMBELIA RIBES BURM. F.

Synonyms: *Embelia glandulifera* Wight; *Samara ribes* Kurz.

Classification:

Kingdom:	Plantae
Subkingdom:	Phanerogamea
Class:	Dicotyledonae
Subclass:	Gamopetalae
Series:	Bicarpellatae
Order:	Gentianales
Family:	Primulaceae
Genus:	*Embelia*
Species:	*Embelia ribes*

Common names: Bengali: Vidang, Bidanga, Bhaibirrung; **English:** False black pepper; **Gujrati:**Bavadang; **Hindi:** Vaividanga; **Kannada:**Vayubaliga, Amogha; **Malayalam:**Vijhala,Tirit ticanni; **Marathi:** Bavidanga, Ambti; **Sanskrit:** Vidangaka, Amogha, Chitrabija; **Tamil:** Vilanga, Kattukodi; **Telugu:** Vayu vidangalu, Potosul.

Distribution: It is widely distributed throughout India. It is also present in countries such as Singapore, Sri Lanka, Indonesia, Malaysia and Africa. In India, it grows mostly in Arunachal Pradesh, Assam, Maharashtra, Andhra Pradesh, Karnataka, Tamil Nadu and Kerala. Commonly seen up to a height of 1500 m.

Description: Woody shrub, almost a climber, with flexible and terete branches, bark with lenticels, about 40–60 cm in girth. Leaves are simple, alternate and elliptic-lanceolate, coriaceous gland dotted and obtusely acuminate, about 6–10 cm long and 2–4 cm broad, entire and shiny above. Flowers are small and globular, greenish yellow and pinkish white. The fruits are like black pepper and reddish-brown to black in colour, wrinkled, a short pedicel often present, and 1-seeded, seeds covered with a brittle pericarp with thin membrane. The root is brownish grey in colour with hairy reddish rootlets. It is called 'false black pepper' because it mimics black pepper in appearance and nature.

Flowering: February.

Fruiting: March–April.

Chemical constituents: The principal constituents of this herb are embelin, christembine, homo-embelin, vilangine, quercitol, tannins, volatile oil, resins and fatty acids. The plant also contains potassium embelate.

Used in: Ayurveda.

Parts used: Fruits, leaves and root bark.

Uses: The root bark is acrid, astringent, anthelmintic, antifertility, antiestrogenic, carminative, digestive, laxative, soothing, stimulant, stomachic and thermogenic. It is used in treating abdominal disorders, constipation, flatulence, fungal infections, headache, hemorrhoids, intestinal parasites, indigestion, lung diseases, mouth ulcers, obesity, piles, pneumonia, skin diseases, sore throat, toothaches and worms. Its decoction is useful in insanity and heart diseases. The leaves possess astringent, demulcent, depurative and thermogenic properties. They are useful in skin diseases and leprosy.

The fruits are acrid, astringent, anthelmintic, bitter, brain tonic, carminative, contraceptive, diuretic, febrifuge, laxative, stimulant and thermogenic. They are useful in asthma, colic, dyspepsia, flatulence, leprosy, nervous debility, tumors, fever, general debility and skin diseases. The fruit paste is used as mouthwash to combat cavities, and also applied externally for skin-related problems. The root decoction is taken for treating insanity and heart diseases. The fruit pulp is used as a purgative. The fresh juice is used as a cooling, diuretic and laxative drink. The dried berries are useful in treating intestinal worms and piles.

Methods of propagation: Through seeds. The seeds are directly sown in the field during June–July with a spacing of 1.0 × 1.0 m.

EMBLICA OFFICINALIS GAERTN.

Synonym: *Phyllanthus emblica* L.

Classification:

Kingdom:	Plantae
Subkingdom:	Phanerogamea
Class:	Dicoyledonae
Subclass:	Monochlamydae
Series:	Unisexuales
Order:	Geraniales
Family:	Euphorbiaceae
Genus:	*Emblica*
Species:	*Emblica officinalis*

Common names: Bengali: Akla, Amlaki; **English:** Gooseberry; **Gujarat:** Amali, Ambala; **Hindi:** Amla; **Kannada:** Bettada nelli; **Malayalam:** Nellikka; **Sanskrit:** Amla, Amlaki; **Tamil:** Nelli; **Telugu:** Usirikai, Amalkamu.

Distribution: The plant occurs throughout tropical and sub-tropical India; it was found to be abundant (and practically unexploited) in deciduous forests of Madhya Pradesh.

Description: A small or middle-sized deciduous tree with smooth, greenish grey, exfoliating bark. Leaves small, 10–13 mm long, 2–3 mm broad, very closely set in pinnate fashion; branchlets look rather feathery. Male and female flowers borne on same tree, flowers pale green, usually in small dense clusters below the leaves; male flowers small, numerous, on short stalks; female flowers also small, fewer. Fruit 1.5–2.5 cm diameter, fleshy, roundish, rather indistinctly marked into 6 lobes, pale green or yellowish; seeds 6, fruits of cultivated forms are larger.

Flowering: March–May.

Fruiting: March–May.

Chemical constituents: The fruits are the richest source of vitamin C (ascorbic acid, 2–3 %). They also contain gallotannins, carbohydrates, phenolic acids, alkaloids, pectin and minerals.

Used in: Folk, traditional, Ayurveda, Siddha and Unani medicine; homeopathy.

Parts used: Fruit, flowers, root, bark and leaves.

Uses: The fresh or dried fruits of this tree constitute the drug. The fruits are one of the 3 constituents of the well-known Indian preparation Triphala (the other two constituents being *Bahera – Terminala bellirica* and *Harra – T. chebula*). *Triphala* is used as laxative and in treatment of enlarged liver, piles, stomach complaints, pain in eyes, etc. The fruits are a good liver tonic; the raw fruits are cooling and mildly laxative. The fermented liquor made from the fruits is considered useful in indigestion, anaemia, jaundice, certain heart complaints and colds; it also promotes urination. The dried fruits are prescribed in Indian medicine. The flowers, roots and bark of the tree are also medicinal. The seeds are reported to cure asthma, stomach disorders and peptic ulcers. *Emblica* also exhibits anti-tumour, hypotensive and immunomodulatory properties. The extracts are anticytotoxic, antigenotoxic, anticlastogenic and antimutagenic in nature.

Methods of propagation: This plant can be propagated through seeds as well as by vegetative means. It grows in a wide range of soils including mildly alkaline, shallow, gravelly soils, but prefers deep, moist and loamy to alluvial soils.

- The ripe fruits are collected during January and dried in the sun.
- The fleshy portion of the fruit is removed and soaked in water for 24 hours. It is then wrapped in a wet gunny cloth till roots emerge. Soaking the seeds in hot water for 5 minutes hastens the germination. The seeds are viable for about a month. Fresh seeds are immediately sown due to their low viability on storage. Sowing at the time of commencement of rainy season will give the best results.
- Seeds germinate in about 3–4 days and there is about 70–80 % germination.
- For vegetative propagation, layering and budding are practised. For budding, shield budding has been found to be most practical and commercially viable method. It is important that proper mother plants are selected and that buds are selected from a branch that has a good number of female flowers.
- The success rate in grafting is in the range of 60–70 %, while in the case of softwood grafting it is about 42 %.
- The advantage of the vegetative method of propagation lies in the fact that vegetatively propagated plants start fruiting in 5–6 years, while seedling trees take at least 10 years to fruit.

Photo: Dr R.S. Modi

EUPHORBIA HIRTA L.

Synonyms: *Euphorbia capitata* Lam.; *Euphorbia nodiflora* Steud.; *Euphorbia pilulifera* Linn.; *Chamaesyce pilulifera* Linn.

Classification:

Kingdom:	Plantae
Subkingdom:	Phanerogamea
Class:	Dicotyledonae
Subclass:	Monochlamidae
Series:	Unisexuales
Order:	Malpighiales
Family:	Euphorbiaceae
Genus:	*Euphorbia*
Species:	*Euphorbia hirta*

Common names: English: Asthma herb, Snake weed; **Gujrati:** Dudeli; **Hindi:** Bada dudhi; **Kannada**: Akkigida, Kempuneneyakk; **Marathi:** Dudnali, Govardhana; **Malayalam:** Nelapali; **Sanskrit:** Amampatchairaisi, Barokherui, Dugadhika; **Tamil:** Amumpatchaiyarissi; **Telugu:** Baidarie, Nanabala, Reddinanabrolu.

Distribution: This is distributed throughout the hotter parts of India, found in waste places and along the road sides.

Description: It is a slender-stemmed, branched, hairy annual plant. It is about 40 cm across and reddish purple in colour. Leaves opposite, 1–3 cm long, elliptic to oblong acute, dark green above and pale below, blotched with purple in the middle. Inflorescence a cyathium; single ones are paired in axillary clusters. Male flowers 4–6, ebracteate; female florets pendulous, stigma bifid from the base. The fruit is a capsule, yellow, 3-celled and hairy, about 1–2 mm in diameter. Seeds 3 per fruit, brown, angular and wrinkled.

Flowering: Throughout the year.

Fruiting: Throughout the year.

Chemical constituents: Numerous compounds are reported to be present in this plant, such as afzelin, myricetin, rutin, quercetin and euphorbin A, B and C. The plant also contains β-amyrin, kaempferol and gallic and protocatechuic acids.

Used in: Ayurveda and folk medicine.

Parts used: The whole plant.

Uses: It is used in the treatment of diarrhea, dysentery and intestinal parasitosis. Also used in the treatment of asthma, bronchitis, hay fever and conjunctivitis. The roots are used for snake bites. A decoction of the herb is used as gargle to treat thrush. A decoction of the dry herb is used to treat skin diseases. It is also reported to have hypotensive and tonic properties.

Methods of propagation: Through existing root stalks.

Photo: Prof Srinath Rao

EVOLVULUS ALSINOIDES (L.)

Synonyms: *Evolvulus alsinoides* (L.); *L. var. angustifolius Torr.*; *Evolvulus alsinoides* (L.); *L. var. debilis (Kunth) van Ooststr.*

Classification:

Kingdom:	Plantae
Subkingdom:	Phanerogamea
Class:	Dicotyledonae
Subclass:	Gamopetalae
Series:	Bicarpellatae
Order:	Polemoniales
Family:	Convolvulaceae
Genus:	*Evolvulus*
Species:	*Evolvulus alsinoides*

Common names: Hindi: Sankhapushpi, Vishnukranta, Shyamakranta; **Kannada:** Vishnukranti, Shankha vaelu, Vishnugranthi soppu; **Malayalam:** Vishnukranthi; **Marathi:** Sankhapushpi, Vishnukranti; **Sanskrit:** Vishnukranta, Laghu vishnukrantha; **Tamil:** Vishnukranthi, Vishnukarandi; **Telugu:** Vishnukraanthamu, Nalla Vishnukraantha.

Distribution: Throughout India. Open-grassy places, preferably well-drained soil.

Description: It is a prostrate, much-branched herb with small woody rootstock. The stem is spreading, wiry, hairy, more than 30 cm long. The leaves are numerous, oblong-elliptic or lanceolate, 1.0–3.5 × 0.3–1.0 cm, obtuse at apex, densely covered with appressed silky hairs, petioles short.

Flowers are light blue, solitary, axillary; bracts lanceolate. Sepals are 4, lanceolate, densely silky. Corolla is funnel-shaped. They are pollinated by insects. Seeds are dispersed by wind. Capsules are globose, 3–4 mm long, 4-valved. Seeds are usually 4, glabrous.

Flowering: July–November.

Fruiting: July–November.

Chemical constituents: The prevailing compounds were Squalene, 2-Hydroxy-3[(9e)-9-Octadecenoyl, Oleic acid, Octadecanoic acid, 1-(+)-Ascorbic acid 2,6dihexadecanoate and Caryophyllene.

Used in: Ayurveda and Unani medicine.

Parts used: The whole plant and leaves.

Uses: It promotes intelligence, enhances memory, reduces stress, anxiety, depression and confusion; induces peaceful sleep. It is one of the psychotropic drugs that cure nervous debility and dementia. It gives relief in epilepsy and fits. It detoxifies the body and helps in digestive impairment due to nervousness. It has aphrodisiac and laxative properties and strengthens the reproductive system. The paste of the whole plant is applied externally for cuts and ulcers to stop bleeding. The leaves are made into cigarettes and smoked to get relief from asthma and chronic bronchitis and the leaf paste is applied for the growth of hairs. The paste of the plant is applied on the affected teeth to cure toothache. The root powder is used to improve libido.

Methods of propagation: Through seeds.

Photo: Prof Srinath Rao

FAGONIA CRETICA LINN.

Synonyms: *Fagonia arabica* Linn.; *Fagonia schweinfurthii* Hadidi; *Fagonia indica var. schweinfurthii* Hadidi.

Classification:

Kingdom:	Plantae
Subkingdom:	Phanerogamea
Class:	Dicotyledonae
Subclass:	Polypetalae
Series:	Disciflorae
Order:	Geraniales
Family:	Zygophyllaceae
Genus:	*Fagonia*
Species:	*Fagonia cretica*

Common names: Bengali: Duralabha; **English:** Khorasan thorn; **Gujarati:** Dhamaaso; **Hindi:** Dhamasa, Damah, Ustarkhar; **Malayalam:** Kodittuva; **Marathi:** Dhamasa; **Sanskrit:** Dhanvayas, Duralabaha; **Tamil:** Tulganari; **Telugu:** Chittigava, Gilaregati.

Distribution: The plant is found in the arid areas of India, Pakistan, Africa and parts of Europe.

Description: A small, much-branched spiny under shrub covered with glandular hairs. The plant is about 30 cm tall. Leaves: compound, cauline, opposite, 1–3 foliate leaflets, linear, sessile, with 2 pairs of sharp stipular spines; leaflets linear, acute, almost sessile. Flowers: solitary, pedicellate, purplish pink.

The wood of the stem is white, very hard and covered with a ragged light-brown bark which becomes shiny and mucilaginous when moistened.

The fruits are glandular, pubescent, 5-partiate, schizocarp. Seeds: flat, compressed, ovoid. Roots are cylindrical, tough and woody and externally light yellow to light brown in colour.

Flowering: Throughout the year.

Fruiting: Throughout the year.

Chemical constituents: The aerial parts of the plant contain 6 saponins. Saponin A and B are based on nahagenin; saponin C is based on 21α, 22β-dihydroxy nahagenin. The plant also contains diterpenes, fagonone, chinovic acid, ceryl alcohol, β-sitosterol, fagonin, fagogenin, betulin and oleanolic acid. Flavonoids such as ternatin, herbacetin, gossypetin are also present. Phenolic acids such as p-hydroxybenzoic, vanillic, syringic, p-coumaric and ferulic acids and an alkaloid, hermin, are found in the plant.

Used in: Ayurveda and Unani medicine.

Uses: The plant is antitumor, antioxidant, analgesic, astringent, febrifuge and prophylactic against smallpox agents (Puri and Bhandari, 2014). It is a blood purifier and cures fevers, asthma, diarrhea and dermatitis. It also possesses stimulant and laxative properties. A 20 gm minimum daily dose taken in the form of a herbal tea for a period of 6 months has the ability to keep cancer in remission. It starts attacking and killing cancerous cells within 24–72 hours of intake of the medicine. It is also found to be very effective against the side effects of chemotherapy as the action of the medicine is limited only to the cancerous cells and not extended to the normal cells.

Methods of propagation: Through seeds.

FOENICULUM VULGARE MILL.

Synonyms: *Anethium dulce* DC.; *Anethum foeniculum* L.; *Foeniculum officinale* All.

Classification:

Kingdom:	Plantae
Subkingdom:	Phanerogamea
Class:	Dicotyledonae
Subclass:	Polypetalae
Series:	Calyciflorae
Order:	Umbellales
Family:	Umbelliferae (Apiaceae)
Genus:	*Foeniculum*
Species:	*Foeniculum vulgare*

Common names: Bengali: Mouri; **English:** Fennel, Sweet cumin; **Gujarati:** Variyali; **Hindi:** Badi saunf; **Kannada**: Soump; **Marathi:** Soump; **Telugu:** Soump.

Distribution: This plant is a resident of Southern Europe and Asia. The *vulgare* variety is cultivated chiefly in Russia, Romania, Hungary, Germany, France, Italy, India, Japan, Argentina, and the USA. It is grown on a small scale in Indian states such as Karnataka, Maharashtra, Uttar Pradesh, Punjab, Bihar, Jammu and Kashmir.

Description: It is a hardy, perennial herb with yellow flowers and feathery green leaves. It grows to a height of up to 2.5 m with hollow stems. Leaves pinnately decompound, petiolate, grow to 40 cm; they are finely dissected with the last section filiform (like thread), approximately 0.5 mm wide. Inflorescence: compound umbel. Flower terminals are produced in compound umbels. Stamens 5, epigynous, ovary inferior, crowned by a disk. The fruit is a cremocarp, oblong, carpophore 2-partite, dry seeds 4–10 mm long (https://en.wikipedia.org/wiki/Fennel).

Flowering: August–September.

Fruiting: August–September.

Chemical constituents: Phenols, phenolic glycosides and volatile aroma compounds such as anethole, estragole, limonene, safrole, α-pinene, β-pinene, camphene, β-myrcene and fenchone. The phenolics present in the plant are apigenin and acids such as caffeic, chlorogenic, neochlorogenic, cinnamic, ferulic, gallic and p-cocid, rosmaric acids, as well as quercetin

(Al-Snafi, 2018).

Used in: Ayurveda and folk medicine.

Parts used: Fruits.

Uses: Widely used as digestive and diuretic. It treats asthma edema, anxiety, depression and gastrointestinal disorders. It, helps lower cholesterol levels and augment the production of breast milk. Its seeds are used to lower blood pressure, reduce asthma symptoms and water retention, purify blood and improve eyesight.

Methods of cultivation: Through seeds.

Photo: Prof Srinath Rao.

GARCINIA CAMBOGIA (GAERTN.) DESR.

Synonyms: *Garcinia gummi-gutta*; *Garcinia hanburyi*.

Classification:

Kingdom:	Plantae
Subkingdom:	Phanerogamea
Class:	Dicotyledonae
Subclass:	Polypetalae
Series:	Thalamiflorae
Order:	Guttiferales
Family:	Gutiiferae (Clusiaceae)
Genus:	*Garcinia*
Species:	*Garcinia cambogia*

Common names: **English:** Malabar tamarind; **Hindi:** Hila; **Kannada:** Upagimara; **Malayalam:** Pinaru; **Sanskrit:** Vrikshaml, Kankusta; **Tamil:** Kadumpuli; **Telugu:** Simachinta.

Distribution: It is a tropical tree from India, Sri Lanka and Malaysia. In India it is found in the evergreen forests of the Western Ghats at an altitude of 6000 feet.

Description: It is a small to medium tree that grows to a height of 20 m with a rounded crown and horizontal or drooping branches. The leaves of the tree are dark green, shiny, ovate in shape, ranging from 2½ to 3½ in in length and from 1 in to 1½ in in breadth. The tree bears small green fruits that are ovoid in shape and have a diameter of 1–1.5 in, yellow or red, with 6–8 grooves. The tree flowers during the hot season and the fruits ripen during the rainy season. The fruit resembles a small yellow or reddish pumpkin, or it may have a unique purple color. The seeds are surrounded by a succulent aril part.

Flowering: January–March.

Fruiting: April–August.

Chemical constituents: The rind of the fruit contains hydroxycitric acid, which has many useful properties. The root contains the xanthone garbogiol. The bark of the stem contains benzophenones such as garcinol and isogarcinol. The compounds present are carbogiol, garcinol and gamma-aminobutyric acid (Semwal et al., 2015).

Used in: Ayurveda and traditional medicine.

Parts used: Fruit rind.

Uses: Anthelmintic, anti-diabetic, anti-obesity and anti-ulcerogenic activity. It is used in treating bowel complaints, bilious affections, constipation, cancer and delayed menstruation. It is also employed as a demulcent and to treat edema and rheumatism. It is an ingredient in weight loss supplements. The fruit rind exhibits anti-obesity activity by reducing food intake and body fat (Semwal et al., 2015; Hemshekhar et al., 2011).

Methods of cultivation: Monsoon is the time for propagation and transplantation (Tharachand Selvaraj and Avadhani, 2013).

GLORIOSA SUPERBA L.

Synonyms: *Eugone superba*; *Gloriosa simplex*; *Gloria abyssinica.*

Classification:

Kingdom:	Plantae
Class:	Monocotyledonae
Subclass:	Liliidae
Series:	Coronarieae
Order:	Alismatales
Family:	Liliaceae
Genus:	*Gloriosa*
Species:	*Gloriosa superba*

Common names: English: Supper lily, Tiger lily; **Hindi:** Karihari; **Kannada:** Agnishike, Huliyugu; **Malayalam:** Malattamara; **Sanskrit:** Langali, Agnishikha; **Tamil:** Kanvali poo, Kandal; **Telugu:** Kalappa gadda, Agnishikha.

Distribution: A Paleotropic species found in Africa, Madagascar, India, Indo-China and Malesia. Within India, it is found throughout the tropical region, in moist, deciduous forests.

Description: It is a climbing herb with a root stock of arched solid fleshy cylindrical bifurcately branched tubers pointed at both ends. Annual stems arise from the angles of tubers. Leaves: simple, alternate or whorled, sessile, ovate-lanceolate, with a tendril like a spiral tip. Flowers: large, axillary, solitary and showy, perianth segments 6; petals linear with wavy margins, basal half yellowish and upper half red, stamens 6, ovary glabrous, 3-celled. Fruit: a capsule, large, 6–8 cm long with 3 equal lobes, green when young and black when mature, coriaceous and septicidal. Seeds 50–150 per capsule, pale orange attached to the suture, oval in shape.

Flowering: September–October.

Fruiting: January–February.

Chemical constituents: The roots contain colchicines (0.37 %) and 7 colchicine derivatives; sitosterol, β- and γ-lumicolchicine and glycyrrhizin. Seeds and flowers also contain glorisines, colchicines and their derivatives (Banu and Nagarajan, 2012).

Used in: Ayurveda, Tibetan, Unani, Siddha and folk medicine.

Parts used: Seeds, roots and tubers.

Uses: The root powder is used to treat arthritis, boils, piles, for the extraction of foreign bodies, for easy delivery, baldness, scrofula and ear diseases. The rhizomes are used as tonic, anti-periodic, anthelminthic and against snake bites. They are also purgative, cholagogue, and used in rheumatic fever and skin diseases (http://agritech.tnau.ac.in/banking/PDF/Medicinal%20Aromatic_Medicinal%20&%20Aromatic%20Plants.pdf). The tubers are boiled with sesame oil and applied on joints and areas affected with arthritis to reduce pain. Seeds are used for rheumatic pain and as a muscle relaxant. Root paste is effective against paralysis, rheumatism and snake and insect bites (Banu and Nagarajan, 2012).

Methods of propagation:

This is propagated through seeds and tubers. It prefers red or black loamy soils with good water-holding capacity and good drainage.

- Seeds are collected during October. They are soaked in water for 24 hours and sown in seedbeds. The seeds germinate within 20–30 days.
- For vegetative propagation, cut the tubers into divisions with at least 2 eye buds and weighing not less than 50–60 gm.
- Plant the tubers horizontally, finger deep in furrows opened 2 feet apart. The ideal planting season is during March–April.
- The dormant tubers start sprouting from the month of May till August.

Photo: Dr Kotresh

GLYCOSMIS PENTAPHYLLA (RETZ.) DC.

Synonyms: *Glycosmis arborea* (Roxb.) A. DC.; *Limonia pentaphylla* Ritz.

Classification:

Kingdom:	Plantae
Subkingdom:	Phanerogamea
Class:	Magnoliopsida
Subclass	Rosidae
Series:	Not available
Order:	Sapindales
Family:	Rutaceae
Genus:	Glycosmis Correa
Species:	Glycosmis pentaphylla (Retz.) A. DC

Common names: English: Ash sheora, Orange berry; **Hindi:** Ban nimbu; **Kannada:** Guroda gida; **Malayalam:** Kuttippannel; **Marathi:** Kirmira; **Sanskrit:** Asvasakhotah; **Tamil:** Kulu Pannai; **Telugu:** Golugu.

Distribution: Indo-Malaysia. Throughout India the plant is found in various states such as Assam, Arunachal, Meghalaya, Nagaland and Mizoram, as well as in the Western Ghats,

Undergrowth in forests up to 1400 m.

Description: Large shrubs to small trees up to 4 m tall. Bark brown, smooth. Young branchlets subterete, glabrous. Leaves compound, imparipinnate (rarely 1- foliolate), alternate, spiral; rachis pulvinate, terete, glabrous; petiole 0.2 cm long, planoconvex in cross section, glabrous; leaflets 3–5, alternate, 9–16.5 × 3.5–5 cm, oblong or elliptic, apex shortly acuminate with blunt tip or obtuse to acute, base acute, margin crenulate, glandular punctate, glabrous; secondary nerves 713 pairs, looped near margin. Inflorescence axillary panicles, flowers small, white, sessile, – Sepals 5(4), 1–1.5 mm long, ovate-acute, ciliate on margins. Petals 5(4), white, 4–5 × 2–2.5 mm, obovate, surface gland-dotted. Stamens 8–10, longer and shorter filaments; longer ones about 3 mm long, shorter about 2 mm long. Disc prominent, annular. Ovary 2–2.5 mm across, ovoid, 5-celled; stigma flat or obscurely lobed. Berry 1–1.2 cm in diameter, ovoid, white turning pink. Berry globose, apiculate, up to 2 cm across; seeds 2–3.

Flowering: February–April.

Fruiting: May onwards.

Chemical constituents: Some of the major classes of compounds reported from this plant include terpenoids, amides, imides, alkaloids, coumarins and flavonoids (Sreejith et al., 2012). Air-dried leaves yielded two furoquinoline bases, kokusaginine and skimmianine. Other alkaloids reported from the leaves include glycosine, arborine, glycosminine, arborinine (major), glycosamine, glycorine, glycosmicine, g-fagarine triterpenes, arbinol and isoarbinol. The roots contain carbazoles, glycozolines, gltcozolidinearborine and skimmianine. The flowers are known to contain glycophymine, glycosolone and glycophymoline (Sreejith et al., 2012). Vanillic, syringic and ferulic acids are the other compounds present in the leaves. Alkaloids and diglycoside acyl esters have been isolated from the root and bark of this plant. Its stem has also been reported as a rich source of flavonoids, naphthoquinone and acridone alkaloids (Ito et al., 1999; Wang et al., 2006).

Parts used: Leaves, stem and roots.

Uses: Various parts of this plant are used during confinement, in intestinal ailments, anaemia and jaundice. It was found to possess anti-pyretic, antimicrobial, anthelmintic, anti-inflammatory and cytotoxic properties (Rahaman et al., 2013; Rahamani et al., 2010; Jeeshna et al., 2009). *Glycosmis pentaphylla* was found to exhibit significant anti-tumor activity (Ito et al., 1999) (Quader et al., 1999).

Methods of cultivation: Through seeds.

Photo: C.T. Chandralekha

GLYCYRRHIZA GLABRA L.

Synonyms: *Glycyrrhiza echinata* Lepech.; *Glycyrrhiza hirsuta* L.

Classification:

Kingdom:	Plantae
Subkingdom:	Phanerogamea
Class:	Dicotyledonae
Subclass:	Polypetalae
Series:	Calyciflorae
Order:	Fabales (Rosales)
Family:	Leguminosae (Fabaceae)
Subfamily:	Papilionoidae
Genus:	*Glycyrrhiza*
Species:	*Glycyrrhiza glabra*

Common names: Bengali: Jashtimodhu; **English:** Liquorice; **Hindi:** Mulethi; **Kannada:** Jeshtamadhu, Yashtimadhkam; **Malayalam:** Irathimadhuram; **Marathi:** Jestha madha; **Sanskrit:** Madhuyashti, Yashtimadhu,; **Tamil:** Atimadhuram; **Telugu:** Yashtimadhkam.

Distribution: Liquorice is native to the Mediterranean region, Southern Europe and Central and Western Asia. In India, it grows in Punjab, Jammu and Kashmir. Semi-arid areas of the Haryana, Rajasthan and Gujarat states are suitable for the cultivation of liquorice.

Description: A tall erect hairy herb or under shrub up to about 1.5 m high. The word

Glycyrrhiza is of Greek origin, meaning 'sweet', and glabra means 'smooth', which refers to

the smooth fruit of the species. Leaves: pinnately compound, 7–15 cm long with 9–17 leaflets. Flower: lilac or light violet, small, 1–1.2 cm long, borne in slender axillary spikes which are equal to, or slightly longer than, the leaves. Fruit: a pod, 1–3 cm long, flat, densely covered all over with small spinous outgrowths. Rhizome is soft, flexible and fibrous, light yellow, with a characteristic sweet taste. It has long cylindrical burrowing root stock and horizontal creeping stolons which reach 1.5–1.8 m in length.

Flowering: June–July.

Fruiting: July–September.

Chemical constituents: The roots contain saponin, glycrrhizin – which is 60 times sweeter than cane sugar (Kaur et al., 2018) – and glycyrrhizinic acid. Other compounds in the roots include liquirtin, isoliquertin, liquiritigenin and rhamnoliquirilin. Glucoliquiritin apioside, prenyllicoflavone-A have been isolated from the root (The Wealth of India, 2005).

Used in: Ayurveda, Tibetan folk and traditional medicine; homeopathy.

Parts used: Roots and underground stems.

Uses: The dried roots and underground stems of this plant constitute the drug. Liquorice is believed to be useful in cough, sore throat, bronchitis, abdominal pains, consumption and epilepsy. For sore throat and cough, a small piece of liquorice is just chewed or sucked. It promotes urination, and is demulcent, antibacterial and antihepatotoxic. Due to its sweet taste and demulcent properties, it is largely used in making syrups and to counter the bitter taste of medicines in mixtures or pills; it is also used in treating asthma, bronchitis, peptic ulcer, arthritis and allergic complaints. Liquorice powder, mixed with butter (ghee) and honey, is applied on cuts and wounds. The leaves of the plant, if applied as a poultice, are believed to be useful in scalds of the heads. All these properties are attributed to the presence of glycyrrhizin. Glycyrrhetinic acid exhibits antitumor activity and is cytostatic. It also induces phenotypic reversion, i.e. the cancer cells get converted into normal cells. Roots are also used in coughs, epilepsy, flatulence, gout, hyperdipsia, leucorrhoea, skin diseases, sore throat, sexual debility and tonsillitis (Kaur et al., 2018).

Methods of propagation: Through seeds, rhizome cuttings. Rhizomes of 20–30 cm in length and 15–20 mm in diameter seem to be the most effective. Stolon pieces are also used.

GMELINA ARBOREA ROXB.

Synonym: *Gmelina rheedii* Hook.

Classification:

Kingdom:	Plantae
Subkingdom:	Phanerogamea
Division:	Phanerogams
Class:	Dicotyledones
Subclass:	Gamopetalae
Series:	Bicarpellatae
Order:	Lamiales
Family:	Verbenaceae
Genus:	*Gmelina*
Species:	*Gmelina arborea*

Common names: Bengali: Gumbar, Gamari, Gamargachha; **English:** Ganhar, White teak; **Gujarati:** Shewan, Seevan; **Hindi:** Gumhar; **Kannada:** Shivani, Kasmiri mara; **Malyalam:** Kumilu, Kumpil, Kumizhu, Kumbi; **Marathi:** Shivan; **Sanskrit:** Gambhari, Kasmari; **Tamil:** Gomadi, Kumadi, Umi-thekku; **Telugu:** Gumar teak, Gummadi.

Distribution: Found in tropical regions with high rainfall, in Western Ghats, West Bengal and Assam (wet tropical) and in the tropical medium-rainfall areas of the Indo-Gangetic Plains and the peninsula (moist tropical).

Description: Moderate-sized, unarmed, deciduous tree. In well-thinned plantations under optimum conditions it may reach heights of about 30 m and stem diameter over 60 cm, but it is more usually about 20 m with a clear bole of 6–9 m. In the open, however, it develops low, heavy branches, a wide crown, and a strong bark, greyish yellow, rather corky. Branchlets and young parts clothed with fine white mealy pubescence. Leaves: opposite decussate, 4–8 ×3–6 in, broadly ovate, acuminate, entire, glabrous above when mature, stellately fulvous-tomentose beneath, base cordate or sometimes truncate and shortly cuneate. Petioles 2–3 in long, cylindric, puberulous, glandular at the top. Flowers appearing with or sometimes before young leaves, usually in small cymes of about 3 flowers arranged along the branches of densely fulvous-hairy panicle reaching 12 in long; buds clavate, angular; bracts linear-lanceolate. Calyx broadly campanulate, densely fulvous-hairy, teeth 5, small, triangular, acute. Corolla brownish yellow, densely hairy outside, 5-lobed, 2-lipped; upper lip rather more than 3–8 in long, deeply divided into 2 oblong, obtuse lobes; lower lobe nearly 1 in long, 3-lobed, the middle lobe projecting forward, ovate, sub-obtuse, with irregularly crenulate margin, much longer and broader than the obovate rounded lateral lobes. Fruit is a drupe, 1.8–2.5 cm long, ovoid or pyriform, smooth, orange-yellow when ripe (http://www.flowersofindia.net/catalog/slides/Gamhar.html)

Flowering: January–April.

Fruiting: May–June.

Chemical constituents: The roots and heartwood of this plant are reported to contain gmelinol, hentriacontanol, n-octacosanol and β-sitosterol. The heartwood gives ceryl alcohol, cluytyl ferulate, lignans and arboreol. The leaves yield luteolin, apigenin, quercetin, hentriacontanol, β-sitosterol, quercetagetin and other flavones. The fruits contain butyric acid, tartaric acid and saccharine substances. The roots contain sesquiterpenoid and apiosylskimmin, a coumarin known as umbelliferone-7-apiosyl glucoside, and gmelofuran (Kumari et al., 2017).

Used in: Ayurveda.

Parts used: Bark, fruits, flowers, leaves and root; the whole plant.

Uses: The root and bark of this plant are claimed to be stomachic, galactagogue, laxative and anthelmintic. They improve appetite and are useful against hallucination, piles, abdominal pains, burning sensations, fevers, *tridosha* and urinary discharge. Leaf paste is applied to relieve headache and juice is used as wash for ulcers. The leaves are also demulcent, antigonorrhoeic and bechic. Flowers are sweet, cooling, bitter, acrid and astringent. They are useful in leprosy and blood diseases. In Ayurveda, it has been observed that the fruit is acrid, sour, bitter, sweet, cooling, diuretic tonic, aphrodisiac, astringent to the bowels, promotes growth of hairs, useful in *vata*, thirst, anaemia, leprosy, ulcers and vaginal discharge. The plant is recommended for the treatment of snake bite and scorpion sting. In snake bite, a decoction of the root and bark is given internally (Kumari et al., 2007).

Methods of propagation: Normally, the seed is collected from the ground and de-pulped and the stones are dried. For quick germination, the seeds should be soaked for 48 hours. The seeds germinate within 20–50 days under ideal conditions; the average rate for a healthy seed lot is 60 %. Trees can be raised easily by transplanting, which is carried out in the rainy season, or by direct sowing in lines; the latter has proved to be more successful in some instances. Large cuttings planted during the rainy season do well.

Source: http://en.wikipedia.org/wiki/File:Gmelina_arborea_flower_

GYMNEMA SYLVESTRE (RITZ.) R. BR.

Synnonyms: *Asclepias geminate* Roxb.; *Gymnima humile* Decne.; *Gymnema parviflolium* Wall.; *Gymnema germinatum* R.Br.; *Periploca sylvestris* Retz.

Classification:

Kingdom:	Plantae
Subkingdom:	Phanerogamea
Class:	Dicotyledonae
Subclass:	Gamopetalae
Series:	Bicapellatae
Order:	Gentianales
Family:	Asclepiadaceae
Genus:	*Gymnema*
Species:	*Gymnema sylvestre*

Common names: English: Australian cow plant, Periploca of the woods; **Gujarati:** Kavali, Gudmar; **Hindi:** Gudmar, Merasimgi; **Kannada:** Kadhasige, Kodapathre; **Malayalam:** Cakkarakkolli; **Marathi**: Bedakicha phala, Kavali; **Sanskrit:** Madhunasini, Mesasrngi; **Tamil:** Sirukurumkay; **Telugu:** Podapatra, Vodapatte.

Distribution: Globally, the species is distributed in the tropical regions of Africa and Asia, and in Central and Peninsular India. Common in the Western Ghats.

Description: It is a large, much-branched, woody twiner. Leaves: opposite elliptic, narrow-tipped, base narrow, leaves are smooth above and densely velvety beneath. Flowers: small, yellow, in sessile or umbel-like pedunculate cymes, bell-shaped. Corona: single with 5 fleshy scales (http://www.flowersofindia.net/catalog/slides/Gurmar.html).

Flowering: October–January.

Fruiting: March–May.

Chemical constituents: The leaves contain gymnemic acid, gurmarin and gymnamine. The stem contains conduritol A, 1-heptadecanol, stigmasterol, stigmasterol glucoside, 1-quercitol, 1-octadecanol, lupeol cinnamate. Other phytoconstituents include anthraquinones, flavones, hentriacontane, pentatriacontane, phytin, formic acid, butyric acid, lupeol, β-amyrin-related glycosides and stigmasterol (Tiwari et al., 2014; Zhen et al., 2008).

Used in: Ayurveda, folk, Tibetan, Unani and Siddha medicine; homeopathy.

Parts used: The whole plant. The most used part are the leaves.

Uses: The leaves and root are used in the form of powder, paste and decoction to treat inflammation of glands, enlargement of spleen, indigestion, constipation, jaundice, piles, sinusitis, cough, respiratory diseases, feeling of heaviness in the head, urinary stones, malarial fevers. The leaves and stem are hypoglycemic and, therefore, used to treat diabetes. They are diuretic and stimulate the cardiovascular system.

Methods of propagation: This plant can be propagated through seeds, cuttings and apical shoots. The plant prefers humus-rich loamy soils with good shade.

- Seeds are collected during February–April. They are directly collected into polybags during June. The potting mixture for polybags is prepared by mixing 2 parts of FYM/compost, 1 part sand and 1 part clay.

- Within 7–10 days, about 40 % seed germination can be observed.
- For vegetative propagation, semi-hardwood cuttings 6–8 inc long are taken from the apical shoots and planted in poly bags. When sufficient rooting is observed they can be transplanted in the field. Rooting initiates in about 6 days and the chance of success is 60 %.

Photo: Dr Kotresh

HEDYCHIUM CORONARIUM (KOENIG) KUNTZE.

Synonyms: *Amomum filiforme* Hunter ex Ridl.; *Gandasullum coronarium* (Koenig) Kuntze.; *Gandasullum lingulatum* (hassk.) Kuntze.; *Hedychium chrysoleucum* Hook.; *Hedychium lingulatum* Hassk.; *Hedychium.*

Classification:

Kingdom:	Plantae
Subkingdom:	Phanerogamea
Class:	Monocotyledons
Subclass:	As per bentham and hooker classification
Order:	Zingiberales
Family:	Zingiberaceae
Genus:	*Hedychium*
Species:	*Hedychium coronarium*

Common names: English: Common ginger lily, Butterfly common lilly; **Hindi:** Dollan champa; **Kannada**: Suruli sugandhi; **Malayalam:** Saugandhikam, Elipoochedi; **Marathi:** Sontaka; **Tamil:** Chantikantam; **Telugu:** Kichchiligadda, Vasanthi.

Distribution: It is native to the Eastern Himalayas region of India (Sikkim), Nepal and Bhutan through northernmost Myanmar. It is commonly cultivated in the warm, temperate and sub-tropical regions of the world and in China as an ornamental.

Description: It is a stout, strongly aromatic, perennial plant growing 1–2.5 m tall from a fleshy rhizome that is 25–50 mm in diameter. Leaves large, oblong to lanceolate, 16–60 cm long ×5–10 cm wide, glabrate; ligule 2 cm long. Flowers in spike 15 cm long and 7 cm wide, about 2–4 in each bract, calyx 4 cm long, glabrous. Corolla tube 8–10 cm long, narrow, glabrous, lobes 3–4 cm long, lip 6 × 6 cm, lobes emarginate, white in colour. Anthers 1–1.5 cm long, filaments shorter than lip. The flowers have a delicious scent.

Flowering: August–December.

Fruiting: August–December.

Chemical constituents: Eucalyptol and β-trans-ocimenone are the major components present in the plant's rhizome and flower. Caryophyllene and α-Phellandrene are the other components present.

Used in: Folk medicine.

Parts used: Rhizome.

Uses: The seed is aromatic, carminative and stomachic. The root is antirheumatic, excitant and tonic. The ground rhizome is used as a febrifuge. A decoction is used to treat pain in the chest and arms. An essential oil from the roots is carminative and has anthelmintic indications. A decoction of the basal part of the stem is used for gargling in the treatment of tonsillitis or, alternatively, a part of the stem may be chewed. The chewed stem is applied to infected nostrils. An infusion of the leaves is used to treat abdominal complaints. The plant is commonly used for treating eye infections and rheumatism. It is also employed as an anthelmintic and mild tranquillizer (Parida et al., 2019).

Methods of cultivation: By means of rhizome.

Photo: Prof Srinath Rao

HELICTERES ISORA L.

Synonyms: *Helicteres brevior* Mill.; *Helicteres chrysocalyx* Miq. ex Mast. *Helicteres Grewifolia* DC.; Isora corylifolia Schott & Endl.

Classification:

Kingdom:	Plantae
Subkingdom:	Phanerogamea
Class:	Dicotyledonae
Subclass:	Polypetalae
Series:	Thalamiflorae
Order:	Malvales
Family:	Sterculiaceae
Genus:	*Helicteres*
Species:	*Helicteres isora*

Common names: Bengali: Atmora; **English:** East Indian screw tree; **Gujarati:** Mriga Shinga, Maradashingh; **Hindi:** Jonkaphal, Marorphali, Mrigashinga; **Kannada:** Yedmuri; **Malayalam:** Valampiri; **Marathi:** Murudsheng; Kevani, Varkati; **Sanskrit:** Mriga-singha; **Tamail:** Vadampiri; **Telugu:** Valambiri.

Distribution: Found in Australia, Philippines, Malaya and Nepal. Distributed throughout Eastern, Western, Northern and Southern India.

Description: It is a sub-deciduous shrub or a small tree with stem 1–2 in in diameter. It reaches a height of 5–10 feet. The plant is covered with a grey bark and the young shoots have stellar hairs covering their entire surface. Leaves are simple, alternate, bifarious, 8–10 cm long and 4–5 cm broad, oblong ovate, shortly acuminate, lamina covered with stellar hairs on both surfaces, margins serrate, stipules short, petiole 6–9 mm long. Flowers are solitary or in clusters of 2–4, red in colour, 2.5–3.5 cm long, distinctly bilabiate, petals red turning pale blue when old. Fruits greenish brown, beaked, cylindrical, spirally twisted when ripe (hence the name 'screw tree').

Flowering: July–December.

Fruiting: July–December.

Chemical constituents: The roots contain betulic acid, daucosterol, β-sitosterol, oleanolic acid, cucurbitacin B and isocucurbitacin B. (Wenhao et al., 1991; Ben et al., 1985; Bean et al., 1985).

Used in: Ayurveda.

Parts used: Stem bark and roots.

Uses: The roots and stem bark are considered to be expectorant, demulcent, astringent and anti-galactagogue. Bark and root juice are used in diarrhea, dysentery, scabies, biliousness and gripping of the bowels. Fried pods are given to children to kill intestinal worms. The plant has insulin-sensitizing and hypolipidemic properties and has the potential for use in the treatment of type-2 diabetes. Traditionally, the root juice is claimed to be useful in diabetes, emphysema and snake bite (Kirtikar and Basu, 1995; Singh et al., 1984) (Qu et al., 1991).

Photo: Prof Srinath Rao

HELIOTROPIUM INDICUM LINN.

Synonyms: *Heliotropium africanum* Schum. & Thonn.; *Heliotropium anisophyllum* Beauv.; *Heliotropium foetidum* Salisb.; *Tiardium indicum* (L.) Lehm.

Classification:

Kingdom:	Plantae
Subkingdom:	Phanerogamea
Class:	Dicotyledonae
Subclass:	Gamopetalae
Series:	Bicarpellatae
Order:	Polemoniales
Family:	Boraginaceae
Genus:	*Heliotropium*
Species:	*Hilotropium indicum*

Common names: Bengali: Hasti-sura; **English:** Scorpion weed; **Hindi:** Siriyari, Hatishura; **Kannada:** Chelukondi gida; **Malayalam:** Thekkada; **Marathi:** Bhurundi; **Tamil:** Thel-Kodukku; **Telugu:** Nagadanthi.

Distribution: It is a common weed in waste places and settled areas. It is native to Asia.

Description: This plant is an annual, erect, branched herb slightly woody at the base that can grow to a height of about 30–50 cm. It has a hairy stem, bearing alternating ovate to oblong-ovate leaves, distinctly petiolate, petioles 5 cm long. Inflorescence terminal, tip coiled, axis up to 20 cm long. The flowers are arranged on only one side of the inflorescence. It has small white flowers with green calyx. Calyx lobes linear-lanceolate, ciliate, 2 mm long; corolla-tube 4–5 mm long; petals rounded; stamens 5, borne on a corolla tube; style terminal; ovary 4-lobed. Fruit is 3.5 mm long, ovoid, ribbed, separating into 2 nutlets each 2-celled, 3–3.5 mm long.

Flowering: Throughout the year.

Fruiting: Throughout the year.

Chemical constituents: Contains major pyrrolizidine alkaloid (heliotrine, indicine N-oxide) tannins. Indicine, AC-indicine, indicinine, indicinine-N-oxide, lupeol, rapnone and estradiol (Guhabakshi, 1999).

Used in: Traditional and folk medicine.

Parts used: The whole plant.

Uses: The juice extracted from the pounded leaves of the plants is used to cure wounds, skin ulcers and furuncles. The juice is also used as an eye drop for conjunctivitis. The pounded leaves are used as poultice. The whole plant is used in high fever, throat infection, ulcer fever, gonorrhea, localized inflammation, rheumatism, ringworm, ulcers, wounds. Aphrodisiac, astringent, bitter, expectorant.

Photo: Dr Kotresh

HEMIDESMUS INDICUS (L.) R. Br. ex SCHULT.

Synonym: *Periploca indica* L.

Classification:

Kingdom:	Plantae
Subkingdom:	Phanerogamea
Class:	Dicotyledonae
Subclass:	Gamopetalae
Series:	Bicapellatae
Order:	Gentianales
Family:	Asclepiadaceae
Genus:	*Hemidesmus*
Species:	*Hemidesmus indicus*

Common names: English: Indian sarsaparilla; **Hindi:** Salsa, Anantmul; **Kannada:** Sogdeberingida; **Malayalam:** Nannari; **Sanskrit:** Nagajihya; **Tamil:** Saribam, Nannari; **Telugu:** Sugandipala, Muttavapulagamu.

Distribution: The plant occurs almost throughout India.

Description: A perennial twine or creeper, root stock woody, fragrant; stems slender, hairless; leaves vary greatly in shape and size; they are 5–10 cm long, but their breadth varies from 5 to about 4 cm, dark green, often with whitish blotches, pale or whitish, hairy on lower surface. Flowers: very small, greenish, in small, compact cluster. Fruits 10–15 cm long, green, narrow, cylindric, pointed at tip, in pairs. Seeds: small, black, with a tuft of white hairs at top. All parts of the plant have white milky juice.

Flowering: October–January.

Fruiting: October–January.

Chemical constituents: Air-dried roots contain essential oils containing p-methoxy salicylic aldehyde as the major constituent. The aroma of the drug is attributed to this aldehyde. The other constituents present in the roots are β-sitosterol, lupeol, tetracyclic triterpene alcohols, coumarino-lignoids, hemidesmine and steroids such as hemidesmol, hemidosterol 1 and 2.

Used in: Ayurveda, Unani, Siddha and folk medicine; homeopathy.

Parts used: Roots, leaves and stem.

Uses: The dried roots of the plants constitute the drug. The drug is an alternative, anti-diarrheic, anti-leucorrhoeic, antirheumatic, anti-syphilitic, aphrodisiac, depurative, diaphoretic, diuretic and febrifuge remedy. It helps in curing bronchitis, cough, dyspepsia, dysentery, dysuria and blood diseases. It improves the quality and quantity of spermatozoa. The drug is also used to treat anemia, debility, fever, poisoning, skin diseases, thirst and vomiting (Anonymous, 2001; Chakrabortty and Choudhary, 2014).

Methods of propagation: This plant is propagated through seeds and root cuttings. It prefers garden loam soil. Fresh seeds can be collected any time of the year (since fruiting is observed throughout the year). Seed are soaked in water for 24 hours and sown in raised beds; seeds germinate within a fortnight. The rate of germination is about 60 %. For vegetative propagation, plant root cuttings 4 in long in poly bags or raised beds. They sprout in 15–20 days with 40 % success. Transplant the seedlings to polythene bags when they are 4-leaved.

Photo: Dr Kotresh

HEMIGRAPHIS COLORATA (BLUME) H. G. HALLIER F.

Synonyms: *Hemigraphis alternata* (Burm. f.) t. Anders.; *Ruellia colorata* Blume.

Classification:

Kingdom:	Plantae
Subkingdom:	Phanerogamea
Class:	Dicotyledonae
Subclass:	Gamopetalae
Series:	Bicapellatae
Order:	Personales
Family:	Acanthaceae
Genus:	*Hemigraphis*
Species:	*Hemigraphis colorata*

Common names: English: Aluminium plant, Metal leaf, Red flame ivy; **Hindi: Kannada:** ; **Malayalam:** Murikootti, Murian pacha.

Distribution: Native of Java. Now widely distributed in cultivation.

Description: It is a prostrate herb with spreading rooting stems. Leaves are opposite, ovate to cordate, bearing well-defined veins, 4.5–8 cm long, 3.5–4.5 cm wide, with a sub-obtuse tip and a rounded and heart-shaped base with toothed scalloped margins. Upper surface of leaves is dark purple throughout. Flowers are white, borne in terminal spikes up to 7 cm long, with large imbricate bracts. Calyx is green and 5-parted. Corolla is tubular, cylindric below, swollen above, 10–15 mm long and 5-lobed. Capsule is linear (Reshma, et al., 2018).

Flowering: All seasons.

Fruiting: All seasons.

Chemical constituents: Extracts of leaves and stems yielded phenol, carbohydrates, steroids, saponins, coumarins, tannins, proteins, carboxylic acids, flavonoids, xanthoproteins and alkaloids. The leaves of the plant contain flavonoids, polyphenols and tannins along with high potassium and low sodium levels; stem contains saponins and tannins, while roots contain flavonoids and polyphenols (Reshma, et al., 2018).

Used in: Folck medicine.

Parts used: Leaves; the whole plant.

Uses: Leaves consumed as a cure for gallstones and used as diuretic. Leaves used in treatment of bloody dysentery and hemorrhoids and as decoction for curing excessive menstruation. Paste used externally for skin complaints and wounds in order to stop bleeding and promote healing.

Paste of leaves applied for wound healing; also used for anemia. Sap of leaf buds squeezed in water and drunk as contraceptive and to induce sterility. This is an ornamental plant that can remove harmful volatile organic compounds to improve the quality of indoor air (Devi Priya 2013).

Methods of propagation: Through stem cuttings.

Photo: Prof Srinath Rao

HIBISCUS ROSA-SINENSIS (L.)

Synonyms: *Hibiscus arnotti* Griff. ex Mast.; *Hibiscus rosiflorus* Stokes; *Hibiscus boryanus* DC.

Classification:

Kingdom:	Plantae
Subkingdom:	Phanerogamea
Class:	Dicotyledonae
Subclass:	Polypetalae
Series:	Thalamiflorae
Order:	Malvales
Family:	Malvaceae
Genus:	*Hibiscus*
Species:	*Hibiscus rosa-sinensis*

Common names: Bengali: Jaba; **English:** China rose, Shoe flower; **Hindi:** Gudhal, Jasum; **Kannada:** Dasvala; **Malayalam:** Chemparati, Tavilamma; **Marathi:** Jaswanda; **Tamil:** Chembarutti; **Telugu:** Mandarum.

Distribution: Native to East Asia. Cultivated extensively in all parts of India.

Description: This is a bushy, evergreen shrub growing 8–10 feet tall and 3–5 feet wide. Leaves: simple opposite, petiolate and glossy. The leaf shape is ovate, the tip is acute and the margin is serrate/coarsely toothed; venation is reticulate. The stem is aerial, erect, green, cylindrical and free; lateral stipules are present. Flowers solitary, 8–10 cm in diameter, brilliant red, pink or white in colour, regular, actinomorphic, bracteate or ebracteate, pedicellate, dichlamydeous, pentamerous, with prominent orange-tipped red anthers. Calyx is green, about 2 cm long, with ovate lobes. Petals are red, orange or rose-white, obovate, entire, rounded tip, imbricate. Stamens form a long staminal tube enclosing the entire style of the pistil and protruding out of the corolla. Ovary is hypogynous, superior 5-celled, styles 5, fused below. Fruits are capsules, loculicidally 5-valved, but rarely formed in cultivation. The root is a branched tap root. *Hibiscus rosa-sinensis* is the national flower of Malaysia.

Flowering: Throughout the year.

Fruiting: Throughout the year.

Chemical constituents: *Hibiscus* contains linolenic and hyaluronic acids which stimulate hair growth and strength. The plant contains high levels of vitamin C. Flowers contain hisbiscetin. polyphenols, flavonoids, anthocyanins, polysaccharides, diglucoside and vitamins such as thiamine, riboflavin, ascorbic acid and niacin, which promote wound healing and are immune-modulating. Leaf extracts yielded tannins, anthraquinones, cardiac glycosides and alkaloids. The stem and leaves contain stigmasterol, β-sitosterol and taraxeryl acetate (Pearline, et al., 2015).

Used in: Ayurveda, Siddha and folk medicine.

Parts used: Leaves, stem, flowers.

Uses: Water mixed with an extract of *Hibiscus* stem and flowers has been traditionally used as contraceptive and abortifacient, to ease menstrual cramps, prevent excessive bleeding during periods and help in childbirth. A preparation from the leaves is used to treat postpartum relapse sickness, boils, sores and inflammations. It is also good for hair growth and helps hair become shiny and healthy. It moisturizes the hair and prevents conditions such as dandruff and itchy scalp. It also prevents hair breakage. The juice of the leaves is considered beneficial in blackening the hair. Moreover, the leaves are emollient, having moisturizing properties. The juice from the leaves is considered beneficial in gonorrhea and alopecia. It is used to boost metabolism and fight fatigue. The flowers are reported to possess anti-fertility properties and are used to cure epilepsy. The extracts of the flower have been long used in skin care. The flower also has anti-inflammatory properties that help prevent acne and increase the elasticity of the skin; it is used in removing wrinkles, marks and lines. The flower extract has strong antibacterial effects on human pathogens. It is also a natural moisturizer that has astringent properties, which helps prevent many kinds of skin diseases such as eczema. It exfoliates the skin by removing dirt, impurities and dead cells, thus keeping it soft, smooth and fresh-looking. *Hibiscus* is found to be effective in lowering blood pressure and highly beneficial for curing cardiovascular problems. It helps reduce the levels of cholesterol in the blood and prevents the accumulation of fat in the blood vessels; it is also useful in relieving stress. Various parts of this plant have shown potential results in reducing blood sugar and lipid levels as well as preventing tumors and cancer. *Hibiscus* also enhances brain activity and cognition. It protects from age-related amnesia, dementia, other cognitive disorders and reperfusion injury in ischemia. *Hibiscus* helps soothe hot flushes. The flowers of *Hibiscus* are used in several countries to treat menstruation-related problems. In Bangladesh and India, a decoction of *Hibiscus* flowers is used to regulate the menstrual cycle. Hot-water mixed with an extract of the flowers and bark is used in China to treat irregular menstruation. According to traditional medicine, *Hibiscus* can be used to treat several liver disorders. The antioxidants in this plant have been found to protect the liver, reduce inflammation and also boost liver health. The leaves are used in the treatment of diarrhea. A decoction prepared from the leaves is used to relieve pain and treat headaches and fevers. The leaves offer mild laxative properties. The buds work as astringent and help in treating urinary tract and vaginal disorders. They are also used in preventing intestinal worms (Gauthaman et al., 2006) (Maganha et al., 2009).

(Jadhav, et al., 2009; Pearline, et al., 2015; https://www.bimbima.com/hair-care/medicinal-use-of-gudhal-or-hibiscus-rosa/1825/).

Methods of propagation: Through stem cuttings.

Photo: Prof Srinath Rao

HOLARRHENA ANTIDYSENTERICA (ROTH) WALL EX A. DC.

Synonym: *Chonemorpha antidysenterica* (Roth) G. Don.; *Holarrhena codag* G. Don.; *Holarrhena pubescens* Wall.

Classification:

Kingdom:	Plantae
Subkingdom:	Phanerogamea
Class:	Dicotyledons
Subclass:	Gamopetalae
Series:	Bicarpallatae
Order:	Gentianales
Family:	Apocynaceae
Genus:	*Holarrhena*
Species:	*Holorrhena antidysenterica*

Common names: Bengali: Kutaja; **English:** Conessi bark, Kurchi bark, Bitter oleander; **Hindi:** Kurchi, Karchi; **Kannada:** Koodsalo, Beppale; **Malayalam:** Kodagapala; **Marathi:** Indrajav; **Sanskrit:** Kutaja, Kalinga; **Tamil:** Veppalei, Kodagapalei; **Telugu:** Pala kodaga.

Distribution: This is a native of India, found throughout the country.

Description: It is a small tree. Leaves: broad, elliptical. Flowers: white, borne in terminal corymbose cymes. Fruits: follicles, in pairs of 2, long, cylindrical.

Flowering: April–July.

Fruiting: August–October.

Chemical constituents: The bark of this tree contains steroidal alkaloids such as conessine, 7-α-hydroxy conessine, holonamine, regholarrenine A-F, pubescine, norholadiene, pubescimine, kurchinin and kurchinine. The root bark contains holacetine. The leaves contain alkaloids such as kurchiphylline, kurchiphyllamine, kurchaline, holadysine and holadysamine. Also present are kaempferol and quercetin.

Used in: Ayurveda, Tibetan and folk medicine; homeopathy.

Parts used: Bark, seeds and leaves.

Uses: The bark and seeds are used in dropsy and dysentery, stomach disorders, abdominal and glandular tumors. Fruits are anti-cancerous and hypoglycemic and used to regulate menstruation. The trunk bark possesses amoebicidal properties. A bath containing a decoction of bark cures scabies.

Methods of propagation: The plant is propagated through seeds.

- Collect fresh seeds and soak in cold water for 24 hours.
- Sow them in poly bags or raised beds.
- 80 % of the seeds germinate in 7–10 days.

Photo: Dr Kotresh

HOLOSTEMMA ADA-KODIEN (ROXB. K.) SCHUM.

Synonyms: *Asclepias annularia* Roxb.; *Sarcostemma annulare* Roth.; *H. rheedei* Wall.; *H. rheedianum* Spreng.; *H. fragrans* Wall.

Classification:

Kingdom:	Plantae
Subkingdom:	Phanerogamea
Class:	Dicotyledons
Subclass:	Gamopetalae
Series:	Bicarpellate
Order:	Gentianales
Family:	Asclepiadaceae
Genus:	*Holostemma*
Species:	*Holostemma ada-kodien*

Common names: English: Holostemma creeper; **Gujarati:** Kharner; **Hindi:** Chhirvel; **Kannada: Malayalam:** Adakodien; **Marathi:** Duduri; **Sanskrit:** Jivanti, Arakapushpi; **Tamil:** Palaykirai; **Telugu:** Palagurugu.

Distribution: This species is distributed in Sri Lanka, India, Myanmar and China. Within India, it is distributed in Peninsular India and tropical rain forests. It grows over hedges and in open forests, especially on the lower slopes of hills.

Description: Stems twining, branched; the bark is deeply cracked. Latex is milky white. Leaves opposite, decussate, lamina 2.1–15.7 × 1.8–15.5 cm, ovate-oblong, caudate or aristate at apex, cordate with incision up to 1.8 cm deep at base with rounded lobes, more or less pubescent along vein beneath, lateral veins 5–7 pairs, arched; petioles terete, 1–6 cm long and 1–2 mm in diameter. Flowers: very fragrant, greenish yellow in colour, born in axils, in umbellate cymes, many-flowered; peduncles terete, 0.6–4 cm long, about 1.5 mm in diameter, glabrous; bracts 1.5–2 × 0.5 mm linear, apex acute, grooved in the centre from inner side, glabrous; pedicels terete, 1–4 cm long, about 1 mm in diameter, glabrous. Calyx 5-lobed, divided up to base, lobes 5–7 × 4.5–6 mm, broadly ovate, saucer-shaped, apex obtuse, margin papery, ciliate, glabrous, glandular within. Corollas rotate, about 1.5 × 1–1.5 mm, ovate-oblong, apex obtuse, leathery with papery margin, glabrous. Corona staminal, uniseriate, annular ring, arising from the base of the staminal column, 3.5–4 mm high, 7–7.5 mm across, truncate. Stamens 5, about 8 mm long; pollinia 5, pollen masses solitary in each anther cell, yellow, waxy, attached by light-brown caudicles about 1195.2 mm long and 158.4 mm wide to dark-brown corpuscles about 734.4 mm long and 216 mm wide. Gynostegium about 11 mm long. Follicles paired or usually solitary, about 11 cm long and 3.5 cm wide, ovoid, thick, apex acute, glabrous; seeds many, ovate, base-rounded, flat, brown with light-brown margin; about 0.5 mm wide; coma silky white, 2–4 cm long.

Flowering: April–October.

Fruiting: April–October.

Chemical constituents: Ethanol extract of roots contains 6 amino acids, viz. alanine, glycine, serine, aspartic acid, threonine and valine, and benzene extract contains α-amyrin, lupeol and a sitosterol.

Used in: Ayurveda and folk medicine.

Parts used: Tubers.

Uses: The root is the most accepted source of Jivanti and is considered as an important *rasayana* drug capable of maintaining youthful vigour and strength. It has cooling, alternative, tonic and laxative properties and is also an astringent for the bowels. The root is made into a paste and applied to the eyes in case of ophthalmic issues; it is also used for scalding in gonorrhea. The root and leaves of this plant are used in the form of powder and juice to treat spider-poisoning and fever with burning sensation. A decoction of the root mixed with ghee removes burning sensation and fever. It also cures ulcers, biliousness, diseases of the blood, worms, itching and vesicular calculi. In case of defects of vision, the leaves are used as a vegetable. The root paste along with milk is used in diabetes. The dried roots are used to cure spermatorrhea.

Mode of propagation: Through seeds.

Photo: Prof Srinath Rao

HYOSCYAMUS NIGER L.

Synonyms: *Hyoscyamus auriculatus* Ten.; *Hyoscyamus pictus* Roth.; *Hyoscyamus vulgaris* Lej.

Classification:

Kingdom:	Plantae
Subkingdom:	Phanerogamea
Class:	Dicotyledonae
Subclass:	Gamopetalae
Series:	Bicarpellatae
Order:	Polemoniales
Family:	Solanaceae
Genus:	*Hyoscyamus*
Species:	*Hyoscyamus niger*

Common names: English: Henbane, Common henbane; **Hindi:** Khurasani; **Kannada:** Kurasanivadaki; **Malayalam:** Khurasaanee, Paarasika, Yavaani; **Sanskrit:** Dipya, Parasikaya; **Tamil:** Kurasaniyomam; **Telugu:** Khurashanivamam.

Distribution: The plant occurs in the Western Himalayas from Kashmir to Garhwal at altitudes of about 1500–3000 m, commonly in waste places near habitations. It has been cultivated in several regions of India, such as Kashmir, Punjab, Uttar Pradesh, Maharashtra and the Nilgiri Hills.

Description: An annual or biennial bad-smelling herb, densely covered with glandular hairs; stem up to 1 m high. Lower basal leaves 15–20 cm, margins toothed. Upper leaves smaller and divided into many segments. Flowers 2–3 cm in diameter, pale green, streaked with purple, some borne solitary at the place of branching of stems, others in long, terminal spikes. Fruits 1.3 cm in diameter, globose.

Chemical constituents: The dried leaves contain alkaloids, mostly hyoscyamine and scopolamine. Compared to the leaf, the petiole yields more alkaloids.

Used in: Homoeopathy, Ayurveda, naturopathy, modern and traditional Medicine.

Parts used: Seeds.

Uses: The dried leaves, flowers and flowering tops are collected soon after the flowering of the plant constitute the drug. The drug has similar properties as Belladonna *(Atropa belladonna)*. It is useful for relieving certain painful spasmodic conditions of the muscles, in nervous irritations such as hysteria, to treat cough, etc. It also dilates the pupil of the eye. (It is said that Queen Cleopatra used extracts of henbane to expand her pupils and appear more attractive.) The seeds of the plant also have medicinal properties; they are usually pasted and applied locally on pains.

Methods of propagation: It is propagated through seeds and vegetative method.

Other species: *Hyoscyamus muticus* L. (Egyptian henbane) is being cultivated in Kashmir and other places. The alkaloid content of this plant is higher than that of Indian henbane; it also has a more intense narcotic property. The leaves of Egyptian henbane are smoked for intoxication.

IMPATIENS BALSAMINA (L.)

Synonyms: *Balsamina angustifolia* Bl.; *Balsamina balsamina* Huth.

Classification:

Kingdom:	Plantae
Subkingdom:	Phanerogamea
Class:	Dicotyledonae
Subclass:	Gamopetalae
Series:	Heteromerae
Order:	Ericales
Family:	Balsaminaceae
Genus:	*Impatiens*
Species:	*Impatiens balsamina*

Common names: English: Garden balsam; **Hindi:** Gulmehendi; **Kannada**: Karnamamdala; **Malayalam:** Mechingam; **Sanskrit:** Dushparijati; **Tamil:** Kacit-tumpai; **Telugu:** Gulivinda.

Distribution: It is probably native to India and Myanmar; cultivated in India.

Description: Erect annual herbs up to 60 cm tall; stems simple or branched. Leaves alternate, 2.5–9 × 0.8–2.2 cm, lanceolate-elliptic, base narrowed, margins serrate, apex acute to acuminate; petioles 1–4 cm long, glandular. Flowers solitary or 2–3 in axillary fascicles, rose or white; pedicels up to 1.5 cm long. Lateral sepals 2–3 mm long, ovate; lip 1–1.6 cm long, cymbiform, petaloid; spur 1.2–2.2 cm long, incurved. Standard petal about 1 cm long, ovate; wings about 2 cm long, obovate, deeply notched; auricles ovate-rounded. Capsules 1–1.5 cm long, ellipsoid, tomentose. Seeds globose, tubercled.

Flowering: March–October.

Fruiting: March–October.

Chemical constituents: Naphthoquinones lawsone, or hennotannic acid, and lawsone methyl ether and methylene-3,3'-bilawsone are some of the active compounds in the leaves of this plant. It also contains kaempferol and several derivatives. Baccharane glycosides have been found in Chinese herbal remedies made from the seeds (Bot 2004) (Hua et al., 2001).

Used in: Folk, Ayurveda, Siddha, Unani, traditional and Chinese medicine.

Parts used: Leaves, flowers,

Uses: Antibacterial, antimicrobial, antifungal, anti-cancer and analgesic properties. The plant is cathartic, diuretic and emetic. It is used in the treatment of pains in the joints. The crushed leaves are used as a poultice on wounds and skin inflammations (Sakunphueak and Panichayupakaranant 2010).

Methods of cultivation: Through seeds.

Photo: Prof Srinath Rao.

INDIGOFERA TINCTORIA LINN.

Synonym: *Indigofera summatrana* Gaertn,

Classification:

Kingdom:	Plantae
Subkingdom:	Phanerogamea
Class:	Dicotyledonae
Subclass:	Polypetalae
Series:	Calyciflorae
Order:	Rosales (Fabales)
Family:	Leguminosae (Fabaceae)
Genus:	*Indigofera*
Species:	*Indigofera tinctoria*

Common names: Bengali: Nil; **English:** Indian indigo; **Hindi:** Gouli; **Kannada:** Nili; **Marathi:** Kattukasturi; **Sanskrit:** Nilini; **Tamil:** Averi; **Telugu:** Aviri, Nili.

Distribution: This plant is distributed in South and South-east Asia and tropical Africa and has been introduced in tropical America. In India, it is found almost throughout and cultivated in many parts.

Description: A branched shrub which grows up to 2 m high. Stems and branches are green; branchlets silvery, pubescent. Leaves are alternate, stipulate, imparipinnate, with 7–13 leaflets which are elliptic-oblong, membranous, 1.7 × 0.9 cm, shortly mucronate, pale green or bluish. Flowers are small, rose-coloured, in axillary racemes. Calyx 5-cleft, gamosepalous; corolla papilionaceous;

stamens diadelphous; ovary sessile with a short, incurved style ending in a capitate stigma. Pods are linear, cylindrical, 2–5 cm long, deflexed, with 8–12 seeds.

Flowering: July–September.

Fruiting: September–October.

Chemical constituents: Glabretephrin, semiglabrin, pseudosemiglabrin, flavonal glycoside, indigotin and indigo 2,2-bisindole alkaloid.

Used in: Ayurveda.

Parts used: Leaves, flowers, tender stems and roots.

Uses: Neelini-Indigo or Indigofera tinctoria is used for the treatment of fever, liver and spleen disorders, rheumatoid arthritis, gout, grey hairs etc. The leaves promote hair growth. The plant is deobstruent, alterative, antitoxic, anti-asthmatic and antiepileptic. The aerial part is hypoglycemic, CNS depressant and antitoxic. The leaves, flowers and tender shoots are considered to be cooling, demulcent and alterative. The leaf is anti-inflammatory. The root and stem are laxative, expectorant, anti-tumorous, febrifuge, anti-cephalalgic and anthelmintic. They are used as an antidote for snake bite and to promote hair growth. Indirubin is antineoplastic and has toxicity. Nili is antitoxic, purgative and laxative.

Methods of propagation: Through seeds.

Photo: Prof Srinath Rao

INULA RACEMOSA HOOK. F.

Synonym: *Inula royleana* C. B. Clarke.

Classification:

Kingdom:	Plantae
Subkingdom:	Phanerogamea
Class:	Dicotyledonae
Subclass:	Gamopetalae
Series:	Inferae
Order:	Asterales
Family:	Compositae (Asteraceae)
Genus:	*Inula*
Species:	*Inula racemosa.*

Common Names: English: Inula; **Gujarati:** Pohkarmul; **Hindi:** Puskarmul; **Kannada:** Rasnabheda; **Malayalam:** Puskaramulum; **Marathi:** Pushakar; **Sanskrit:** Pushkarmoola, Kusthabheda; **Tamil:** Puskaramulum; **Telugu:** Pushkarmul.

Distribution: This species is distributed in the Hindu Kush Himalayan region across Afghanistan, Pakistan, India and Nepal. Within India, it is found in Uttar Pradesh, Jammu, Kashmir and Himachal Pradesh at an altitude range of 2800–3200 m.

Description: Stout, erect her6, up to 1.75 m tall, branched, furrowed; young parts densely hairy. Leaves elliptic-lanceolate to obovate, short, acuminate at apex, amplexicaul at base, woolly beneath, upper surface pubescent. Heads yellow, many in racemes, 4–5 cm across, stout; peduncles hairy. Involucral bracts multi-seriate, densely hairy; outer ovate, 1.5 × 0.8 cm, tips recurved, densely woolly outside, tubercled within. Ray florets ligulate, 2–3 cm long; ligules around 1.5 cm long, hyaline, conspicuously nerved. Disc florets around 1.5 cm long. Achenes 0.2 cm long. Pappus up to 8 mm long.

Flowering: July–September.

Fruiting: September and October.

Chemical constituents: Root contains sesquiterpene lactones such as alantolactone, isoalantolactone and dihydroisolantolactone (Rastogi and Mehrotra 1960–1969). A germacranolide-inunolide is reported from the root oil (Rastogi and Mehrotra 1960–1969). The roots also contain β-sitosterol, daucosterol, isoalantolactone, octadecanoic acid and D-mannitol inuline, as well as an essential oil containing alantolactone. (-)dammar-20,24-dien-3β-yl acetate was isolated from the roots. Aplotaxene and phenylacetonitrile are reported in the essential oil.

Used in: Ayurvedic medicine.

Parts used: Roots.

Uses: The roots are antispasmodic anthelmintic, expectorant, resolvent, diuretic, beta-blocking, hypotensive, antianginal, antimicrobial. The seeds are aphrodisiac. This plant is also used to improve cardiovascular conditions and ischemic heart diseases, especially combined with *Commiphora mukul.* It also lowers diastolic blood pressure, decreases cholesterol and treats asthma, bronchial disease and rheumatism. The rhizome is sweet, bitter and acrid in taste with a natural potency. It acts as antiseptic, antibacterial, antifungal, anti-inflammatory, analgesic and mild diuretic. It is used in the treatment of contagious fevers as well as in cough, hiccup, bronchial asthma, indigestion, flatulence, anorexia and fever. Externally, the paste of its roots is used to dress wounds and ulcers as the her0 possesses antiseptic properties. It is also used to boost appetite.

Methods of propagation: Through seeds.

IONIDIUM SUFFRUTICOSUM GING.

Synonyms: *Hybanthus enneaspermus* Muell.

Classification:

Kingdom:	Plantae
Subkingdom:	Phanerogamea
Class:	Dicotyledonae
Subclass:	Polypetalae
Series:	Thalamiflorae
Order:	Parietales
Family:	Violaceae
Genus:	*Ionidium*
Species:	*Ionidium suffruticosum*

Common names: Bengali: Nunbora; **English:** Pink lady's slipper, spade flower; **Hindi:** Ratanpurus; **Kannada:** Charaata; **Malayalam:** Nelam-parenda; **Marathi:** Rathanparas; **Sanskrit:** Rathnapurusha; **Tamil:** Camaracum, Purusharathnam; **Telegu:** Ratnapurusha.

Distribution: This is a seasonal perennial herb, widely distributed in Africa, Madagascar, Sri Lanka, China, New Guinea, tropical Australia and India. Within India, it is found in the warmer parts from Delhi to Bengal, and southwards throughout the Deccan Peninsula. In nature, the plants appear for only a few months from June to September (monsoon season). The roots and the few basal stem stocks remaining in the soil are regenerated only during the rainy season, and soon after the plants dry out and disappear.

Description: A small perennial herb about 60 cm in height. Leaves: simple alternate, linear to lanceolate, 2–3 cm long, margins recurved, stipule acuminate, 1–4 mm long, margins serrate with nectariferous glands. Flowers: zygomorphic, 8–10 mm across, solitary, pink to purple, spade-shaped, sepals 5, 3–4 mm long, lanceolate, lower ones clawed. Corolla: lower petal broad, spade-shaped, with deep purple veins, upper petals linear-oblong, 3–4 mm long. Fruit: a capsule, 4–9 mm long. Seeds: 5–12, ribbed.

Flowering: January–July.

Fruiting: October–November.

Chemical constituents: The plant extract contains an alkaloid, aurantiamide acetate, β-sitosterol and isoarborinol. The nectar contains high amounts of amino acids including valine, leucine and glutamic acid. The major phytoconstituents are -Docosadienoic acid (20.90 %) and Cedran-diol, which possess many biological properties.

Used In: Ayurveda, folk and Siddha medicine.

Parts used: Root, leaves and fruit.

Uses: The herb is used as diuretic, demulcent and tonic. The root is diuretic and is used in urinary and bowel complaints of children. The root is also administrated as an infusion in gonorrhea. The decoction from the leaves and tender stalks is demulcent. Dried powdered leaves are used in asthma. The fruit is used to treat scorpion sting. An infusion of the plant is given in case of cholera. A Siddha medicine called *Orithazhthamarai choornam* is given in a dose of 2–4 gm twice a day with milk for gonorrheal diseases in ladies and leucorrhoea. It increases sexual power or libido and improves the quality of semen. In women, during parturition, it enhances milk secretion.

Methods of propagation: Propagated thorough seeds.

1. The samoolam is crushed and squeezed well in water. This extract is used to wash the eyes in conditions such as itching, pain, conjunctivitis, etc.
2. The samoolam of this plant is crushed well in water and used as a hair application in case of dandruff.
3. This herb is a Kayakalpa. Regular intake acts as a tonic and relieves ulcers and headaches when externally applied.
4. An infusion or decoction of the roots is given for urinary infections (http://siddham.in/ionidium-suffruticosum).

Photo: Prof Srinath Rao

IPOMOEA OBSCURA (L.) KER GAWL.

Synonyms: *Convolvulus obscures* L.; *Convolvulus subangularis* Buch. Ham. ex Steud.; *Ipomea luteola* R. Br.; *Ipomea acutifolia* A. Rich.

Classification:

Kingdom:	Plantae
Subkingdom:	Phanerogamea
Class:	Dicotyledonae
Subclass:	Gamopetalae
Series:	Bicarpellatae
Order:	Polemoniales
Family:	Convolvulaceae
Genus:	*Ipomoea*
Species:	*Ipomoea obscura*

Common names: Bengali: Chhagalkuri; **English:** Obscure morning glory; **Gujarati:** Gumbadvel; **Hindi:** Pan bel; **Kannada:** Bilichita bogari, Bokadi mooguthi balli; **Malayalam:** Cherutali, Tirutali; **Marathi:** Pilibonvari, Bokadi, Pungali; **Sanskrit:** Laksmana, Vachagandha; **Tamil:** Chirutali, Kuruguttali, Sirudali; **Telugu:** Nalla kokkita, Golla jiddu aaku.

Distribution: Widespread in tropical and South Africa, through tropical Asia to Northern Australia and Fiji. Present in grasslands, thickets, hedges, open forests, waste ground, road sides and as a weed in cultivated fields.

Description: It is an annual or perennial herb with slender, twining or prostrate stems up to 3 m long and 2 cm in diameter. Leaves simple alternate, 3–9 cm long, heart-shaped or oval-shaped, with long and glabrous petioles. Flowers solitary or grouped, funnel-shaped, white or pale yellow with a darker yellow 5-pointed star shape and a dark-purple throat. Corolla with 5 petals, shallow apiculate lobes, 2–2.5 cm long, white or yellow with purple centre. Sepals 5, pubescent, egg-shaped, hairless to lightly hairy. Stamens 5, villous at base. Ovary about 1.5 mm long, conical; stigma capitate. Capsule 6–8 mm across, sub-globose, apically beaked; seeds 5 mm long, ovoid, black, thinly pubescent.

Flowering: August–October.

Fruiting: September–November.

Chemical constituents: Seed extract contains indole alkaloids, such as ipobscurines B–D, calystegins B1, B2, B3. B4 and C1. The most common biologically active constituents in these plants are ergoline alkaloids, indolizidine alkaloids, nortropane alkaloids, phenolics compounds, coumarins, norisoprenoids, diterpene, isocoumarin, benzenoids, flavonoids, anthocyanosides, glycolipids, lignan and triterpenes (Meira et al., 2012).

Used in: Traditional and folk medicine.

Parts used: The whole plant.

Uses: The plant possesses antimicrobial, analgesic, spasmolitic, spasmogenic, hypotensive, psychotomimetic and anti-cancer activities. The leaf sap is used to treat fits of insanity. A paste of the leaves, combined with the leaves of *Argyreia mollis* and alcohol, is applied to open sores and pustules. The dried and powdered leaves are used to treat aphthae. A root decoction is orally given against dysentery.

Methods of propagation: Through seeds, cuttings.

Photo: Prof Srinath Rao

KAEMPFERIA ROTUNDA L.

Synonyms: *Kaempferia bhucampac* Jones; *Kaempferia longa* Jacq.; *Kaempferia versicolor* Salisb.; *Zerumbet zeylanacia* Garsault.

Classification:

Kingdom:	Plantae
Subkingdom:	Phanerogamea
Class:	Monocotyledons
Subclass:	Zingiberidae
Order:	Zingiberales
Family:	Zingiberaceae
Genus:	*Kaempferia*
Species:	*Kaempferia rotunda*

Common names: English: Indian crocus, Peacock ginger; **Hindi:** Abhuyicampa; **Kannada:** Nelasampiga; **Malayalam:** Chengazhuneerkizhengu; **Marathi:** Bhuichampa; **Sanskrit:** Bhumicampaka, Bhucampaka; **Telugu:** Bhucampakamu.

Distribution: The plant is distributed in the tropics and sub-tropics of Asia and Africa. It grows wild in the forests of South India in wet and humid places. It is also cultivated.

Description: It is an aromatic herb with tuberous root stalk and very short stem. Leaves are simple, few, erect, oblong or ovate lanceolate, acuminate, 30 cm long, 10 cm wide, variegated, green above and tinged with purple below. Flowers are fragrant, white, tip purple or lilac, arranged in crowded spikes. Rhizome: subglobose, tuberous, from which many roots bearing small oblong or rounded tubers arise.

Flowering: March–May.

Fruiting: March–May.

Chemical constituents: The tubers contain crotepoxide, β-sitosterol and essential oils.

Used in: Ayurveda and traditional medicine.

Parts used: Rhizome.

Uses: The tubers are widely used as a local application for tumors, swellings and wounds. They are also given in gastric complaints. They help to remove blood clots and other purulent matter from the body. The juice of the tubers is given in dropsical affections of hands and feet, and of effusions in joints. The juice causes salivation and vomiting. It is a component in many Ayurvedic preparations such as Chyavanaprasam, Asokarishtam, Baladthatryaditailam and Kalyanaka Ghritham. It also improves the complexion and cures burning sensations, mental disorders and insomnia. The drug Hallakam is prepared from this plant and is used in the form of powder or as an ointment and applied to cure wounds and bruises as well as to reduce swellings.

Methods of propagation: Through rhizome.

LAGENARIA SICERARIA (MOLINA) STANDLEY

Synonyms: *Cucurbita siceraria* Molina; *Lagenaria vulgaris Ser*
L. leucantha (Duchesne) Rusby; *Cucurbita leucantha* Duchesne in Lam.

Classification:

Kingdom:	Plantae
Subkingdom:	Phanerogamea
Class:	Dicotyledonae
Subclass:	Polypetalae
Series:	Calyciflorae
Order:	Passiflorales
Family:	Cucurbitaceae
Genus:	*Lagenaria*
Species:	*Lagenaria siceraria*

Common names: English: Bottle gourd; **Gujrati:** Dudhi; **Hindi:** Lauki; **Kannada:** Kumbalkai; **Malayalam:** Bellashora; **Marathi:** Dudhya; **Sanskrit:** Brihatphala; **Tamil:** Shorakkai; **Telugu:** Sorakaya.

Distribution: The plant is cultivated throughout India.

Description: It is a climbing or trailing herb softly pubescent, stems 5-angled. Tendrils bifid. Leaves soft, 15 cm in diameter, 3–5-lobed, acuminate at apex, hairy on both surfaces, basal sinus broad and deep; petioles 6–20 cm long, hollow. Flowers: white, large, solitary. Male flowers longer than female flowers; calyx tube campanulate, 5-teeth, narrow; petals 5, free, hairy on both surfaces. Female flowers: calyx and corolla as in male. Fruit: Large, bottle-shaped, thick or woody. Seeds compressed, white, horizontal with a marginal groove or apex bidentate, smooth.

Flowering: June–February.

Fruiting: June–February.

Chemical constituents: 2 triterpenoids-22-deoxocucurbitacin D and 22-deoxoisocucurbitacin D isolated from fruits (Rastogi and Mehrotra, 1960–1969). Palmitic palmitoleic (0.81 %), stearic (6.7 %), oleic (9.67 %) and linoleic acids (40.42 %) isolated from seed oil (Rastogi and Mehrotra, 1970–1979) Rhamnose, fructose, glucose, galactose, sucrose, raffinose and stachyose isolated from seeds (Rastogi and Mehrotra, 1970–1979). Protein (44.68 %) and nitrogen (7.15 %) determined in defatted seed meal; all essential amino acids except tryptophan also present. Avenasterol, codisterol, clerosterol, isofucosterol, stigmasterol, campesterol, sitosterol, spinasterol, 25(27)-dehydroporiferasterol, 25(27)-dehydrofungisterol, 25(27)-dehydrochondrillasterol, 24β-ethy-(25)-dehydrolathosterol, 24-methyllathosterol and 22-dihydrospinasterol identified in seeds. Palmitic, stearic, oleic and linoleic acids are present in seed oil.

Used in: Traditional, folk, Ayurvedic and Unani medicine.

Parts used: Fruit.

Uses: The fruit is traditionally used for its cardioprotective, cardiotonic, general tonic and aphrodisiac properties. It is also used in treatment of various allergic and inflammatory disorders such as bronchial asthma, rhinitis, bronchitis and rheumatism. Various extracts of the fruit of *Lagenaria siceraria* were found to have anti-inflammatory, analgesic, hepatoprotective, antihyperlipidemic, diuretic and antibacterial activities (Shirwaikar and Sreenivasan 1996).

Methods of cultivation: Through seeds.

Photo: Prof Srinath Rao

LEPIDIUM SATIVUM L.

Synonyms: *Lepia sativa* (L.) Desv.; *Lepidium hortensis* Forssk.; *Lepidium spinescens* DC.

Classification:

Kingdom:	Plantae
Subkingdom:	Phanerogamea
Class:	Dicotyledonae
Subclass:	Polypetalae
Series:	Thalamiflorae
Order:	Parietals
Family:	Cruciferae (Brassicaceae)
Genus:	*Lepidium*
Species:	*Lepidium sativum*

Common names: English: Garden cress; **Hindi:** Chandrasur, Hurf, Halim; **Kannada:** Alavibija; **Malayalam:** Asali; **Sanskrit:** Asalika; **Tamil:** Alivirai; **Telugu:** Aditya.

Distribution: This species is native to the Mediterranean region, but has been introduced, naturalized and cultivated throughout the world. Within India, it is cultivated as a salad plant.

Description: An erect glabrous annual herb which grows to 15–45 cm in height. The leaves are entire, the upper ones are sessile whereas the lower ones are petiolate. The flowers are white in colour, small, in long racemes. The fruits are small, pods, obovate, with 2 seeds per pod. The seeds are brownish red in colour and become slimy when soaked in water.

Flowering: December–March.

Fruiting: December–March.

Chemical constituents: The seeds contain 20–25 % yellowish semidrying oil and the major fatty acid in them is α-linolenic acid (32–34 %). Garden cress oil (GCO) has a balanced amount of polyunsaturated fatty acids (PUFA) (46.8 %) and monounsaturated fatty acids (MUFA) (37.6 %) and also contains natural antioxidants, viz. tocopherols and carotenoids, which protect the oil from rancidity. 7 imidazole alkaloids, lepidine B, C, D, E, and F (dimeric), and two new monomeric alkaloids, semilepidinoside A and B, sinapic acid and sinapine were reported in the seeds of *L. sativum*.

Used in: Traditional, modern and Ayurvedic medicine.

Parts used: Seeds and leaves.

Uses: The seeds of this plant have great medicinal value. They are useful in abdominal diseases, hiccough, eye disorders, skin diseases, diarrhea, menorrhagia, asthma and joint pains. The seeds are also galactagogue and emmenagogue. In *The Wealth of India*, this plant has been mentioned as a rubefacient, laxative, tonic, aphrodisiac, lactogenic and diuretic. In Gujrat, it is used as an effective remedy for *sandhivata*, *amavata* and *katishula*. The leaves are diuretic and stimulant.

Methods of propagation: Through seeds.

Seed can be sown in nursery beds and seedlings transplanted when 2–3 leaves have emerged.

Source: https://educalingo.com/en/dic-en/garden-cress

LEPTADENIA RETICULATA

Synonyms: *Asclepias tuberosa*; *Asclepias volubilis*; *Cynanchum asthmaticum*; *Cynanchum ovatum*; *Cynanchum reticulatum*; *Gymnema aurantiaca*; *Leptadenia appendiculata*; *Leptadenia imberbis*; *Secamone canescens.*

Classification:

Kingdom:	Plantae
Subkingdom:	Phanerogamea
Class:	Dicotyledonae
Subclass:	Gamopetalae
Series:	Bicarpellatae
Order:	Gentianales
Family:	Asclepiadaceae
Genus:	*Leptadenia*
Species:	*Leptadenia reticulata*

Common names: Hindi: Jivanti; **Kannada:** Jivanti; **Marathi:** Shinguti; **Sanskrit:** Jiwanti; **Tamil:** Plalkudai; **Telugu:** Tammudu.

Distribution: Distributed in sub-Himalayan tracts of India, mainly in Punjab, Uttar Pradesh and throughout the Deccan Peninsula up to 900 m. Found abundantly in forests and fields.

Description: A much-branched twining shrub; stems are cylindrical, occasionally bent at places, 5–10 cm long, 0.5–2.5 cm in diameter, surface rough, longitudinally ridged, wrinkled and furrowed, transversely cracked, with vertically elongated lenticels at places. Fracture short and splintery. The roots are externally rough, white or buff-colored with longitudinal ridges and furrows. In the transverse section, the wide cork, lignified stone cell layers and medullary rays can be seen. Leaves are ovate to cordate, entire, acute, subacute to mucronate, coriaceous above, finely pubescent below. Flowers in cymes, greenish yellow; fruit a follicle.

Flowering: November–December.

Fruiting: Fruits do not appear.

Chemical constitution: This plant contains α-amyrin, β-amyrin, ferulic acid, luteolin, diosmetin, rutin, β-sitosterol, stigmasterol, hentriacontanol, a triterpene alcohol, simiarenol, and apigenin. Pregnane, glycosides, reticulin, deniculatin and leptaculatin were isolated from the aerial parts which on hydrolysis give calogenin and tocopherols 4–5 (Krishna et al., 1975).

Used in: Traditional medicine.

Part used: The whole plant.

Uses: The plant is galactagogue, cooling, nutritive, aphrodisiac, stimulant and diuretic. It is also an eye tonic. Useful to cure eye diseases, seminal debility, general weakness, cough, dyspnea, fever, asthma, constipation, sore throat and gonorrhea. Extracts from the roots and leaves of the plant act as antibacterial and antifungal agents useful in various skin diseases. The plant promotes health and vigour, improves voice and balances the three doshas – *vata*, *pitta* and *kapha*. It also cures hematemesis, emaciation, and burning sensations Sastry et al., 1985.

Methods of propagation: Rooted semi-wooded stem cuttings from six-month-old to one-year-old plants having three to four nodes at pre flowering stage are generally used for planting. Seeds may also be collected in November for raising the plants.

Photo: Dr R.S. Modi

LEUCAS CEPHALOTES (ROTH) SPRENG.

Synonyms: *Leucas capitata* Desf.; *Phlomis cephalotes* Roth.

Classification:

Kingdom:	Plantae
Subkingdom:	Phanerogamea
Class:	Dicotyledonae
Subclass:	Gamopetalae
Series:	Bicarpellatae
Order:	Lamiales
Family:	Labiatae (Lamiaceae)
Genus:	*Leucas*
Species:	*Leucas cephalotes*

Common names: Bengali: Darunaphula, Hulkasha; **Gujarati:** Kulnnphul; **Hindi:** Goma madhupati; **Kannada:** Dronapushpi; **Malayalam:** Thumbe; **Marathi:** Tumba, Bahuphul; **Sanskrit:** Dronapushpi, Chitrapathrika, Chitrakshup; **Tamil:** Thumbai; **Telugu:** Thummichittu.

Distribution: This plant is found throughout India and in the plains of Bangladesh, Southeast Asian countries, Mauritius, Java and Philippines.

Description: Annual, branched, erect plant, reaches up to a height of 15–60 cm; leaves sub-sessile, simple, opposite decussate, linear, obtuse, pubescent, up to 8 cm long and 1.25 cm broad, with entire or crenate margin, petiole 2.5–6 mm long; flowers white, sessile, small, in dense terminal or axillary whorls; 30–40 flowers per whorl, bracts numerous, 6 mm long, linear, acute, bristle-tipped, ciliate with long slender hairs; sepal cup about 1.5–2 cm, tubular, slightly curved, nerved, hairy, corolla bi-lipped, upper lip bearded, lower lip longer than upper. Nutlets are narrow, oblong, about 3 mm long and 2 mm broad.

Flowering: July–October.

Fruiting: July–October.

Chemical constituents: The whole plant contains alkaloids, glycoside, sitosterol and flavonoids. It also contains, β-sitosterol and its glycosides, oleanolic acid, stigmasterol, pillion, gonzalitosin-I, tricin and anisofolin.

Used in: Ayurveda, Siddha, traditional medicine.

Parts used: Leaves, flowers.

Uses: It has antimicrobial, insecticidal, fever-reducing, larvicidal and inflammation-reducing properties. It is useful in skin diseases. In malarial fever, the leaves juice is given. A decoction made from the whole plant is used for curing fever. The juice removes toxins from the body. The juice of the leaves is applied externally for skin diseases, scorpion sting, snake bite and swelling. It is also put in the nostrils as drops to treat cough, congestion, blocked nose, headaches due to cough, sinusitis, migraine and phlegm. A few drops of the juice of the whole plant are put in the nostrils in case of snake bite. A flower decoction is taken orally to quench excessive thirst.

Methods of propagation: Through seeds.

Photo: Prof Srinath Rao

LIMONIA ACIDISSIMA L.

Synonyms: *Feronia elephantum* Correa; *Hesperethusa crenulata* L.; *Naringi crenulata*; *Feronia limonia*; *Schinus limonia* L.

Classification:

Kingdom:	Plantae
Subkingdom:	Phanerogamea
Class:	Dicotyledonae
Subclass:	Polypetalae
Series:	Disciflorae
Order:	Geraniales
Family:	Rutaceae
Genus:	*Limonia*
Species:	*Limonia acidissima*

Common names: Bengali: Koth Bel, Kapittha; **English:** Wood apple, Elephant apple; **Gujarati:** Kotha, Kothi; **Hindi:** Kaitha, Kath Bel; **Kannada:** Baelada Hannina; **Malayalam:** Vilam Kai; **Marathi:** Kavant; **Sanskrit:** Kapittha; **Tamil:** Vilam palam, Kavittam; **Telugu:** Vellaga pandu.

Distribution: This is the only species within the monotypic genus *Limonia*. It is native to the Indo-Malaya zone, Bangladesh, India, Pakistan, Sri Lanka and the Indo-Chinese region east of Java.

Description: It is a large tree growing to 9 m tall, with rough, spiny bark. The leaves are pinnate, with 5–7 leaflets, each leaflet 25–35 mm long and 10–20 mm broad, with a citrus scent when crushed. The fruit is a berry 5–9 cm in diameter, and may be sweet or sour. It has a very hard rind which can be difficult to crack open, and contains sticky brown pulp and small white seeds.

Flowering: January–March.

Fruiting: April onwards.

Chemical constituents: Phytochemical analysis of Limonia acidissima plant parts showed the presence of alkaloids, flavonoids, phenols, terpenoids, tannins,fats steroids, saponins, glycosides, gum, mucilage and fixed oils. Leaves contain stigmasterol, psoralen,bergapten, orientin, vitedin, saponarin, tannins and an essential oil. Seeds contain fixed oil. carbohydrates, proteins and amino acids. Roots contain feronia lactone, geranylum belliferone, bargapten, osthol, isopimpinellin, marmesin and marmin (Pratima and Rekha 2014; Zarga 1986) 2-quinolone, 2-hydroxyquinoline, N-acetyl-N methyltryptamine tanakine and tanakamine from the stem bark. Coumarins including 6-methoxy-7-geranyloxycoumarin, aurapten, bergapten, dihydrosuberenol, fernolin, isopimpinellin, marmesin, osthenol, osthol, and xanthotoxin are obtained from root and root bark.

Used in: Ayurveda, Unani and traditional medicine (Zarga 1986).

Parts used: Fruit pulp, stem bark.

Uses: This plant has many properties such as antioxidant, tyrosinase inhibitory, anti-inflammatory and antibacterial. The fruit pulp of the plant has been reported in traditional medicine as a curative for various ailments such as diarrhea, pruritis, impotence, dysentery, heart disease, vomiting and anorexia, and has also been used for the treatment of asthma and tumors, and as a liver tonic.

Methods of propagation: Through seeds.

Photo: Vedavyas Kulkarni

LYCIUM BARBARUM L.

Synonyms: *Boberella halimifolia* (Mill.) E. H. L. Krause; *Jasminoides flaccidum* Moench; *Lycium halimifolium* Mill.; *Lycium turbinatum* Veill.; *Lycium vulgare* Dunal; *Teremis elliptica* Raf.

Classification:

Kingdom:	Plantae
Subkingdom:	Phanerogamea
Class:	Dicotyledonae
Subclass:	Gamopetalae
Series:	Bicarpellateae
Order:	Pleomoniales
Family:	Solanaceae
Genus:	*Lycium*
Species:	*Lycium barbarum*

Common names: English: Wolfberry, Goji berry; **Kannada:** Uruvalu; **Tamil:** Sottai kala, Naavithar mooli.

Distribution: Native to Southeast Europe and Asia. In India it grows in the mountainous region of Himalayan Kashmir.

Description: It is a deciduous woody perennial plant, growing 1–3 m high. Leaves: either in an alternating arrangement or in bundles of up to 3, lanceolate. Leaves 7–8 cm wide × 3–4cm broad with blunted or round tips. Flowers born in groups of 1–3 in the leaf axils. The calyx consists of bell-shaped or tubular sepals forming short, triangular lobes. The corollas are light purple, 9–14 mm wide with 5 lobes shorter than the tube. The stamens are structured with filaments longer than the anthers. The anthers are longitudinally dehiscent. Fruit: a berry, ellipsoid, orange-red, 1–2 cm deep. The number of seeds in each berry varies widely based on cultivar and fruit size, containing 10–60 tiny yellow seeds that are compressed with a curved embryo.

Flowering: June–September.

Fruiting: August–October.

Chemical constituents: Wolfberries contain many nutrients and phytochemicals, including polysaccharides, β-carotene, zeaxanthin, betaine, cerebroside, β-sitosterol, atropine and p-coumaric, ascorbic, linoleic and α-linolenic acids.

Used in: Homeopathy, Ayurveda and Chinese medicine.

Parts used: Fruit and root bark.

Uses: This plant might have potential benefits against cardiovascular and inflammatory diseases. It is used to treat diabetes, high blood pressure, fever, malaria, cancer, blood circulation problems, sexual problems (impotence), dizziness, ringing in the ears (tinnitus), macular degeneration and glaucoma. It is more frequently used as antiaging, cancer preventive and immune-system enhancer. The root bark is used as a cooling agent to clear heat and lower consumptive fever.

Methods of propagation: Through seeds.

Photo: https://www.google.co.in/search?q=Lycium barbarum&tbm

MELIA AZEDARACH L.

Synonyms: *Melia sempervirens* Sw.; *Melia bukayun* Royle.

Classification:

Kingdom:	Plantae
Subkingdom:	Phanerogamea
Class:	Dicotyledonae
Subclass:	Polypetalae
Series:	Disciflorae
Order:	Geraniales
Family:	Meliaceae
Genus:	*Melia*
Species:	*Melia azedarach*

Common names: Bengali: Ghoranim, Mahanim, Gora-Nim; **English:** Bead tree, Chinaberry, Indian lilac, Persian lilac; **Gujarati**: Bakan limado, Bakai nimbu; **Kannada:** Kadu Bevu; **Malayalam:** Malaveppu; **Marathi:** Bakana nimb; **Sanskrit:** Dreka maha nimba, Ramyaka; **Tamil:** Malaivembu, Mallay vembu; **Telugu:** Turakavepa, Taraka vepa.

Distribution: *Melia azedarach* is found throughout India. It grows naturally in the sub-Himalayan tracts up to 1800 m. It is cultivated and naturalized throughout India.

Description: It is a deciduous tree reaching up to 45 m in height. It has a spreading crown and sparsely branched limbs. The bark is smooth, greenish-brown when young and turns grey and fissured with age. The leaves are up to 50 cm long, alternate, long-petioled, 2–3 times compound (odd-pinnate); the leaflets are dark green above and lighter green below, with serrate margins. The flowers are small and fragrant, with 5 pale purple or lilac petals, growing in clusters. Sepals 5-lobed, 1 cm long; petals 5-lobed, 0.9 cm long, pubescent; staminal tube deep-purple/blue, 0.5 cm long, 1 cm across. The fruit is a drupe, marble-sized, light yellow at maturity, hanging on the tree all winter, and gradually becoming wrinkled and almost white. The bark is thin, about 0.2–0.6 cm thick; outer surface black and rough, slightly fissured, exfoliating in small slightly woody pieces, light and dark grey to greyish black in colour.

Flowering: April–May.

Fruiting: April–May.

Chemical constituents: Quercitrin, rutin, tetranortriterpenoids, salanin and vilasinin are present in the leaves. The heartwood contains bakayanin, a lactone, and bakalactone.

The oil extracted from the seeds contains butyric, valeric, stearic, palmitic, oleic, linoleic, arachidic, liquoceric and unsaturated resinous acids. The bark of the tree contains an alkaloid – paraisine. The inner bark is extremely bitter and nauseous but not astringent and contains yellowish-white resin. The outer bark contains tannins.

Used in: Ayurveda, Unani and f.olk medicine

Parts used: Root, stem bark, flower, leaf, root, bark, fruit, seed, gum and sap.

Uses: It is used traditionally to treat various ailments such as piles, mouth ulcers, skin problems, dandruff, gout, inflammation, etc. A decoction of the leaf is anthelmintic, astringent and emetic. Boiled leaves are applied on the affected area to get relief from gout and arthritis. Leaf juice is applied to relieve dandruff and cure eczema, ringworm infections, itching and burns on the skin. It is also given orally to cure irregular and heavy menses. Leaf decoction is given to relieve hernia. The root and the bark are used as an anthelmintic, vermifuge, cathartic and emetic as well as for intermittent fevers and dysentery. The dried stem bark is used to cure asthma, emesis, gastroenteritis, giddiness, increased frequency and turbidity of urine, nausea, skin diseases and vertigo. An infusion of the bark is effective against ascariasis. Bark decoction is given to cure kidney problems, increased serum creatinine and increased uric acid. A poultice of the bark is applied to cure leprosy and scrofulous ulcers. A paste made from the flowers is applied to hair to kill lice (Grover and Yadav 2004).

Methods of propagation: Through seeds.

Photo: Prof Srinath Rao

MOMORDICA CHARANTIA L. SP. PL.

Synonyms: *Momordica elegans* Salisb.; *Momordica chinensis* Hereng.; *Momordica indica* L.; *Cucumis argyi* H. Lev.; *Cucumis intermedius* M. Roem.

Classification:

Kingdom:	Plantae
Subkingdom:	Phanerogamea
Class:	Dicotyledonae
Subclass:	Polypetalae
Series:	Calyciflorae
Order:	Passiflorales
Family:	Cucurbitaceae
Genus:	*Momordica*
Species:	*M. charantia*

Common names: Bengali: Karala; **English:** Bittergourd; **Gujarati:** Karalu; **Hindi:** Karela; **Kannada:** Hagalakayi; **Malayalam:** Kaypa; **Marathi:** Karale; **Sanskrit:** Karavelaka; **Tamil:** Pavakkay, Iracavalli; **Telugu:** Kakarakayi.

Distribution: It is a tropical plant widely cultivated in Asia, India, East Africa and South America.

Description: This herbaceous, tendril-bearing vine grows up to 10–15 feet in length. It is branched. The slender, green stems are usually slightly 5-angled or ridged, and carry unbranched tendrils in the leaf axils. The plant bears simple, alternate leaves 6–10 cm across, with 3–7 deeply separated lobes. Each plant bears separate yellow male and female flowers, both borne in the leaf axils. The flowers occur singly in the upper leaf axils on 2–10 cm long stalks with a small leaf-like bract towards the base. Male flowers have a slender basal swelling which is continuous with the base of the sepal tube, ending in 5 blunt sepals. There are 5 oval yellow petals 10–20 cm long, and 5 central stamens. Female flowers are similar to the male flowers but have a distinct warty swelling well below the base of the sepal tube and 3 stigmas. The pendulous cylindrical fruits are egg-shaped/oblong, 10–20 cm long and covered with longitudinal ridges and warts. The fruit has a distinct warty exterior and an oblong shape. It is hollow in cross-section, with a thin layer of flesh surrounding a central seed cavity filled with large, flat seeds and pith. Bright-red arils enclose the seeds, which adhere to the inside of the fruit. Seeds woody, flattened, 5–9 mm long, withfinely pitted surfaces.

Flowering: June–July.

Fruiting: September–November.

Chemical constituents: The main constituents of this plant which are responsible for its anti-diabetic effects are triterpene and proteid, steroid, alkaloid, inorganic, lipid and phenolic compounds. Several glycosides have been isolated from the *M. charantia* stem and fruit and are grouped under the genera of cucurbitane-type triterpenoids. *M. charantia* consists of chemical constituents including alkaloids, charantin, charine, cryptoxanthin, cucurbitins, cucurbitacins, cucurbitanes, cycloartenols, and diosgenin. This plant is one of the most commonly used vegetables containing polypeptide-p and is employed to control diabetes naturally. Amino acids present are aspartic acid, serine, glutamic acid, threonine, alanine, g-amino butyric acid, pipecolic acid and luteolin. Fatty acids present are lauric, linoleic, linolenic, myristic, palmitic, palmitoleic, stearic and oleic acids (Joseph and Jini 2013) (Liu et al., 2009).

Used in: Ayurveda and traditional medicine.

Parts used: The whole plant, especially the fruits.

Uses: In traditional Indian medicine, different parts of the plant are used as treatments for diabetes (particularly polypeptide-p, an insulin analogue), and as stomachic, laxative, anti-bilious, emetic and anthelmintic agents, as well as for the treatment of cough, respiratory diseases, skin diseases, wounds, ulcer, gout, and rheumatism. *Momordica charantia* has a number of purported uses, including cancer prevention, treatment of diabetes, fever, HIV and AIDS. Bitter melon tea, also known as gohyah tea, is a herbal tea made from an infusion of dried slices of bitter melon. It is sold as a medicinal tea (Kumar et al., 2010).

Methods of propagation: Through seeds.

Photo: Prof Srinath Rao

MOMORDICA CYMBALARIA (HOOK.) FENZL.

Synonyms: *Luffa tuberosa* (Roxb.); *Momordica tuberosa* (Roxb.).

Classification:

Kingdom:	Plantae
Subkingdom:	Phanerogamea
Class:	Dicotyledonae
Subclass:	Polypetalae
Series:	Calyciflorae
Order:	Passiflorales
Family:	Cucurbitaceae
Genus:	*Momordica*
Species:	*Momordica cymbalaria*

Common names: Hindi: Kakrol; **Kannada**: Karchikai; **Malayalam:** Kattupaval; **Marathi:** Kadavanchi; **Sanskrit:** Kuduhunchi; **Tamil:** Athalakkai; **Telugu:** Kasarakayee

Distribution: Originating in tropical regions of India and Southeast Asia. Found in the Indian states of Andhra Pradesh, Karnataka, Madhya Pradesh, Maharashtra and Tamil Nadu.

Description: The plant is a climbing annual or perennial herb with slender, scandent, branched, deeply striate stem. The leaves are orbicular-reniform in outline, deeply cordate at the base, 5–7 angled. The flowers are white or yellow, male flowers in short racemes, female flowers solitary, long-peduncled. The fruits are 20–25 mm long, pyriform, with 8 sharp ridges, 24 mm × 15 mm, attenuated at the apex and with the base narrowed into a curved peduncle, which is fleshy, dark green and ribbed (Figure 1b). The seeds are 4.6 mm long, ovoid-shaped, smooth and shiny. Flowers are unisexual. The male flower peduncle is 5–30 mm long, filiform, puberulous, ebracteate, with 2–5 flowers in racemes with a pale-yellow corolla and 2 stamens for each flower. The female flower is solitary on a peduncle of 28 mm in length. The roots are tuberous and perennial.

Chemical constituents: *Momordica* has bioactive compounds such as phenolic acids, flavonoids, carotenoids, cucurbitane triterpenoids, and phytosterol. It is a rich source of vitamin C, fibres and β-carotene as well as iron and calcium.

Used in: Ayurveda and traditional medicine.

Parts used: Fruits.

Uses: Anti-diabetic, hypoglycemic, hepatoprotective, cardio-protective, anti-ulcer, nephroprotective, anti-cancer and anti-diarrheal properties.

Methods of cultivation: Propagated through tuberous roots. The tubers sprout after the onset of monsoon and undergo dormancy during the summer.

Photo: Dr R.S. Modi

MOMORDICA DIOICA ROXB. EX WILLD.

Synonyms: *Momordica balsamina*, Wall; *Momordica wallichii*, Roem.

Classification:

Kingdom:	Plantae
Subkingdom:	Phanerogamea
Class:	Dicotyledonae
Subclass:	Polypetalae
Series:	Calyciflorae
Order:	Passiflorales
Family:	Cucurbitaceae
Genus:	*Momordica*
Species:	*Momordica dioica*

Common names: Bengali: Ban-karela; **English:** Teasel gourd; **Hindi:** Jangli karela; **Kannada:** Gidda haagalu; **Malayalam:** Ben-pavel; **Marathi:** Banzakartholi; **Sanskrit:** Vishakantakini, Sarpadamani; **Tamil:** Aegarvalli; **Telugu:** Angaakakara.

Distribution: This species is globally distributed in Indo-Malaysia. Within India, it has been found almost throughout in open areas and in plateaus.

Description: Climbing herbs with tuberous roots; tendrils simple. Leaves 4–8 × 3–6 cm, broadly ovate, deeply 3–5-lobed, base cordate, margin denticulate, apex acute, membranous; petiole up to 2.5 cm. Flowers dioecious, solitary, axillary. Male flowers with long peduncles with foliaceous bracts at apex; peduncle 4–6 cm long. Calyx tube about 1 cm long, lobes linear, villous. Petals about 2.5 × 1 cm, oblong, obtuse, yellow. Stamens 3, inserted at the mouth of calyx tube; 1 1-celled, 2 2-celled; staminodes of 2–3 glands. Ovary oblong with many horizontal ovules. Fruit about 3 × 2.2 cm, ovoid, softly echinate. Seeds 6–7 × 5–6 mm, ovoid, emarginate, pale yellow.

Flowering: July–December.

Fruiting: July–December.

Chemical constituents: Root contains traces of alkaloids. Fruit contains ascorbic acid. Seed kernel contains semi-drying oil.

Used in: Ayurveda, folk and Siddha medicine.

Parts used: The whole plant.

Uses: It is traditionally used as astringent, febrifuge, antiseptic, anthelmintic and spermicidal remedy. Also used in bleeding piles, urinary infections and as a sedative. Studies indicate that it possesses antioxidant, hepatoprotective, antibacterial, anti-inflammatory, anti-lipid peroxidative, hypoglycemic and analgesic properties. Roots are used in bleeding piles, bowel affections and urinary complaints. The root of the male creeper is used in ulcers caused by snake bites and the root of the female plant is used as an expectorant.

Methods of propagation: Through seeds or tuberous roots. The tubers remain dormant in winter and sprout in summer. The fleshy roots may remain in the field or be dug out for storing. The plant can also be propagated by shoot cuttings of 20–25 cm. Long shoots are put into the sand medium for rooting. They root within 2 weeks. By this process, a greater number of female plants is obtained.

MORINGA OLEIFERA LAM.

Synonym: *Moringa pterygosperma* Gaertn.

Classification:

Kingdom:	Plantae
Subkingdom:	Phanerogamea
Class:	Dicotyledonae
Subclass:	Polypetalae
Series:	Disciflorae
Order:	Sapindales
Family:	Moringaceae
Genus:	*Moringa*
Species:	*Moringa oleifera*

Common names: Bengali: Sajina; **English:** Drumstick tree, Horse radish tree; **Hindi:** Sahijan, Mungana; **Kannada:** Nuggekaigida, Murunga; **Malayalam:** Murinna; **Marathi:** Sheraga; **Sanskrit:** Sigruh, Sobhanjanah; **Tamil:** Murunkai; **Telugu:** Munaga, Mulaga.

Distribution: Found all over India, growing wild or being cultivated.

Description: A medium-sized soft-wooded tree with pungent roots. Leaves: compound, tripinnate, with articulated rachis 30–60 cm long. Flowers: bisexual, oblique, stalked, small, white, borne in axillary or terminal panicles, sepals refluxed, petals spathulate, stamens yellow. Fruit: a capsule, long, pendulous, cylindrical, 9-ribbed, constricted between seeds. Seeds: trigonous, broadly winged at angles.

Flowering: July–August.

Fruiting: August–November.

Chemical constituents: Leaves contain kaempferol, acacetin, vitexin and vanillic, syringic and melilotic acids. The root bark contains 3 alkaloids: moringine, moringinine and spirochin.

Used in: Ayurveda, Siddha, folk and traditional medicine.

Parts used: Pods, seeds, flowers, leaves, roots and bark.

Uses: All parts of the plant are useful. Roots and root bark are used in fever, asthma, rheumatism, inflammation, epilepsy and hysteria. Leaves are galactagogue, useful in eye diseases, fainting fits, spasmodic problems of bowels and hysteria. Roots, leaves, fruits and seeds are aphrodisiac, cardiotonic and anthelmintic. Also used in enlargement of spleen and joints, syphilis and rheumatism. Gum is used as an antiseptic.

Methods of propagation: Through seeds and stem cuttings. Seeds have no dormancy periods and can be planted as soon as they are mature. They have longevity of up to 6 months. They are soaked in cold or tepid water for 24 hours and directly sown in the month of June. In India, the plant is propagated by planting limb cuttings 1–2 m long, from June to August, preferably. It can also be propagated through stem cuttings. Cuttings 45–100 cm long, 4–10 cm wide should be taken from the woody parts of the previous year's branches. They can be then planted in a nursery or in the field. They take about 15–20 days to root.

Photo: Prof Srinath Rao. Fruits

MOLLUGO NUDICALIS LAM.

Synonyms: *Mollugo bellidifolia* (Poir.) Ser.; *Pharnaceum bellidifolium* Poir.; *Pharnaceum spathulatum* Swartz.; *Lampetia nudicalis* (Lam.) Raf.

Classification:

Kingdom:	Plantae
Subkingdom:	Phanerogamea
Class:	Dicotyledonae
Subclass:	Polypetalae
Series:	Calyciflorae
Order:	Ficoidales
Family:	Aizoaceae
Genus:	*Mullugo*
Species:	*Moollugo nudicalis*

Common names: English: Daisy-leaved chickweed, Naked-stem chickweed; **Kannada:** Murukalu gida; **Malayalam:** Ullukkumarunna; **Tamil:** Parpaadagam, Parpatakam;

Distribution: South America, tropical Africa, Ceylon, India and Pakistan. Found in open places such as waste areas, road sides, river beds, grassland and cultivated land.

Description: Annual up to 20 cm tall, rosette habit, stem reduced, only the inflorescence is erect, highly branched. Leaves basal, 1.5–6.0 cm long, 5–15 mm broad, spathulate to oblanceolate, glabrous, base attenuate, margin entire, apex obtuse, retuse, petiole about 1 cm. Flowers in dichasial cymes; scapes slender, glabrous; involucres bracts ovate-oblong, 1–2 mm long, scarious; pedicels up to 1.4 cm long. Sepals oblong, 2.5–3.0 mm long, mucronate, imbricate, persistent. Stamens 3–5, filaments about 1 mm long, persistent, anthers less than 1 mm long, subglobose. Ovary 3-lobed, subglobose, 1.5 mm long. Stigmas 3, short, curved. Fruit a capsule as long as the sepals, subglobose, dehiscing by 3 valves. Seeds less than 1 mm, muricate, minutely strophiolate, black and shiny.

Flowering: June–December.

Fruiting: June–December.

Chemical constituents: The plant is rich in flavonoids and related phenolic compounds. It also contains cyanogenic glycosides, terpenes and saponins.

Used in: Folk, Siddha and traditional medicine.

Parts used: The whole plant.

Uses: The whole herb is vermifuge, pectoral, a cure to whooping cough, cold and jaundice. It is useful mainly in preventing small ringworms in the stomach, to control diabetes and to increase the life energy of blood cells. It is also used to treat wounds, cough, fever, and body pain. A decoction of the whole plant is taken for treating gonorrhea. Crushed leaves and stems mixed with garlic are given to typhoid patients.

Methods of cultivation: Through seeds.

Photo: Prof Srinath Rao

MURRAYA KOENIGII (L.) SPR.

Synonyms: *Bergera koenigii* Linn.; *Chalcas koenigii.*

Classification:

Kingdom:	Plantae
Subkingdom:	Phanerogamea
Class:	Dicotyledonae
Subclass:	Polypetalae
Series:	Disciflorae
Order:	Geraniales
Family:	Rutaceae
Genus:	*Murraya*
Species:	*Murraya koenigii*

Common names: Bengali: Karipata; **English:** Curry leaf tree; **Hindi:** Kurry patta; **Kannada:** Karibevu; **Malayalam:** Kariveppu, Karivepu; **Marathi:** Karipat; **Sanskrit:** Alakavhaya; **Tamil:** Karivempu; **Telugu:** Karepaku.

Distribution: Found almost throughout India and the Andaman Islands up to an altitude of 1500 m. It is commonly found in forests, often as gregarious under growth. It is much cultivated for its aromatic leaves and for ornament throughout India. The species grows naturally in the Konkan, Western Ghats, Deccan, Northern Circars and along the foothills of Kumoan to Sikkim, up to 5000 feet. It is widely cultivated in home gardens.

Description: Aromatic, more or less deciduous shrub or a small tree, up to 6 m in height and 15–40 cm in diameter. Bark dark brown or almost black; leaves imparipinnate: leaflets 9–25. ovate, lanceolate or somewhat rhomboid, irregularly crenate, dentate, acuminate, obtuse or acute, base usually oblique, almost glabrous above, pubescent beneath, gland-dotted, strongly aromatic. The leaves have a slightly pungent, bitter and feebly acidulous taste. They retain their flavour even when moderately dried. The inflorescence is terminal, pedunculate, many white-flowered, compact, corymbiform, cymose panicle; berries subglobose or ellipsoid, purplish black when ripe, 2-seeded.

Flowering: April–August.

Fruiting: September–December.

Chemical constituents: The leaves of the plant are extensively employed as flavouring in curries and chutneys. The analysis of the leaves shows the presence of proteins, fats, carbohydrates, fibres and mineral matter such as calcium; phosphorus, iron, carotene (as vitamin A), nicotinic acid, vitamin C thiamine and riboflavin. The leaves are a fair source of vitamin A; they are also a rich source of calcium. The free amino acids present in the leaves are: asparagine, glycine, serine, aspartic acid, glutamic acid, threonine, alanine, proline, tyrosine, tryptophan, gamma aminobutyric acid, phenylalanine, leucine, isoleucine and traces of ornithine, lysine, arginine and histidine. The leaves also contain crystalline glucoside, koenigin, and a resin. Extracts of the stem contain hydroxymellein, vanillin, octacosanoic acid, β-sitosterol and β-sitostenone, Girinimbin is found in the bark

Used in: Ayurveda, Siddha, Unani, folk and traditional medicine.

Parts used: Leaves, root, bark, fruits, wood.

Uses: Curry leaf is used in many Ayurvedic and Unani medical prescriptions.

The leaves and roots are slightly bitter, acrid, cooling, alexeric, anthelmintic and analgesic. They are reported to cure piles and allay heat of the body, thirst, inflammation and itching. Bark and roots are used as stimulants. The green leaves are said to be eaten raw for the treatment of dysentery. An infusion of toasted leaves is used to stop vomiting. The powdered leaf is used to aid in the healing of fresh cuts, and a decoction of the leaves is drunk for dropsy. The tribal peoples of India, especially the Munda, use its ground bark for a drink, and they rub the bark on the bitten limb as a snake-bite remedy. The root is eaten and rubbed on for body aches.

The leaves, root and bark are considered tonic, stomachic and carminative. The leaves are used internally in dysentery and diarrhea, and also for checking vomiting; they are applied externally to bruises and eruptions. The juice of the root is taken to relive pain associated with the kidneys. The bark of the root is a stimulant, used externally to cure eruptions and venomous bites. The juice of the root is given to relieve renal pains. The essential oil is used in soaps and in the perfume industry.

Methods of propagation:

Propagation through seeds,

- Within 3–4 days of collection of fruits, the seeds should be pulled and sown in nursery beds or poly bags filled with soil and farmyard manure.
- If sowing is delayed, there will be drastic deterioration in the germination of the seeds, which germinate freely under partial shade. A spacing of 15–20 feet has been suggested for planting seedlings.
- Red sandy loam with good drainage will be ideal for normal and fleshy growth which will result in better leaf yield. To a certain extent, the plant can tolerate drought also.
- Raising seedlings in a nursery and planting them in the main field is the normal practice. For raising seedlings, well-ripe fruits should be harvested from high-yielding mother trees.
- Nursery beds are prepared in a well-drained area since water stagnation will severely affect the germination. Beds of 1 m × 1 m in size and 30 cm in height are formed. Adequate farmyard manure should be applied before preparing the beds.
- Lines are drawn at a spacing of 10 cm and seeds are sown in these lines. The seeds will germinate in about 3 weeks' time. Seedlings can also be raised in polythene bags.

- 1-year-old seedlings are the best suited for planting. In the main field, pits of size 30 × 30 × 30 cm are dug at a spacing of 1.2–1.5 m 1–2 months before planting the seedlings.

Photo: Prof Srinath Rao

MYRISTICA FRAGRANS HOUTT.

Synonyms: *Myristica officinalis* L. f.; *M. moschata* Thunb.; *M. aromatica* Lamk.

Classification:

Kingdom:	Plantae
Subkingdom:	Phanerogamea
Class:	Dicotyledonae
Subclass:	Monochlamidae
Series:	Micrembryeae
Family:	Myristicaceae
Genus:	*Myristica*
Species:	*Myristica fragrans*

Common names: Bengali: Jaiphal; **English:** Nutmeg; **Hindi:** Jaiphal; **Kannada:** Jajeekayi; **Malayalam:** Jatika; **Marathi:** Jayiphal; **Sanskrit:** Jajiphala, Jatiphalam, Jatikosha, Jatipatra; **Tamil:** Jathikai; **Telugu:** Jaaikaaya.

Distribution: This species is a native of the Eastern Moluccas, now cultivated in many tropical countries. In India it is endemic to the Western Ghats of Karnataka and Kerala. It is cultivated in Assam, West Bengal and South India.

Description: It is a dioecious or occasionally monoecious evergreen, aromatic tree. The bark is greyish black; the branches appear in spatially arranged whorls in the main stem and grow outwards. The entire tree has a cone-like appearance. Two types of branches, viz. plagiotropic fruiting branches and orthotropic erect growing branches are present in a tree. Usually male and female trees are separate. Leaves: alternate, simple, exstipulate, elliptic or oblong-lanceolate, coriaceous, small, dark green, shiny above, dull beneath with pointed ends and sebaceous glands on the upper side; aromatic when bruised. Flowers: in umbellate cymes, 8–10 per branch; female trees produce only 1–3 flowers in a branch, creamy yellow and fragrant, glabrescent with sparse, very minute tomentum, pale yellow with a 3-lobed perianth; male flowers have a distinct staminal column formed by the fusion of anthers, 8–10-stamen female flowers with a superior sessile, 1-celled ovary with a single basal ovule. Fruits: yellow, broadly pyriform or globose, 5–8 cm long, splitting open in 2 halves when ripe; the seed, broadly ovoid, albuminous, with a shell-like purplish brown testa and covered by a red, fleshy aril. The dried kernels constitute the nutmeg of commerce. Mace is the dried fibrous aril covering the testa.

Flowering: Throughout the year.

Fruiting: Throughout the year.

Chemical constituents: Nutmeg essential oil is a pale yellow to almost water-white with a fresh, warm, spicy, aromatic odour; it contains about 10 % essential oil, mostly composed of terpene hydrocarbons-pinenes, camphene, p-cymene, sabinene, phellandrene, terpinene, limonene and myrcene (60– 90 %); terpene derivatives such as -linalool, geraniol, terpineol (5–15 %) and phenylpropanes such as -myristicin, elemicin, safrol (2–20 %). Myristicin (methoxy-safrol) is responsible for the hallucinogenic effect of large nutmeg dosages (typically, 1 or more nuts).

Used in: Ayurveda, folk, Tibetan, Unani, Siddha and modern medicine; homeopathy.

Parts used: Dried kernel of the seed.

Uses: Nutmeg is reported to be an expectorant, vermifuge, aphrodisiac, and as a nervine used by psychiatrists. It is used in tonics and electuaries and is recommended for the treatment of inflammations of bladder and urinary tract. The aril of the seeds is cooling, febrifuge and expectorant, and is useful in the vitiated conditions of *vata*, cough, bronchitis, fever and burning sensations. It is also used in folk medicine for the treatment of rheumatism. It has anti-inflammatory, antifungal and antibacterial properties. It has no toxic effect when consumed in normal dosage. Higher dosage has narcotic effects, hence should be taken in prescribed dosage.

Methods of propagation: By vegetative methods such as layering, budding, grafting. In arching, veneer grafting, side grafting and epicotyl grafting using rootstock, about 90 % success rate is observed.

Photo: Prof Srinath Rao

NARDOSTACHYS GRANDIFLORA DC.

Synonym: *Nardostachys jatamansi* DC.

Classification:

Kingdom:	Plantae
Subkingdom:	Phanerogamea
Class:	Dicotyledonae
Subclass:	Gamopetalae
Series:	Inferae
Order:	Asterales
Family:	Valerianaceae
Genus:	*Nardostachys*
Species:	*Nardostachys grandiflora*

Common names: Bengali: Jatamansi; **English:** Jatamansi; **Gujarati:** Ingoriyo; **Hindi:** Jatamansi; **Kannada:** Jatamamshi; **Malayalam:** Jatamashi; **Sanskrit:** Jatamsni; **Tamil:** Jatamashi; **Telugu:** Jatamamshi.

Distribution: The plant occurs in the alpine Himalayas of Nepal and India at 3000–4500 m altitude, as well as in China and Bhutan.

Description: An erect perennial herb, up to about 60 cm tall, rhizomes woody, long, covered with fibers from petioles of withered leaves. Lowers basal leaves up to about 20 cm long; narrowed into the petiole; upper leaves much smaller, almost ovate. Flowers small, rosy, pale pink or blue, in dense cymes; fruit small, covered with minute hair.

Flowering: July–August.

Fruiting: July–August.

Chemical constituents: A sesquiterpene, jatamansone (valeranone) forms the major chemical component of the rhizome. Other sesquiterpenoids present are spirojatamol, patchouli alcohol and norseychelanone. Also present jatamol, jatamansic acid and terpenoid coumarines oroselol and jatamansin.

Used in: Ayurvedic, Unani and traditional medicine.

Parts used: Rhizome.

Uses: The dried rhizomes and roots of this plant are medicinal. Jatamansi is used as tonic: it has antispasmodic and stimulant properties; it is, therefore, useful in the treatment of epilepsy, hysteria, convulsions and palpitation of the heart. It is also used as a laxative and for improving urination, menstruation and digestion.

Methods of propagation: Seed sown as soon as it is ripe. The seed requires light for germination. Plant the seedlings into individual pots when they are large enough to handle and grow them on in the greenhouse for their first winter. Plant out in late spring or early summer.

NIGELLA SATIVA L.

Synonym: *Nigella cretica* Miller; *Nigella indica* Roxb. ex Fleming.

Classification:

Kingdom:	Plantae
Subkingdom:	Phanerogamea
Class:	Dicotyledonae
Subclass:	Polypetalae
Series:	Thalamiflorae
Order:	Ranales
Family:	Rananculaceae
Genus:	*Nigella*
Species:	*Nigella sativa*

Common names: Bengali: Kalijira; **English:** Black cumin, Small fennel; **Gujarati:** Kadu jeeroo; **Hindi:** Kalajira; **Kannada:** Karijirigae; **Malayalam:** Karinchirkam; **Marathi:** Krishanajira; **Punjabi:** Kalonji, Kala jira; **Sanskrit:** Upakunchika; **Tamil:** Karunshirgam, Karunjiragam; **Telugu:** Nallajilakarra.

Distribution: It is occasionally cultivated and often found as a weed of cultivation in Punjab, Himachal Pradesh, Bihar and Assam.

Description: Pretty, small, erect annual herb up to 70 cm tall, with a well-developed yellow-brown taproot and numerous feeder roots. Stem profusely branched, subterete, ribbed, sometimes hollow, puberulous, light to dark green. Leaves alternate, stipules absent; petioles strongly broadened at base, light green, only present in basal leaves, 1–6 cm long, ribbed, puberulous; blade up to 7 × 5 cm in outline, bi-tri- or even multi-pinnately dissected into short, thin, sublinear, divergent, slightly pilose lobes, normally green but sometimes reddish brown.

Flowers terminal, solitary; pedicel 4–8 mm long, puberulous, ribbed; all flower parts inserted on a pale yellow or blue, fleshy, depressed-conical receptacle, about 2 mm in diameter, visible as an orange-brown ring below the carpels in fruit; sepals 5, petaloid, ovate, 13–17 × 6–12 mm, apex obtuse, tapering basally into a 2–3 mm-long claw, papillose to pilose, pale green when young, sometimes partly reddish, inside pale blue/white when older; petals 8 with short claw, blade 2-lobed, lobes enclosing nectar pocket, apex bifid, greenish-white with violet lines, stamens in 6–10 groups of 3–7 stamens each, each group forming a spiraloid line on the receptacle, anthers yellow; pistil-lobed, composed of 3–7 white granular carpels, almost connate at base, forming a compound ovary 4–9 mm long with free stigmas.

Fruit: a capsule, ribbed, oblongoid, tuberculate, 6–16 mm × 5–12 mm, greyish green to brown at maturity, many-seeded, with persistent stigmas.

Flowering: January–December.

Fruiting: January–December.

Chemical constituents: Seeds contain numerous esters of unsaturated fatty acids and terpene alcohols (7 %) and about 0.5 % essential oil. Amongst the components of the essential oil are thymoquinone, thymohydroquinone, p-cymene, α- and β-pinene, further traces of carvacrol, anethole, α -terpineol and d-citronellol. Seed also contains tannins, alkaloids damascenine, nigellone, nigellimine, nigellimine-N-oxide, nigellicene and sterols cholesterol, campesterol, stigmasterol, ß-sitosterol and α-spinasterol (Khan, 1999; Ahmad, et al., 2013; Forouzanfar et al., 2014).

Used in: Modern medicine.

Parts used: Seeds, seed oil and leaves.

Uses: Seeds are acrid, bitter, thermogenic, aromatic, carminative, diuretic, anodyne, antibacterial, anti-inflammatory, deodorant, appetizing, digestive, anthelmintic, constipating, sudorific, febrifuge, stimulant and expectorant. Useful in skin diseases, hemorrhoids, cephalalgia, rheumatic pain, headache, asthma, jaundice, inflammation, fever, paralysis, ophthalmia, halitosis, anorexia, dyspepsia, flatulence, diarrhea, dysentery, cough, amenorrhea, dysmenorrhea, helminthiasis especially tapeworm, strangury, intermittent fevers, agalactia and vitiated conditions of *vata* and *kapha*. Seed and oil possess pharmacological properties such as detergent, seductive, anti-inflammatory, antibacterial. Crushed seeds in vinegar, they are applied in skin disorders such as ringworm, eczema and baldness. Applied as a galactagogue, emmenagogue, vermifuge and aid discharge of pus. Pounded seeds are used against nausea and in parts of North India to induce abortion (Khan, 1999; Ahmad, et al., 2013; Forouzanfar et al., 2014).

Methods of propagation: Through seeds.

- Seeds are broadcasted at a seed rate of 8 kg/ha, as they are very fragile for transplanting, but may also be sown in peat blocks.
- Row spacing 25–40 cm. Planted as a rabi crop in the winter season in the northern plains. Sown during the month of October in North Indian conditions.
- Soil prepared to fine tilth and levelled to make convenient-sized beds. Pre-sowing irrigation should be given for better germination.
- Plant to plant spacing should be 15–20 cm.

NOTHAPODYTES FOETIDA (WIGHT) SLEUMER

Synonyms: *Nothopodytes nimmoniana* (J. Grah.) D. J. Mabberely.; *Mapia foetida* (Wt.) Miers.

Classification:

Kingdom:	Plantae
Subkingdom:	Phanerogamea
Class:	Dicotyledonae
Order:	Icacinales
Family:	Icacinaceae
Genus:	*Nothapodytes*
Species:	*Nothapodytes foetida*

Common names: English: Fetid tree, Stinking tree**; Hindi:** Ghenera; **Kannada**: Durvasane mara; **Malayalam:** Peenari; **Marathi:** Ghanera, Narkya pulippacha; **Tamil** Pillipiccu, Arali.

Distribution: Found in the Western Ghats and in moist, deciduous forests distributed in Maharashtra, Goa, Kerala, Karnataka, Assam and Tamil Nadu in India.

Description: Trees up to 10 m tall. Leaves alternate, ovate-oblong or elliptic-oblong, up to 27 × 10 cm, acute or rounded and unequal at base, acute or acuminate at apex, pubescent beneath; petioles up to 6 cm long. Flowers in terminal, corymbose cymes, yellowish white, fetid; peduncle pubescent. Calyx 5-toothed. Petals 5, pubescent within. Stamens 5, free; filaments flattened and dilated at base. Disk cup-shaped, surrounding the ovary. Ovary silky-tomentose. Drupes ellipsoid or slightly obovoid, about 0.6 cm, blackish purple when ripe. Seed solitary, black.

Flowering: June–September.

Fruiting: June–September.

Chemical constituents: This plant is the major source of a potent alkaloid, namely camptothecin (CPT) and 9-methoxy camptothecin. It also contains 3-ketooctadec-cis-15-enoic, palmitic, stearic, oleic, linoleic and linolenic acids. Other chemical constituents isolated from this plant are acetylcamptothecin, (+)-1- hydroxypinoresinol, Ω-hydroxypropioguaiacone, p-hydroxybenzaldehyde, scopoletin, uracil, thymine, sitosterol and sitosteryl-β-D-glucoside. It also contains palmitic, stearic, oleic and linolenic acids Sheeja et al., 2005.

Used in: Tradational medicine.

Parts used: Seeds, leaves and stem.

Uses: This plant is used in various types of cancer, HIV and malaria. It has antibacterial, anti-inflammatory and antifungal properties, and is an immune modulator. The leaves have anti-inflammatory and antibacterial properties. The stem shows antibacterial activity.

Methods of cultivation: Through seeds. Germination is very poor; therefore the seeds must be treated with acid to improve germination.

Photo: Prof Srinath Rao.

NYCTANTHES ARBOR-TRISTIS (L.)

Synonyms: *Nyctanthes dentate* Blume; *Nyctanthes tristis* Salisb.; *Nyctanthes aculeate* Craib; *Parillium arbor-tristi* (L.) Gaertn.; *Scabrita triflora* L.

Classification:

Kingdom:	Plantae
Subkingdom:	Phanerogamea
Class:	Dicotyledons
Subclass:	Gamopetalae
Series:	Bicarpellatae
Order:	Gentianales
Family:	Oleaceae
Genus:	*Nyctanthes*
Species:	*Nyctanthes arbor-tristis*

Common names: Bengali: Harsingar, Shephalika, Shivuli; **English:** Night jasmine, Coral jasmine; **Gujarati:** Jayaparvati, Harsingar; **Hindi:** Harsingar; **Kannada:** Parijatha; **Malayalam:** Parijatakam; **Marathi:** Parijath; **Sanskrit:** Parijata, Shephali, Shephalika; **Tamil:** Parijata, Paghala; **Telugu:** Parijata.

Distribution: Parijata is native to India. It is also found in Bangladesh and in the Indo-Pak subcontinent as well as in tropical and sub-tropical Southeast Asia.

Description: Small tree growing to about 20–25 feet tall. Leaves simple opposite, with small hairs on the underside. Flowers fragrant, with 5–8-lobed white corolla with an orange red center produced in a cluster of 3–7, open after sunset and fall before sunrise. The fruit is a flat brown (green when young).

Flowering: August–November.

Fruiting: October–December.

Chemical constituents: The leaves and seeds contain iridoid glycosides. The other constituents reported from the leaves are mannitol, β-amyrin, β-sitosterol, hentriacontane, benzoic acid, astragalin, nicotiflorin, oleanolic acid, nyctanthic acid, friedelin and lupeol. The seeds contain polysaccharide glucomannan. Flowers contain nyctanthin.

Used in: Ayurveda.

Parts used: Leaves, flowers and stem bark.

Uses: In Ayurveda, this plant is used for the treatment of many ailments. For medicinal purposes, the leaves, stem, flowers and seeds are used. The leaves are given for treating chronic fever, rheumatism, arthritis, joint pain and obstinate sciatica. For chronic fever, juice made from the fresh leaves with honey is administered orally. The essential oil in the fragrant flowers, which is similar to the oil in jasmine, is used as perfume. The leaves possess anti-arthritic properties. A decoction made from the leaves is recommended to treat sciatica. In addition, it possesses liver-protecting, anti-viral, anti-fungal, analgesic, antipyretic, anti-inflammatory, antispasmodic, hypotensive and respiratory-stimulant activities. The leaves also have anti-leishmanial properties. The flowers, administered orally, tone the stomach, prevent gas formation, are astringent to the bowels, prevent excess bile secretion by the liver, cure cough. The powder of the stem bark is useful in the treatment of rheumatic joint pain and malaria. The stem bark is boiled with dry ginger power and pippali or long pepper and is taken orally to cure malaria.

A saffron-yellow dye obtained from the flowers is used for colouring food. It is an ingredient of some Ayurvedic formulations.

Methods of propagation: Through seeds and stem cuttings.

Photo: Prof Srinath Rao

OCIMUM BASILICUM L.

Synonyms: *O. caryophyllatum* Roxb.; *O. minimum* Linn.; *O. pilosum* Willd.

Classification:

Kingdom:	Plantae
Subkingdom:	Phanerogamea
Class:	Dicotyledonae
Subclass:	Gamopetalae
Series:	Bicarpellatae
Order:	Lamiales
Family:	Labiatae (Lamiaceae)
Genus:	*Ocimum*
Species:	*Ocimum basilicum*

Common names: English: Sweet basil; **Hindi:** Vantulsi; **Malayalam:** Tirunitra; **Sanskrit:** Kharapuspha, Ajgandhikaa

Distribution: Originally native to Tropical Asia, but long cultivated. Its original habitat is obscure.

Description: It is a perennial herb growing to 20–40 cm. Leaves simple, opposite, 6–8 cm long, ovate entire. The flowers are purple-tinged, about 1 cm long, borne in simple terminal racemes.

Calyx 5–7 mm long, upper calyx lobe circular, with wings joined, front 2 teeth of calyx quite distinct. Corolla 7–9 mm long.

Flowering: August–September.

Fruiting: October.

Chemical constituents: Basil is an excellent source of minerals such as calcium, copper, iron, magnesium, manganese and potassium. It is also rich in β-carotene, β-caryophyllene and vitamin A. Basil leaf contains anti-aging properties. Several aroma compounds can be found in different chemotypes of basil, including citral, eugenol, linalool, methyl chavicol, and methyl cinnamate.

Used in: Ayurveda, Siddha and Unani medicine.

Parts used: Leaves, seeds, root.

Uses: Sweet basil has been used for thousands of years as a culinary and medicinal herb. It acts principally on the digestive system, easing flatulence, stomach cramps, colic and indigestion and also nervous system. The leaves and flowering tops are antispasmodic, aromatic, carminative, digestive, galactagogue, stomachic and tonic. They are taken internally in the treatment of abdominal cramps, colds, depression, exhaustion, gastro-enteritis, influenza, insomnia, migraine, nausea and poor digestion. Externally, they are used to treat acne, loss of smell, insect stings, snake bites and skin infections. The mucilaginous seed is given in infusion in the treatment of chronic diarrhea, dysentery and gonorrhea. It is said to remove film and opacity from the eyes. The root is used in the treatment of bowel complaints in children. Extracts from the plant are bactericidal and are also effective against internal parasites. The essential oil is used in aromatherapy (Tilebeni 2011).

Methods of propagation: Through seeds. The seeds are sown in mid- to late spring in a greenhouse. Germination is quick. The seedlings are then picked and transferred into a field when they are large enough to handle.

Photo: Prof Srinath Rao

OCIMUM SANCTUM L.

Synonyms: *Ocimum tenuiflorum* L.; *Ocimum scutellariodes* Willd. ex.

Classification:

Kingdom:	Plantae
Subkingdom:	Phanerogamea
Class:	Dicotyledonae
Subclass:	Gamopetalae
Series:	Bicarpellatae
Order:	Lamiales
Family:	Labiatae (Lamiaceae)
Genus:	*Ocimum*
Species:	*Ocimum sanctum*

Common names: English: Holy basil; **Hindi:** Tulsi; **Kannada:** Vishnu tulsi, Karitulasi; **Malayalam:** Trittavu, Krsinatulsi; **Sanskrit:** Surash, Tulsi, Manjari; **Tamil:** Karuttulaci; **Telugu:** Tulasi.

Distribution: This plant is grown in houses, gardens and temples all over India.

Description: This is a well-known sacred plant of the Indians. It is a much-branched erect herb, up to about 75 cm high, hairy all over. Leaves opposite, about 5 cm long, margins entire or toothed; hairy on upper as well as lower surface, dotted with minute glands, aromatic. Flowers small, purplish or reddish, in small compact clusters on slender spikes. Fruits small; seeds yellowish or reddish.

Flowering: Throught the year.

Fruiting: Throught the year.

Chemical constituents: The leaves yield a volatile oil. Eugenol and caryophyllene are the principal components of the volatile oil. Other constituents are triterenes flavonoids and fatty acid esters.

Used in: Ayurveda, Siddha, Tibetan, folk and modern medicine; naturopathy and homeopathy.

Parts used: The whole plant.

Uses: The leaves and seeds of the plant are medicinal. The oil obtained from the leaves has the property of destroying bacteria and insects. The juice or infusion of the leaves is useful in bronchitis, catarrh and digestive complaints; it is applied locally on ringworm and other skin diseases; it is dropped in the ears to relieve earache. A decoction of leaves is used in Indian homes to cure common colds. Seeds are useful in complaints of the urinary system. Decoction of root is given in malarial fever to bring about sweating. Tulsi is known to possess adapogenic, analgesic, anti-stress, antihypertensive, anti-inflammatory antiulcerogenic, antispasmodic and antitumor properties. It is CNS-depressant, hypoglycemic, immunomodulatory and radioprotective. The plant is rather an essential article in the worship of Hindu gods and goddesses.

Methods of propagation: Through seeds.

Photo: Prof Srinath Rao

OROXYLOM INDICUM L.

Synonyms: *Bignonia indica* L.; *B. pentandra* Lour.; *Spathodea indica* (L.) Pers.

Classification:

Kingdom:	Plantae
Subkingdom:	Phanerogamea
Class:	Dicotyledonae
Subclass:	Gamopetalae
Series:	Bicarpellatae
Order:	Personales
Family:	Bignoniaceae
Genus:	*Oroxylom*
Species:	*Oroxylom indicum*

Common names: Bengali: Nasona; **English:** Indian trumpet flower; **Hindi:** Assarsauna; **Kannada:** Tattuna, Aanemungu; **Malayalam:** Aralu; **Marathi:** Tetu; **Sanskrit:** Shyonaka; **Tamil:** Arulandai; **Telugu:** Mundukaparnamu.

Distribution: Found in the Himalayan foothills, extending to Bhutan

and Southern China, and also in Sri Lanka.

Description: Small trees, up to 12 m tall. Bark brownish grey, soft, often covered by many corky lenticels. Leaves compound, 2–3-pinnate, with opposite pinnae, very large, 90–150 cm long, almost triangular in outline; primary rachis stout, cylindric; secondary and tertiary rachis striate; petiole 5.5 cm long, striate; leaflets 2–4 pairs with terminal odd one, opposite, lamina 7–17 × 3.5–10 cm, ovate or elliptic, apex caudate-acuminate, base rounded or asymmetric, margin entire, chartaceous, glabrous; midrib flat above; 1–2 pairs of secondary nerves arising from base. The large leaf stalks wither and fall off the tree and collect near the base of the trunk, appearing to look like a pile of broken limb bones. Flowers in terminal racemes, large, reddish purple and pinkish yellow within. The tree is a night bloomer and flowers are adapted to natural pollination by bats. They form enormous seed pods that hang down from bare branches. The long fruits curve downward and resemble the wings of a large bird, dangling sickles or swords in the night. Capsule 30–90 cm long, flat, purple brown; seeds many, flattened with broad hyaline papery wings.

Flowering: July–December.

Fruiting: July–December.

Chemical constituents: Isolation of ellagic acid from root wood (Rastogi and Mehrotra, 1990–1994). Four flavonoids – chrysin, baicalein, baicalein-7-O-glucoside and baicalein-7-O-diglucoside (oroxylin B), and one unknown flavonoid have been isolated and purified from the seeds (Chen et al., 2003).

Used in: Ayurvedic system of medicine.

Parts used: Root and stem bark, fruit and seeds.

Uses: The root extract is an antidote to snake bite. The root bark and stem bark possess antiallergic properties and are used in treating allergic disease, urticaria, jaundice, asthma, sore throat, laryngitis, hoarseness, gastralgia, diarrhea, dysentery, infantile erythema and measles. They are also used in dysentery, dyspepsia and epilepsy. The stem bark is used to cure jaundice, oedema and scabies. The seeds are purgative and also used to cure piles. The fruits are acrid, sweet, stomachic, anthelmintic, and good against diseases of the heart and throat, piles and bronchitis.

The plant is also used as an expectorant, improves appetite, and useful against leucoderma.

Methods of propagation: Through seeds.

Photo: Dr Kotresh

PANDANUS FASCICULARIS LAMK.

Synonyms: *Pandanus tectorius* Soland ex Parkinson; *P. odoratissimus* L. F.

Classification:

Kingdom:	Plantae
Subkingdom:	Phanerogamea
Class:	Monocotyledons
Order:	Nudiflorae
Family:	Pandanaceae
Genus:	*Pandanus*
Species:	*P. fascicularis*

Common names: Bengali: Ketaki; **English:** Screw pine; **Gujarati:** Ketaky; **Kannada:** Kedage; **Malayalum:** Kaitha; **Marathi:** Kevada; **Sanskrit:** Ketaki; **Tamil:** Thazhambu; **Telugu:** Mogali.

Distribution: Southern Asia, from Southern India to Taiwan, and from the south of Japan to Indonesia. It grows wildly in the coastal regions of India and the Andaman Islands.

Description: This plant is a branched palm-like shrub with its stem supported by aerial roots. Leaves: glaucous green, 0.9–1.5 m, ensiform, long, lanceolate, acuminate with 3 rows of prickles each on the margins and on midrib beneath. Flowers: dioecious with the female plant producing a non-scented fruit, while the male produces a flower rich in volatiles.

Flowering: April–May.

Fruiting: April–May.

Chemical constitution: Physcion, cirsilineol, n-triacontanol, β-sitosterol, stigmasterol, campesterol, daucosterol and palmitic and stearic acids have been isolated from the rhizomes (Rastogi and Mehrotra, 1985–1989). β-sitosterol, β-sitostenone, stigmasterol, stigmast-4-en and 6-dione were isolated (Rastogi and Mehrotra, 1990–1994a). A new alkaloid – pandamarine – has been isolated from the leaves (Rastogi and Mehrotra, 1990–1994b). Isolation of 3 new piperidine alkaloids – pandamarilactone, pandamarilactone 31 and pandamarilactone 32 – from the leaves (Rastogi and Mehrotra, 1990–1994).

Used in: Ayurveda and traditional medicine.

Parts used: The whole plant.

Uses: The flower extract called 'keorra-ka-arak' is stimulant and antispasmodic. It is are used against headache and rheumatism. The roots are used in the treatment of skin diseases, leprosy, headache, osteoarthritis, otalgia, wounds, ulcers, dyspepsia, colic, fever, diabetes, sterility and general debility. They possess anodyne, antiseptic, aphrodisiac, carminative, deodorant, depurative, emollient, stomachic, sudorific, suppurative, tonic and vulnerary properties. The flowers are also used to flavour food; the essence is used in numerous Indian sweets such as rasgulla, petha (candied gourd or winter squash), ras malai and burfi, especially in Rajasthan. The flower, however, is reputed to attract snakes. The roots act as an antidote against snake bite.

Methods of propagation: Through seeds.

PEGANUM HARMALA L.

Synonyms: *Harmala multifida* All.; *Harmala peganum* Crantz.; *Harmala syriaca* Bubani.

Classification:

Kingdom:	Plantae
Subkingdom:	Phanerogamea
Class:	Dicotyledonae
Subclass:	Polypetalae
Series:	Disciflorae
Order:	Geraniales
Family:	Zygophyllaceae (Nitrariaceae)
Genus:	*Peganum*
Species:	*Peganum harmala*

Common names: English: African rue, Harmal; **Hindi:** Harmal; **Kannada:** Seeme goranti; **Marathi:** Harmala; **Sanskrit:** Harmala, Soma; **Tamil:** Cimai alavanam; **Telgu:** Seemagoranta.

Distribution: This species is globally distributed from Africa through the Middle East, and from Russia to China. It has been introduced in America and Australia. Within India, it has been recorded in Kashmir, Uttar Pradesh, Punjab, Delhi, Bihar, Rajasthan and the Deccan.

Description: It is a bright-green, densely foliaged, perennial, fleshy, succulent shrub which can grow to about 0.3–0.8 m tall. Many-branched stems may spread 4 feet or more. It is almost round and bushy in habit. The roots of the plant can reach a depth of up to 6.1 m, if the soil where it is growing is very dry. Leaves: fleshy, linear, 2 in long, born singly and finely divided into long narrow segments. The flowers are white, about 2.5–3.8 cm in diameter. Sepals: 5 narrow of slightly longer length. Petals: 5, oblong elliptic. Fruit: a capsule, about 1–1.5 cm in diameter, with 3 chambers, carrying more than 50 dark-brown and angular seeds.

Flowering: June–August.

Fruiting: September–October.

Chemical constitution: Harmaline, harmine, harmalol, harman, quinazoline derivatives, vasicine, vasicinone, anthraquinones and fixed oils are reported from the seeds and roots of this plant (Asgarpanah and Ramezanloo, 2012). Harmine and harmaline are reversible inhibitors of RIMA (Massoro, 2002).

Used in: Ayurveda, folk, Unani and Siddha medicine.

Parts used: Leaves.

Uses: *P. harmala* also has antibacterial, antifungal, antiviral, antioxidant, anti-diabetic, antitumor, antileishmanial, insecticidal and cytotoxic activities as well as hepatoprotective and antinociceptive effects. *P. harmala* is a central nervous system stimulant and a reversible inhibitor of MAO-A(RIMA) (Massaro, 2002), a category of anti-depressant. The plant is used as an abortifacient (www.thenook.org).

Photo: http://search.yahoo.com/search?ei=UTF-8&fr=crmas&p=peganum harmala plants

PHYLLANTHUS AMARUS SCHUM. & THONN.

Synonyms: *P. fraternus; P. niruri* L.; *P. scabrellus; P. nanus.*

Classification

Kingdom:	Plantae
Subkingdom:	Phanerogamea
Class:	Dicotyledonae
Subclass:	Monochlamydae
Order:	Unisexuales
Family:	Euphorbiaceae
Genus:	*Phyllanthus*
Species:	*Phyllanthus amarus*

Common names: English: Carry me seed; **Hindi:** Bhiu amala, Jar amala, Hazardana; **Kannada:** Neal nelli; **Malayalam:** Kilanelli; **Marathi:** Bhui avali; **Sanskrit:** Bahupatra; **Tamil:** Keelanelli; **Telugu:** Nela usari.

Distribution: This herb occurs as a weed throughout.

Description: This is a small annual herb, usually 30 cm tall. Leaves: small, ovate/oblong/ elliptic, about 6–8 cm long and distichously arranged on branches; when the plants are picked the feathery leaves fold in completely, closing themselves. Flowers: small, unisexual, male flowers towards the tops and female flowers towards the bases of the branches, hidden below the leaves. Fruit: a 6-lobed capsule about 2 mm. They burst open and the seeds are hurled away.

Flowering: July–October.

Fruiting: July–October.

Chemical constituents: The entire plant forms the drug. It contains lignans –phyllanthin and hypophyllanthin – as major constituents, and also alkaloids such as ent-norsecurinine, sobubbialine and epibubbiline. From the roots, glycoflavones, amarin, geraniin, rutin and quercetin are reported.

Used in: Ayurvedic and Unani medicine.

Parts used: The entire plant.

Uses: This plant is an important plant of Indian Ayurvedic medicine. The herb is known for its hepatoprotective action, which is attributed to the lignans. The drug exhibits antiviral action on hepatitis B. It is also used as hypoglycemic, diuretic, hypotensive, astringent, cooling, febrifuge and antiseptic. The whole plant is used in gonorrhea, menorrhagia and other genital affections. It is useful in gastropathy, diarrhea, dysentery, intermittent fevers, ophthalmopathy, scabies, ulcers and wounds (Patel et al., 2011).

Methods of propagation: This herb is propagated through seeds. It prefers calcareous well-drained, light-textured soils.

- Seeds are sown in well-prepared nursery beds. Sowing in April–May will help obtain a higher germination rate.
- Farmyard manure is applied to the topsoil while preparing the beds.
- The minute seeds are mixed with a sufficient quantity of fine sand to ensure uniform distribution on the nursery bed. A thin layer of soil is placed on the seeds in the nursery beds.
- The beds must be kept sufficiently moist till the seeds germinate.
- Transplant 30–40-day-old seedlings at a distance of half a foot between them.

Photo: Prof Srinath Rao

PHYSALIS MINIMA L.

Synonyms: *Physalis angulata var villosa Botani, Physalis parviflora R.Br. Physalis divaricata* D. Don; *Physalis micrantha* Link; *Physalis lagascae* Roem. & Schult.

Classification:

Kingdom:	Plantae
Subkingdom:	Phanerogamea
Class:	Dicotyledonae
Subclass:	Gamopetalae
Series:	Bicarpellateae
Order:	Pleomoniales
Family:	Solanaceae
Genus:	*Physalis*
Species:	*Physalis minima*

Common name: Bengali: Bantepariya; **English:** Ground cherry; **Gujarati:** Popti; **Hindi:** Rasbhari; **Kannada:** Gaddehannu; **Malayalam:** Notinotta; **Marathi:** Chirboti; **Sanskrit:** Chirapotikaa; **Tamil:** Kupanti; **Telugu:** Kupanti.

Distribution: Afghanistan, Baluchistan, tropical Africa, Australia, Ceylon and throughout India. Commonly found on the bunds of the fields, wastelands, around the houses, on road sides, etc., where the soil is porous and rich in organic matter.

Description: This is a perennial/annual herbaceous plant about 3 feet in height, pubescent. The stem is branched and often reclining. Leaves: alternate, simple, lobed, ovate to cordate, pubescent, delicate, exstipulate, acuminate, with reticulate palmate venation and undulate margins; dorsal surface of the leaves dark green; ventral surface light green; 9–10 cm long and 7–8 cm broad. Flowers: pedicellate with 1.2 cm-long pedicel, hermaphrodite, complete, solitary, small, campanulate, 1.2–1.4 cm in diameter; calyx: gamosepalous, 5-toothed, actinomorphic, green, persistent, downy; corolla: gamopetalous with 5 petals, the petal cup 1.1–1.3 cm long, yellow, with 5 black spots on yellow ground in the middle of the corolla cup; stamens: 5, epipetalous, 6–7 mm long, with a black filament and greenish-yellow anther lobes; style: black, 9 mm long, with a yellowish stigma at the top and a yellowish round ovary at the base. Fruit: a berry, enclosed within the enlarged, 10-ribbed, reticulately veined calyx, which is 4.1 cm long and 2.5 cm broad; berries: stalked (stalk: 2.2 cm long), almost round with a pinhead-sized depression at the end; seeds: globose, Dresden yellow.

Flowering: March–April.

Fruiting: October–November.

Chemical constituents: Withanone, withaferin A, withanolide A, stigmasterol and sitosterol have been reported from leaves, stem and roots. Physalins A, B and X are contained in the aerial parts.

Used in: Ayurveda, Siddha and traditional medicine.

Parts used: The whole plant.

Uses: The plants of *Physalis minima* are bitter, appetizing, tonic, diuretic, laxative and useful in inflammations, enlargement of the spleen and abdominal troubles. The fruit is considered to be a tonic, diuretic and purgative. Antioxidant, anti-inflammatory and anti-cancer properties are also reported.

Methods of propagation: Through seeds.

Photo: Prof Srinath Rao

PIMPINELLA ANISUM L.

Synonyms: *Anisum odoratum*; *Anisum officinal*; *Anisum officinarum*; *Anisum vulgare.*

Classification:

Kingdom:	Plantae
Subkingdom:	Phanerogamea
Class:	Dicotyledonae
Subclass:	Polypetalae
Series:	Calyciflorae
Order:	Umbellales
Family:	Umbelliferae (Apiaceae)
Genus:	*Pimpinella*
Species:	*Pimpina anisum*

Common names: English: Sweet cumin; **Hindi:** Saonf; **Kannada:** Kaadu sompu; **Malayalam:** Perinichirakam; **Marathi:** Badisomp; **Sanskrit:** Karavee; **Tamil:** Shombu; **Telugu:** Gaddisopu.

Distribution: This species is native to the Eastern Mediterranean, and is cultivated in Europe, Russia, Africa, Mexico and South America. Within India, it is grown to a small extent. It is generally confused with *Foeniculum vulgare.*

Description: It is an annual, growing to 1–2 feet in height. The leaves at the base of the plant are simple, 1–5 cm long and shallowly lobed, kidney-shaped, while the leaves higher on the stems are feathery pinnate, divided into numerous leaves. The flowers are hermaphrodite (they have both male and female organs), white, approximately 3 mm in in diameter, produced in dense umbels. The fruit is an oblong dry schizocarp, 3–6 mm long, usually called 'aniseed'.

Flowering: July–August.

Fruiting: August–September.

Chemical constituents: The essential oil comprises 70–90 % anethole, which has an observed estrogenic effect, and the seed is also mildly estrogenic (http://www.pfaf.org). Other studies have demonstrated the presence of eugenol trans-anethole, methylchavicol, anisaldehyde, estragole, coumarins, scopoletin, umbelliferone, estriols, terpene hydrocarbons, polyenes and polyacetylenes as the major compounds of the essential oil from aniseed (Shojil and Fard, 2012).

Other compounds found are methyleugenol, α-cuparene, α-himachalene, β-bisabolene, p-anisaldehyde and cis-anethole (Ozcan and Chalchat, 2006) (Gulcin et al., 2003).

Used in: Ayurveda, Unani, Siddha and modern medicine.

Parts used: Fruits and dried seeds.

Uses: The seed is the part used, generally in the form of an extracted essential oil. It is analgesic, anti-inflammatory, aphrodisiac, carminative, antispasmodic, deobstruent, diuretic, expectorant, emmenagogue and galactagogue. It is also used to cure asthma, diarrhea, flatulence, diseases of heart and ear, headache, kidney stones and oedema. The herb is also a stimulant of sexual drive and of breast-milk production. Moreover, the seeds are employed in traditional medicine as an analgesic in migraine, and for their carminative, disinfectant and diuretic properties(Amin, 2005).

Methods of propagation: Through seeds.

The seeds are sown in early April. 4–5 seeds are sown per pot. They germinate within 3 weeks in a greenhouse; later they can be transferred to a field, with a spacing of 45–60 cm.

PIPER LONGUM L.

Synonym: *Chavica longa* (L.) Karst; *Chavica roxburghi*; *Piper latifolium* Hunter; *Piper roxburghi* (Miq.) J. Presl.

Classification:

Kingdom:	Plantae
Subkingdom:	Phanerogamea
Class:	Dicotyledonae
Subclass:	Monochlamidae
Series:	Microembryae
Family:	Piperaceae
Genus:	*Piper*
Species:	*Piper longum*

Common names: Bengali: Pipul; **English:** Indian long pepper; **Gujarati:** Pipari; **Hindi:** Pipli; **Kannada:** Hippaliballi; **Malayalam:** Pippali; **Marathi:** Pimpli; **Sanskrit:** Pippali, Ushana; **Tamil:** Pippali, Thippli; **Telugu:** Pippallu.

Distribution: It is a native of the Indo-Malayan region. It grows wild in the tropical rain forests of India, and from the Central Himalayas to Assam and the lower hills of Bengal. It is also cultivated in many states of India.

Description: Creeping and rambling aromatic herb, with perennial woody roots. Branchlets erect, glabrous with swollen nodes, roots clasping the nodes, which helps it get attached to the host tree. Leaves: simple, alternate, stipulate, long, elliptic-oblong, cordate, apex acute to acuminate, margin entire, glabrous; lower ones petioled; upper amplexicaul; inflorescence: solitary erect spike with unisexual, small, densely packed flowers, subtended by peltate bracts; male spikes long, slender, yellow; female spikes with short thick stalk, 1.5–2.5 cm in length, stigmas 3–4, lanceolate, spreading, papillose; fruits: berries, long, blackish green, sunk in the fleshy spikes.

Flowering: August–December.

Fruiting: August–December.

Chemical constituents: The major chemical constituents present in piper species are alkaloids, amides, lignans, esters and volatile oils. The fruit contains a large number of alkaloids and related compounds, the most abundant of which is piperine, together with methyl piperine, pipernonaline, piperettine, piperlongumine, refractamide-A, cepharadione-A and B and cepharanone B. The lignans sesamin, pulviatol and fargesin have been isolated from fruits. The essential oil of the fruit is a complex mixture; the 3 major constituents are caryophyllene, pentadecene and bisaboline. Other compounds present are asarinine, fargosine, guineensine, pluriatilol, pipercide and sesamine (Atal et al., 1977).

Used in: Traditional, Ayurvedic and Unani medicine.

Parts used: Fruits and roots.

Uses: The seeds form the drug, which is used to cure cough, dyspnea, ascites, leprosy, diabetes, piles, indigestion, anaemia, thirst, cardiac and spleen disorders. It is carminative, sedative, emollient, demulcent and hematinic. It is also used as an antidote to snake bite and scorpion sting. The root and fruit decoction is used in acute and chronic bronchitis and cough. Root and fruit are used in gout and lumbago.

Methods of propagation: Long pepper can be cultivated successfully in organic-matter-rich fertile, well-drained forest soils. Optimum elevation for its cultivation is 100–1000 m. It is propagated by suckers or rooted vine cuttings. Vine cuttings and suckers are transplanted soon after the setting in of monsoon rains. The best time for raising nursery is during March and April to avoid mealybug attacks on roots. 10 % DP is to be mixed with the potting mixture. Normal irrigation may be given on alternate days. Excess moisture in the nursery can cause diseases. By the end of May, the cuttings will be ready for planting. Long pepper needs heavy manuring. In soils with low fertility, the growth of the plant is very poor. 20 tonnes of cow dung or farmyard manure is required for 1 ha. of land.

Photo: Prof Srinath Rao

PIPER NIGRUM L.

Synonym: *Piper trioicum*; *Piper baccatum*; *Muldera multinervis.*

Classification:

Kingdom:	Plantae
Subkingdom:	Phanerogamea
Class:	Dicotyledonae
Subclass:	Monochlamidae
Series:	Microembryae
Family:	Piperaceae
Genus:	*Piper*
Species:	*Piper nigrum*

Common names: English: Black pepper; **Hindi:** Kalimirch; **Kannada:** Kari menasu; **Malayalam:** Kurumulaku; **Tamil:** Milagu; **Telugu:** Miriyalu.

Distribution: Southwest India and tropical countries. Probably originated in the hills of Southwest India. Also cultivated in warm and moist regions of the country.

Description: A branching, climbing perennial shrub. Branches stout, trailing and rooting at the nodes; leaves entire, 12.5–17.5 × 5.0–12.5 cm, very variable in breadth, sometimes glaucous beneath, base acute, rounded or cordate, equal or unequal; flowers minute, in spikes, usually dioecious, but often the female bears 2 anthers, and the male a pistillode; fruiting spikes very variable in length and robustness, rachis glabrous; fruits: ovoid or globose, bright red when ripe; seeds usually globose, testa thin, albumin had. **Chemical constituents:** Beta-caryophyllene, limonene, β-pinene, -3-carene, sabinene, α-pinene, eugenol, terpinen-4-ol, hedycaryol, β-eudesmol and caryopyllene oxide.

Flowering: March–August.

Fruiting: March–August.

Parts used: Dried unripe fruits.

Used in: Ayurvedic, Unani and Siddha medicine.

Uses: Due to its multidimensional effect on various systems of the body, this plant has been described as antipyretic, diuretic, aphrodisiac, immune-stimulant, antioxidant, hepatoprotective, digestive, rubefacient, counter-irritant, antiseptic and antispasmodic. Moreover, it is known to enhance the bioavailability of food and drugs. Pepper causes sneezing. Some sources say that piperine, a substance present in black pepper, irritates the nostrils, causing the sneezing Piperine present acts as a thermogenic compound. It enhances the thermogenesis of lipids and accelerates energy metabolism in the body; furthermore, it increases the serotonin and β-endorphin production in the brain. Piperine and other components of black pepper may also be helpful in treating vitiligo (Stanely, 2001; Agbore et al., 2006; Meghwal and Goswami, 2012).

Methods of propagation: Through seeds.

Photo: Prof Srinath Rao

PLANTAGO OVATA FORSSK.

Synonyms: *Plantago brunnea* Morris; *Plantago decumbens* Forsk.; *Plantago fastigiata* Morris.

Classification:

Kingdom:	Plantae
Subkingdom:	Phanerogamea
Class:	Dicotyledons
Subclass:	Gamopetalae
Series:	Bicarpellatae
Order:	Polemoniales
Family:	Plantaginaceae
Genus:	*Plantago*
Species:	*Plantago ovata*

Common names: The local names of this plant in different regions and languages of India are only minor variations of the Sanskrit word *ishapgola* or *isabgol*.

Distribution: The plant grows naturally only in restricted areas in Northwestern India, but it is largely cultivated elsewhere.

Description: An almost stemless small herb, covered with dense or soft hairy growth. Leaves 8–25 cm long, very narrow, toothed, 3-nerved. Flowers: minute, borne on ovoid or cylindrical spikes. Fruit: a capsule 8 mm long; its upper half separates like a lid. Seeds: mucilaginous, boat-shaped, with a reddish-brown elongated spot on a convex surface.

Flowering: February–March.

Fruiting: April–May.

Chemical constituents: Seeds contain mucilage (20–30 %), fatty oil and proteins. The mucilage consists of weakly acidic arabinoxylans, xylose, rhamnose and galacturonic acid.

Used in: Ayurveda and Siddha medicine; homeopathy.

Parts used: Seed husk.

Uses: The seeds of this plant (as also of certain other species of the genus) constitute the drug. The useful properties of the seeds are due chiefly to the large amount of mucilage and albuminous matter present in them. *Isabgol* is very useful in several kinds of chronic dysentery, such as of amoebic, bacillary-origin and chronic diarrheas. It is useful as a soothing agent for mucous membranes and in constipation. The seeds should be soaked in water before use, so that they soon get disintegrated in the alimentary canal; otherwise, the whole seeds might cause irritation or even mechanical obstruction in the intestines. The large amount of mucilage in the seeds 'binds' and increases the mass of the stool, and smoothens its passing out. It is also found to be hypocholesterolemic.

Isabgol husk (Hindi: *ishabgul-ki-bhusi)* is the dry seed coat of *Plantago ovata*, obtained by crushing the seeds and separating the husk by winnowing. The husk has the same properties as the seeds; it also has the advantage that there is no risk of mechanical obstruction or irritation in the alimentary canal by the husk. The husk, therefore, is taken without any pre-soaking and is easier to use than the whole seeds.

The embryo oil of the seeds, having 50 % linoleic acid, prevents arteriosclerosis.

Methods of propagation: Through seeds and vegetative method.

Photo: Prof Srinath Rao

PLUMBAGO ROSEA L.

Synonym: *Plumbago indica.*

Classification:

Kingdom:	Plantae
Subkingdom:	Phanerogamea
Class:	Dicotyledons
Subclass:	Gamopetalae
Series:	Heteromerae
Order:	Primulales
Family:	Plumbaginaceae
Genus:	*Plumbago*
Species:	*P. rosea*

Common names: English: Rosy-flowered leadwort, Fire plant; **Hindi:** Lalcitra, Raktacitra; **Kannada:** Kempucitrmula; **Malayalam:** Kotuveli, Chettikkotuveli; **Sanskrit:** Citrkah, Dahanah; **Tamil:** Cenkotiveli, Cittiramulam; **Telugu:** Yerracitramulam.

Distribution: This is a native of Sikkim, cultivated throughout for its medicinal properties.

Description: It is an erect, branched shrub with long tuberous roots. Leaves: ovate-oblong, petiole short. Flowers: rose red, borne in a long terminal spike, calyx glandular, hairy outside. Fruit: a capsule, enclosed in the persistent calyx.

Flowering: December–January.

Fruiting: January–February.

Chemical constituents: The roots contain a naphthoquinone – plumagin – as well as sitosterol and tannins.

Used in: Ayurveda, Siddha and traditional medicine; homeopathy.

Parts used: Roots and seeds.

Uses: The roots are highly valued for their curative action on leucoderma and other skin diseases. They are digestive, stimulant, diuretic, germicidal, vesicant and abortifacient. They are also used for hemorrhoidal inflammation of the anus, diabetes, diarrhea and elephantiasis.

Methods of propagation:

This herb can be propagated either through seeds or stem cuttings. It grows well in intense sunlight and partial shade.

- The freshly collected seeds are soaked in water for 24 hours.
- Seeds are dibbled in polythene bags filled with the rooting medium prepared by mixing equal quantities of sand, red earth and manure.
- About 70 % of seeds germinate within 7–12 days.
- The plant is also propagated through stem cuttings or semi-hardwood cuttings. 4–6in-long cuttings with 3–4 nodes are planted closely on raised beds.

PODOPHYLLUM HEXANDRUM ROYLE

Synonyms: *Podophyllum emodi* Wall. ex Hook. f. & Thomas; *Sinopodophyllum emodi.*

Classification:

Kingdom:	Plantae
Subkingdom:	Phanerogamea
Class:	Dicotyledonae
Subclass:	Polypetalae
Series:	Thalamiflorae
Order:	Ranales
Family:	Berberidaceae
Genus:	*Podophyllum*
Species:	*Podophyllum hexandrum*

Common names: English: Himalayan may apple; **Kannada:** Balukambi, Beelkambi, Belkambi; **Sanskrit:** Laghupatra, Giripatra, Vanavrintaka

Distribution: Himalayan zones at an altitude of 2000–4000 m above mean sea level, in valleys of flowers and in a few pockets in Ladakh.

Description: It is a herbaceous rhizomatous perennial herb, low to the ground, with a few stiff branches. Leaves: lobed, shortly rhizomatous, 15–40 cm. Stems erect, unbranched, glabrous. Leaves usually 2, first deflexed, brownish, petioles up to 20 cm, orbicular-reniform in outline, 6–15 × 3.5–10 cm; palmately divided into 3 ovate lobes, sometimes further lobed; acute, base cuneate, margins serrate, pubescent beneath. Flower solitary, bisexual, erect, white or pale pink, cup-shaped, 2–4 cm across; sepals 3, soon falling; petals 6, stamens 6. Fruit: a large scarlet or reddish berry, ovoid, 2.5–5 cm, with many seeds embedded in pulp.

Chemical constituents: Podophyllin. The rhizome of the plant contains a resin, known generally and commercially as, Indian podophyllum resin, which can be processed to extract podophyllotoxin, or podophyllin, a neurotoxin.

Used in: Traditional and Ayurveda medicine; homeopathy.

Parts used: Rhizomes.

Uses: Antirheumatic, anti-cancer, cholagogue, cytostatic and purgative. The whole plant, but especially the root, is cholagogue, cytostatic and purgative. The plant contains podophyllin, which has an antimicotic effect (it interferes with cell division and can thus prevent the growth of cells). It is, therefore, a possible treatment for cancer, and has been used especially in the treatment of ovarian cancer. However, alopecia is said to be a common side effect of this treatment. This species contains about twice the quantity of active ingredients as compared to *P. peltatum*. The roots contain several important anti-cancer lignans, including podophyllin and berberine. The roots are also antirheumatic. The root is harvested in the autumn and dried for later use; otherwise, the resin is extracted. This plant is highly poisonous and should only be used under the supervision of a qualified practitioner. It should not be prescribed for pregnant women.

Methods of propagation: Propagated through seeds or rhizomes.

PORTULACA OLERACEA L.

Synonyms: *Portulaca consanguinea* Schltdl.; *Portulaca fosbergii* Poelln.; *Portulaca marginata* Kunth.

Classification:

Kingdom:	Plantae
Subkingdom:	Phanerogamea
Class:	Dicotyledons
Subclass:	Polypetalae
Series:	Thalamiflorae
Order:	Caryophyllinea
Family:	Portulacaceae
Genus:	*Portulaca*
Species:	*Portulaca oleracea*

Common names: Bengali: Nunia Sag; **English:** Garden purslane; **Gujarati:** Ghol, Motiloni; **Hindi:** Khursa, Badi-noni, Lunia; **Kannada:** Doodagooni Soopu, Dudagorai; **Malayalam:** Koluppa, Cheriyagolicheera, Kozhuppacheera, Kozhuppa; **Marathi:** Kurfah, Ghola; **Sanskrit:** Lonamala, Lonica, Brihalloni; **Tamil:** Vazhukkaikeerai, Saagaipoondu, Karpakantam; Telugu: Peddhapayilikura, Gangapavilikura.

Distribution: All over the world. It grows in sunny, warm areas, including flowerbeds, cornfields, road sides, gardens and waste places. It isalso cultivated as a vegetable.

Description: Herbaceous annual weed. Stem: almost cylindrical, swollen at the nodes, ribbed, branched, 0.1–0.2 cm in diameter. Root: cylindrical, small, oblique, surface smooth, brownish grey, secondary roots fewer in number, root hairs abundant in upper region, short. Stem: almost cylindrical, swollen at the nodes, ribbed, highly branched, 0.1–0.2 cm in diameter. Leaf simple, sub-sessile, cuneiform, smooth and greenish brown in colour, rounded and truncate at the apex, about 1–2 cm long and 0.3–0.6 cm wide, oblong, spathulate. Flowers: few, bright yellow, at terminal heads, sometimes in axillary clusters of 2–6, subtended by an involucre, 3–4 leaves; sepal 0.25–0.4 cm long; petals obovate, 0.5 cm long, very delicate and soon falling off; stamens 8–12 cm; style 5-6-fid, 0.35–0.4 cm long. Fruit: an ovoid capsule, 0.3 cm long, dehiscing above the base. Seeds: numerous, reniform, black, very minute, dark brown in colour.

Flowering: June–July.

Fruiting: August.

Chemical constituents: It contains two types of betalain alkaloid pigments: reddish betacyanins and yellow betaxanthins. Rich source of pre-vitamins A, C and E, as well as omega-3 fatty acids.

Used in: Ayurveda.

Parts used: The aerial parts.

Uses: The whole plant is used as an anti-diabetic, antioxidant, antiscorbutic, depurative, diuretic, febrifuge, hepatoprotective, refrigerant and spasmodic. It has liver-protective and regenerative properties and is used in the treatment of liver disorders. The juice made from the leaves and plant is particularly effective in the treatment of skin diseases and insect stings. The leaves are prepared as tea and are used in the treatment of stomachaches and headaches. A poultice of leaves is applied to burns and swellings. The seed powder is tonic and expels intestinal worms.

Methods of propagation: Through stem cuttings.

Photo: Prof Srinath Rao

PSORALEA CORYLIFOLIA L.

Synonym: *Cullen corylifolium* (L.) Medik.

Classification:

Kingdom:	Plantae
Subkingdom:	Phanerogamea
Class:	Dicotyledonae
Subclass:	Polypetalae
Series:	Calyciflorae
Order:	Rosales (Fabales)
Family:	Leguminosae (Fabaceae)
Genus:	*Psoralea*
Species:	*Psoralea corylifolia*

Common names: Bengali: Habuch; **English:** Psoralea seeds; **Hindi:** Bavachi; **Sanskrit:** Bakuchi; **Tamil:** Karpokarishi; **Telugu:** Bavanchalu.

Distribution: Found growing wild in fields throughout India.

Description: It is an erect herb. Simple, 3.8 × 2.5–5.0 cm, broadly elliptic, inciso-dentate, rounded and mucronate at apex, white, hairy, nigropunctate, petioles hairy and gland dotted. In dense 10–30 flowered axillary racemes; corolla bluish purple, standard orbicular, clawed. Pod ovoid-oblong, closely pitted, mucronate, black. 1, smooth.

Flowering: August–December.

Fruiting: January–February.

Chemical constituents: Seeds yield furanocoumarins, psoralen and isopsoralen. Roots contain an essential oil, chromonochalcones, and isoflavonoids.

Used in: Ayurveda and traditional medicine.

Parts used: Seeds and fruits.

Uses: Seeds are used as stomachic, deobstruent, in leprosy and leucoderma. According to Ayurveda, root is useful against tooth cavities whereas leaves are good for diarrhea. Fruit is diuretic and causes biliousness. It is useful in the treatment of vomiting, piles, bronchitis, inflammation and anaemia. It improves hair growth and complexion. Seeds are refrigerant, alternative, laxative, antipyretic, anthelmintic, alexiteric and good for heart troubles. Seed oil is used externally in the treatment of elephantiasis. According to the Unani system of medicine, the seed is purgative, stomachic, anthelmintic, vulnerary, stimulant, aphrodisiac and cures blood-related troubles. It is applied externally in the treatment of skin-related troubles. Psoralen derivatives are taken to promote suntanning of the skin and are found to be spasmolytic and vasodilatory.

Methods of propagation: It is propagated through seeds. It prefers well-drained soils and intense light conditions.

Steps to follow:

- Seeds are collected from mature fruits in November–December.
- Seeds are dried and stored. May– June is the right season for sowing.
- Seeds are soaked in hot water for 12–24 hours for good germination.
- Sow these seeds on raised beds. Seeds can also be sown directly in a field.
- Seeds germinate within 7–12 days with a success rate of 65–70 %.

Photo: Dr. R.S. Modi

RAUVOLFIA SERPENTINA (L.) BENTHAM EX KURZ.

Synonym: *Rauvolfia obversa* (Miq.) Baill.; *Rauvolfia trifoliata* (Gaertn.) Baill.; *Ophioxylon album* Gaertn.; *Ophioxylon serpentinum* L.

Classification:

Kingdom:	Plantae
Subkingdom:	Phanerogamea
Class:	Dicotyledonae
Subclass:	Gamopetalae
Series:	Bicarpellatae
Order:	Gentianales
Family:	Apocynacae
Genus:	*Rauvolfia*
Species:	*Rauvolfia serpentina*

Common names: Bengali: Chandra; **Hindi:** Chotachand; **Malayalam:** Chuvannayilpuri; **Marathi:** Harkaya; **Sanskrit:** Sarpagandha, Chandrika; **Tamil:** Chavanda-avalpori; **Telgu:** Patala-gandhi.

Distribution: The plant is native to India and found in almost all parts up to an altitude of about 1000 m; it is more common in the sub-Himalayan regions and in the lower ranges of the Eastern and Western Ghats; it is also collected from several places in the plains of Bengal, Bihar, Uttar Pradesh, Maharashtra, etc. The plant is now being cultivated in several places.

Description: An erect glabrous shrub, 30–75 cm high; roots vertical, tapering, nearly 15 cm thick at the crown and long, with a serpent-like appearance, greenish yellow in colour, extremely bitter in taste. Leaves whorled; 8–20 cm long, elliptic, lanceolate or obovate, gradually tapering into a short petiole. Inflorescence cymose, terminal or axillary. Flowers about 1.5 cm long, petals white or pinkish, peduncle deep red, in small clusters, corolla tubular, 5-lobed. Stamens 5, epipetalous. Carpels 2, connate, style filiform, stigma bifid. Fruit is a drupe, small, round, dark purple or blackish when ripe. Seeds 1–2, ovoid and wrinkled.

Flowering: August–September.

Fruiting: September–November.

Chemical constituents: The roots yield alkaloids, reserpine, rescinnamine and deserpidine. The other constituents of the roots are phytosterol, fatty acids and sugars.

Used in: Folk, traditional, Ayurvedic, Siddha, Unani and Tibetan medicine.

Parts used: Roots and leaves.

Uses: The drug consists of dried roots with bark intact, preferably collected in autumn and from plants of about 3–4 years of age. It is believed that this plant has been known in Indian medicine for about 4000 years. A mention of the plant is found in Charak's work. The roots contain several alkaloids. The chief use of the drug is as a sedative and hypnotic and for reducing blood pressure. The drug is now largely used in insanity; therefore, it is not useful in acute cases; it is more suitable for mild anxiety cases or patients with chronic mental illness. It has a tranquillizing effect. It should not be given to people suffering from bronchitis, asthma or gastric ulcers. The roots of the plant are useful in diseases of bowels and fever. Other species of the genus, such as *R. tetraphylla* L. (syn.) and *R. canescens* L., are also useful in medicine.

Methods of propagation: It is propagated through seeds, stem and root cuttings. It grows on clayey loam or clayey soil with humus and good drainage. Seed germination is very poor.

- Seeds are collected in June–October. They exhibit a short viability of 6 months. Soak the seeds in 5 % common salt solution, discard the floating seeds and select those that sink. Now place these under a moist piece of cloth for 24 hours to soften their hard covering.
- Sow them into seedbeds to about 1 cm deep in lines about a foot apart with a finger's distance between them. The third week of May is the most favourable time for sowing.
- The seedbeds should be kept moist till the seeds germinate. Then carefully transfer the plants to the nursery beds.
- The seeds sprout in 3–8 weeks with 10–50 % germination. Keep in the nursery for 7–8 weeks.
- For vegetative propagation, take 6–9 in-long stem cuttings and plant in the nursery during June. Keep the bed moist until the cuttings root. These can later be transplanted to the main field.
- For propagation through root cuttings, plant thumb-long cuttings during June–July, exposing only 1 cm of the tip above the surface. Within 3 weeks, the roots give out new sprouts. The success rate is about 50 %.

Photo: Prof Srinath Rao

RICINUS COMMUNIS L.

Synonyms: *Ricinus africanus* Mill.; *Ricinus angulatus* Thunb.; *Ricinus amatus* Andr.; *Cataputia major* Ludw.; *Croton spinosa* L.

Classification:

Kingdom:	Plantae
Subkingdom:	Phanerogamea
Class:	Dicotyledonae
Subclass:	Monochlamidae
Series:	Unisexuales
Family:	Euphorbiaceae
Genus:	*Ricinus*
Species:	*Ricinus communis*

Common names: Bengali: Bherenda; **Gujarati:** Diveli; **English:** Castoroil seed; **Hindi:** Arandi; **Kannada:** Haralu, Audala geeda; **Malayalam:** Avanakka; **Marathi:** Arandi; **Sanskrit:** Eranda, Panchangula; **Tamil:** Amanakkam; **Telugu:** Pedda Amadam.

Sometimes the varieties of the plant have different local names, e.g. *bhat-rendi* for the variety with white seeds and *jogia-rendi* for the variety with pale seeds.

In early literature, the tree has been referred to by the names *chitravija, panchangula* and *vatari;* these refer to the shape or properties of the plant and its parts. *Chitravija* refers to the ornamentations on the seeds, and *panchangula* to the 5-nerved, palmate leaves; *vatari* means enemy of rheumatism (*vata*).

Distribution: The plant is largely cultivated on the borders of agricultural fields, gardens, etc., and is also found wild in fields, gardens, near habitations and in waste places.

Description: It is a tall shrub, sometimes becoming tree-like. Leaves: very large, broad, roundish in outline but partly divided into 7 or sometimes 9 lobes, margins toothed. Flowers large, in big terminal bunches. Fruit a prickly capsule, rather marked into 6 parts. Seeds large (9-17mm long) shiny, bean like with variable brownish mottling with a caruncle at the base. One form of this plant is almost perennial with woody stems and large red seeds. This is said to yield oil suitable only for illumination and lubrication. The other form, which is grown as an annual crop, has grey and brown mottled seeds. These yield oil suitable for medicinal uses. There is still another form with purple bronze leaves; it is grown only for ornamental purposes.

Flowering: May–June.

Fruiting: May–June.

Chemical constituents: The seeds contain oil up to 45% and proteins (16%). Enzymes like lipase, alkaloid recinine. Leaves contain recinine and glycosides of kaempferol and quercetin.

Used in: Ayurveda, folk and traditional medicine.

Parts used: Fruits, leaves and seeds.

Uses: The seeds of the plant are used in medicine. They are poisonous and even 2 or 3 seeds can be fatal. Roots and leaves are considered highly effective in rheumatism. They are used for dyspnea, hydrocele, flatulence, piles, leprosy and arthritis. Tender leaves are used in jaundice. The oil obtained from the seeds, called castor oil, is used as a purgative. This oil is best administered with milk or fruit juice. It is also used in ointments as a soothing agent and as an oil vehicle in eye drops.

Castor oil is used in making contraceptive jellies and creams. The tribes of Bastar rub the leaves of the plant on joints to relieve pain. They also pound young leaves and use them as purgative.

A gel prepared from castor oil is useful in dermatosis and protects from eczema and dermatitis.

Methods of propagation: Through seeds and vegetative method.

Photo: Prof. Srinath Rao

RUTA GRAVEOLENS L.

Synonym: No synonym found for this plant.

Classification:

Kingdom:	Plantae
Subkingdom:	Phanerogamea
Class:	Dicotyledonae
Subclass:	Polypetalae
Series:	Disciflorae
Order:	Geraniales
Family:	Rutaceae
Genus:	*Ruta*
Species:	*Ruta graveolens*

Common names: Bengali: Ermul; **English:** Garden rue; **Hindi:** Sadab; **Kannada:** Sadbu; **Malayalam:** Naaga; **Sanskrit:** Soma-lata; **Tamil:** Arvada; **Telugu:** Aruda.

Distribution: A native of the Mediterranean region, now cultivated widely. Found in India in the states of Karnataka, Gujarat, Maharashtra and Madhya Pradesh.

Description: This plant is a strongly scented herb or under shrub. Leaves are petioled, decompound; leaflets spathulate, oblong or linear-oblong, cuneate at base. Flowers in divaricately spreading corymbs, bisexual. Sepals triangular, acute. Petals oblong to obovate, abruptly clawed, margins more or less undulate or denticulate. Capsules obtuse or bluntish lobes; shortly pedicellate. Seeds angled.

Flowering: April.

Fruiting: October.

Chemical constituents: The herb contains a volatile oil, flavonoids and coumarins. The volatile oil (rue oil) is strongly odoriferous and pungent and contains methyl nonyl ketone (80–90 %), methyl heptyl ketone, α-pinene, limonene and cineole. The rutin content in the plant is about 2 %. The coumarins present are imperatorin, isoimperatorin, xanthotoxin, bergaptene and psoralen. The stem and leaves contain alkaloids such as skimmianine, graveolinine and kokusaginine.

Used in: Ayurveda, Unani, traditional and modern medicine; homeopathy.

Parts used: Leaves and arial parts.

Uses: Rue oil is anthelmintic, antispasmodic, antiepileptic, rubefacient and emmenagogue. The herb is used as a poultice in rheumatic joints. The coumarins present in the herb have antispasmolytic activity.

Methods of propagation: Through seeds and vegetative method.

Photo: Dr Kotresh

SCOPARIA DULCIS L.

Synonyms: *Scoparia grandiflora*; *Scoparia ternata.*

Classification:

Kingdom:	Plantae
Subkingdom:	Phanerogamea
Class:	Dicotyledonae
Subclass:	Gamopetalae
Series:	Bicarpellatae
Order:	Scrophulariales
Family:	Scrophulariaceae
Genus:	*Scoparia*
Species:	*Scoparia dulcis*

Common names: Bengali: Bondhonya; **English:** Sweet broomweed**; Hindi:** Mithi patti, Ghod tulsi; **Kannada:** Mruganmhi gida; **Malayalam:** Kallurukki; **Marathi:** Dulas; **Sanskrit:** Jastmadhu; **Tamil:** Sarakkotthini; **Telugu:** Mundari, Madhukam.

Distribution: Throughout India, found along the stream banks and fallows from plains to 400 m.

Description: Branched erect annual herb with wiry stem, about 30 cm tall, leaves simple, narrowly elliptical and stalkless, opposite or in whorls of 3, 3–4 cm in length and 1–1.5 cm in width, margins serrate. Leaf base cuneate, apex acute. Flowers in leaf axil, small, white and hairy. Sepals 4 or 5, imbricate, corolla rotate, 4-fid, throat bearded, lobes subequal, stamens 4, subequal, 4 greenish; ovary green; fruit a capsule, nearly round, septicidal; seeds truncate or 4-angled.

Flowering: Throughout the year.

Fruiting: Throughout the year.

Chemical constituents: The active principles in this plant are scoparinol, scoparic acid, scopadulcic acid, scopadulciol and scopadulin. Scoparic acids A and B, scopadulcic acids A and B, scopadulciol and scopadulin have been shown to contribute to the observed medicinal effects of the plant, which is rich in flavonoids and diterpenoids (Hayashi et al., 1990).

Used in: Traditional and tribal medicine.

Parts used: The whole plant.

Uses: Treatment of wounds, anemia, burns, headache and sickle-cell disease. This is a medicinal plant. It is used to treat diabetes, hypertension and hemorrhoids. It has antihyperglycemic, antimicrobial and antioxidant properties and protects the stomach from ulcers. Scoparinol, an isolate of the plant, was shown to have analgesic, diuretic, and anti-inflammatory properties, as well as to inhibit the replication of herpes simplex virus. It also possesses antitumor, neurotrophic, anti-viral, antimalarial, anti-cancer and anti-diabetic properties. The plant is traditionally used in the treatment of dysentery, earache, fever, gonorrhea, snake bites, jaundice, toothache and warts (Latha et al., 2005) (Li et al., 2004).

Methods of propagation: Through seeds. Seeds can be collected during the months of September–October and sown directly in the soil.

SOLANUM NIGRUM L.

Synonyms: *Solanum americanum* L.; *Solanum incertum* Dunal.

Classification:

Kingdom:	Plantae
Subkingdom:	Phanerogamea
Class:	Dicotyledonae
Subclass:	Gamopetalae
Series:	Bicarpellatae
Order:	Polemoniales
Family:	Solanaceae
Genus:	*Solanum*
Species:	*Solanum nigrum*

Common names: Bengali: ; **English:** Black nightshade; **Hindi:** Mokoi; **Malayalam:** Mulakuthakkali; **Marathi:** Laghukavali; **Sanskrit:** ; **Tamil:** Manatakkali; **Telugu:** Kasaka.

Distribution: It occurs as a weed throughout India.

Description: This is an erect annual herb, with much-branched, slender, glabrous stem. Leaves: ovate/lanceolate. Flowers: small, white in clusters of 3–5, arising from intermodal regions. Fruit: a berry, globose, yellow.

Flowering: March onwards.

Fruiting: Summer.

Chemical constituents: The whole plant and especially the fruits are used as a drug. All parts of the plant contain steroidal alkaloids, of which solanine and solasodine glycosides are the major components. Also present are uttroside A and B and uttronin A, as well as sapogenins such as tigogenin and diosgenin. The leaves yield glycosides of quercetin.

Used in: Ayurveda, Siddha, folk and traditional medicine.

Parts used: The whole plant.

Uses: The whole plant and especially the fruits are used as an expectorant, anti-inflammatory and diuretic. The plant extracts exhibit hepatoprotective and anti-tumour properties.

Methods of propagation:

This herb is easily propagated through seeds. It grows on all types of soils, but prefers loamy soils.

- Seeds are collected from ripe fruits dried under shade and stored in room conditions. Seeds do not lose their viability during storage at room temperature.
- Since the seeds are small and sometimes it is difficult to separate them from the fruits, the entire fruit may also be sown. However, care should be taken not to sow them deep.
- Fruits/seeds are sown in rows or in flat beds after mixing them with fine soil for better distribution.

Photo: Prof Srinath Rao

SOLANUM TORVUM

Synonym: *Solanum ficifolium* auctt. Non Jacquem. Roxb.

Classification:

Kingdom:	Plantae
Subkingdom:	Phanerogamea
Class:	Dicotyledonae
Subclass:	Gamopetalae
Series:	Bicarpellatae
Order:	Solanales
Family:	Solanaceae
Genus:	*Solanum*

Species: *Solanum torvum*

Common names: English: Turkey berry; **Hindi:** Bhankatiya; **Kannada:** Sundekkayi, Kadu badane kaye; **Malayalum:** Anachunda; **Marathi:** Marang; **Sanskrit:** Brihati; **Tamil:** Sundaikkai; **Telugu:** Kottuvastu.

Distribution: A plant of the moist tropics, where it is found at elevations up to 1600 m.

Description: Evergreen shrub growing up to 100–300 cm tall. The stems are armed with stout, straight or lightly curved reddish or pale yellow prickles. The leaves occur singly along the stems and are broadly ovate to elliptical in shape and 10–15 cm long, alternate, with prickles along the midvein, and range from unlobed to 7 broad, blunt lobes. Both surfaces are covered with very fine stellate hairs; the upper surface is darker than the lower; the leaves have finely hairy petioles, 1–5 cm long. The inflorescence is a dense, compact, branched, extra axillary panicle cyme, consisting of 50–100 flowers at the ends of the branches; flowers are small, pale white in colour, Black nightshade, bearing glandular or stellate hairs, pedicle about 1 cm long. Sepals 5 about 0.5 cm long and hairy. Corolla about 1 cm long, petals 5, white to cream, star-shaped, each about 1 cm long; stamens 5, elongated, yellow with a central stigma. Fruits: small yellow berries with 210–220 seeds. Seeds are ovate to broadly elliptic, or nearly circular, infrequently C-shaped in outline. The root system consists of a deep, strong, woody taproot with numerous woody laterals. The stems are branched above, densely covered with fine stellate hairs and are scattered with broad-based, hooked prickles, 3–7 mm long.

Flowering: April–May.

Fruiting: Throughout the year.

Chemical constituents: The aerial parts of *S. torvum* are rich sources of steroids and saponins. Solanolides, yamogenin, solanolactosides A and B and torvosides are reported. Leaves contain neochlorogenin, solagenin, tetratriacontanoic acid, sitosterol, stigmasterol and campesterol. Fruits contain chlorogenone and neochlorogenone. The root is a source of steroidal glycosides such as torvosides A–G (Badola et al., 1993) (Cuervo et al., 1991).

Used in: Ayurveda and traditional medicine.

Parts used: Leaves and fruits.

Uses: The juice of the plant is used to treat fevers, coughs, asthma, chest ailments, sore throats, rheumatism, dropsy, stomachaches and gonorrhea. The stem and root have anti- tumour, antibacterial, anti-viral and anti-inflammatory properties. The leaves are dried and ground to powder: this is used as a medicine for diabetes. The juice of the flowers, with salt added, is used as eyedrops and to treat cuts, wounds and skin diseases. An infusion of the leaves and fruits is used as a treatment for bush yaws and sores. The fruit is diuretic and it is used in the treatment of malaria, stomachaches and problems with the spleen. A decoction of the root is used to treat venereal diseases. The roots are boiled and mixed with lime juice to make a beverage that is employed as a treatment for malaria. The juice of the roots is used to treat vomiting caused by weakness (Yousafa et al., 2013).

Methods of propagation: Seeds are sown in a nursery and seedlings are transplanted after 5–6 weeks at a spacing of 1 m. Branch cuttings are also used for propagation. Semi-hardwood cuttings 12–15 cm long, collected from fresh shoots and with their leaves removed, will produce roots and new shoots in 3–4 weeks.

Photo: Prof Srinath Rao

SOLANUM TRILOBATUM L.

Synonyms: *Solanum canaranum* Miq.; *Solanum griffithi* (C. B. Clarke) Kuntze; *Solanum procumbens* Lou.; *Solanum sarmentosum* Nees.

Classification:

Kingdom:	Plantae
Subkingdom:	Phanerogamea
Class:	Dicotyledonae
Subclass:	Gamopetalae
Series:	Bicarpellatae
Order:	Polemoniales
Family:	Solanaceae
Genus:	*Solanum*
Species:	*Solanum trilobatum*

Common names: Bengali: Kshudra agnidamani, Nabhi ankuri; **English:** Thai nightshade, Climbing brinjal; **Hindi:** Kantakari lata; **Kannada:** Kakamunji, Ambusondeballi; **Malayalam:** Tutavalam, Padarchunda; **Marathi:** Mothiringnee, Thoodalam; **Sanskrit:** Agnidamini, Alarka, Achuda, Vallikantakarika; **Tamil:** Thuduvalai, Nittidam, Sandunayattan; **Telugu:** Alarkapatramu, Kondavuchinta, Mullumusti.

Distribution: This species is globally distributed in Indo-Malaysia. Within India, it has been recorded in Peninsular India along the highway and waste lands; more commonly available in Southern India.

Description: It is a thorny, perennial herb, woody at the base. It is about 1–2 m in height. The plant is highly branched and prickly; prickles curved, broad-based, yellowish and numerous along the stems. The leaves are rounded-ovate in outline, obtusely 3–5-lobed, 2–7 cm long, 1–4 cm wide, glabrous or slightly stellate, with a few prickles along the petiole and midrib. Inflorescences extra-axillary, peduncle short, 3–9-flowered. Flowers are purplish blue pedicels 1–2 cm long, widely divergent. Calyx 3 mm long, with narrow teeth, sparsely stellate. Corolla 12–16 mm long, deeply lobed, stellate-pubescent outside, purple, reflexed; stamens protruding, anthers 7 mm long, yellow. Fruit: a berry, globose, purple black, about 1 cm in diameter. Seeds compressed, 3 mm diameter.

Flowering: August–September.

Fruiting: October–December.

Chemical constituents: Leaves contain cardiac glycoside, flavonoids, phytosterols, saponins, steroidals, tannins, terpenoids and alkaloids such as soladunalinidine and tomatidine. It also contains compounds such as sobatum, β-solamarin, solasodine and solain.

Used in: Ayurveda, Siddha and folk medicine.

Parts used: Stems, leaves, fruits and flowers.

Uses: The plant is an effective antibacterial, anticancer, antifungal and antioxidant. The leaves are used to cure asthma, deafness, emphysema, indigestion and tuberculosis. They increase fertility in males by treating watery semen; they are also used to cure tuberculosis and lung congestion. The fruits are used to cure cold, constipation, flatulence and rheumatism. The seeds are used as vermifuge.

Methods of propagation: Through seeds.

SOLANUM XANTHOCARPUM SCHRAD. & WENDL.

Synonyms: *S. surattense* Burm. L.; *S. virginianum.*

Classification:

Kingdom:	Plantae
Subkingdom:	Phanerogamea
Class:	Dicotyledonae
Subclass:	Gamopetalae
Series:	Bicarpellatae
Order:	Polemoniales
Family:	Solanaceae
Genus:	*Solanum*
Species:	*Solanum xanthocarpum*

Common names: Bengali: Kantakari; **English:** Yellow-berried nightshade; **Gujarati:** Bhoyaringani; **Hindi:** Kateli; **Kannada:** Chickchundi, Kantakari; **Malayalam:** Kandankattiri; **Marathi:** Bhuiringani; **Sanskrit**: Kantakari; **Tamil:** Kandankattiri; **Telugu:** Pinnamulaka, Nelamulaka, Vankuda.

Distribution: Common in Bangladesh, Nepal, Sri Lanka and India. Very commonly found throughout India in plains from seashore to hills up to 1000 m high.

Description: It is a very prickly perennial herb with woody base. Stem much-branched; younger ones clothed with dense, stellate and tomentose hairs. Prickles are compressed, straight, glabrous and shining, often 1–3 cm long. Leaves 5–8 cm long, 3–5 cm broad, ovate or elliptic, sinuate or sub-pinnatifid, obtuse or subacute, stellately hairy on both sides, armed on the midrib and often on the nerves with long yellow sharp prickles. Petiole is long, stellately hairy and prickly. Flowers are in cymes or sometimes solitary. Calyx tube is short, globose; lobes linear-lanceolate, acute, densely hairy and prickly. Corolla purple, lobes deltoid, acute, and hairy outside. Anther filament is long, glabrous; anthers open by a pore. Ovary is ovoid and glabrous. Fruit is a berry, yellow, 1.5 cm in diameter, blotched green and surrounded by enlarged calyx. Seeds are glabrous.

Flowering: February–March.

Fruiting: April–June.

Chemical constituents: Steroidal alkaloid solasodine is the principal alkaloid. Alcoholic extracts of the plant contain fatty and resinous substances. Solasonine and solanocarpidine are present in fruits – solasodine is reported to possess anti-spermatogenic activity. The unsaponifiable matter of fruits contains 2 sterols, one of which is carpesterol (Gupta and Dutta, 1936). Saponins and flavonoids are also reported.

Used in: Ayurveda.

Parts used: Dried roots; the whole plant.

Uses: According to Ayurveda, it is used for treating fever, lumbago, piles (especially bleeding piles), bronchitis, asthma, thirst, urinary and heart disease, and also facilitates the treatment of gastrointestinal disorders such as constipation and flatulence, because of its laxative and carminative properties. The root paste is utilized by the Mukundara tribes of Rajasthan for the treatment of hernia. Stem, flowers and fruits are bitter and carminative. They are prescribed for relief of burning sensations in the feet accompanied by vesicular watery eruptions. Leaves are applied locally to relieve pain.

Photo: Prof Srinath Rao

SPHAERANTHUS INDICUS L.

Synonyms: *Sphaeranthus hirtus* Willd.; *Sphaeranthus mollis* Roxb.

Classification:

Kingdom:	Plantae
Subkingdom:	Phanerogamea
Class:	Dicotyledonae
Subclass:	Gamopetalae
Series:	Inferae
Order:	Asterales
Family:	Compositae (Asteraceae)
Genus:	*Sphaeranthus*
Species:	*Sphaeranthus indicus*

Common names: Bengali: Belaunja; **English:** East Indian globe thistle; **Hindi:** Gorakmundi; **Kannada:** Bodu kadlesoppu; **Malayalam:** Adakumaniyam; **Marathi:** Gorakhamundi; **Sanskrit:** Mahashvranika; **Tamil:** Anantai; **Telugu:** Bodasaramu.

Distribution: It grows in rice fields, dry waste places and cultivated lands in tropical parts of India. It is distributed throughout India, Sri Lanka, Africa and Australia from sea level to 1200 m altitude.

Description: The herb is much-branched, strongly scented, and annual erect with branched tapering roots. Stems are cylindrical with toothed wings. Leaves are sessile, decurrent, 2–7 cm long, 1–1.5 cm wide, obovate-oblong, rounded or subacute, glandular-hairy, spinous-serrate or dentate, narrowed at the base and greenish brown in color. Flowers are borne in terminal, solitary, globose clusters of heads. Heads of flowers are purple, bracts are short, slender and acuminate. In each head, the outer flowers are female, few or many, fertile, the central flowers bisexual, fertile or sterile, involucre narrow, bracts paleaceous, spathulate, acute, ciliate; receptacle small, naked. Corollas of female flowers are purple, slender, tubular, minutely 2– 3-toothed; corollas of hermaphrodite flowers are purplish white, tubular or funnel-shaped, 4–5-toothed; anther base sagittate, auricles acute or tailed, style armed, filiform, sometimes connate. Fruits are oblong and have compressed achenes in which pappus is absent. Odor of herb is slightly aromatic but disappears on long storage.

Flowering: January–April.

Fruiting: January–April.

Chemical constituents: The air-dried roots and rhizome contain steroids, fats, oils, proteins, amino acids, tannins, phenols, carbohydrates, saponins and alkaloids (Shirwaikar et al., 2006; Ambavade et al., 2006). A sesquiterpene lactone, 7α-hydroxyfrullanolide, was isolated from the plant. The powdered drug contains stigmasterol, β-sitosterol, hentriacontane, n-triacontanol, sesquiterpene lactone, sesquiterpene glycoside, sphaeranthanolide, flavone and isoflavone glycosides (Ambavade et al., 2006; Jha et al., 2010; Jha et al., 2009).

Parts used: The entire plant.

Used in: Ayurveda, folk, Tibetan, Unani and Siddha medicine.

Uses: The juice expressed from the plant is used in vitiated conditions of *vata*, epilepsy, hemicranias, jaundice, hepatopathy and gastropathy. A paste of the herb mixed with oil is good for pruritus and painful swelling. The roots are bitter, acrid, sweet, thermogenic, diuretic, expectorant, febrifuge and stomachic. They are useful in strangury, diabetes, leprosy, fever, pectoralgia, cough, gastropathy, hernia, haemorrhoids, helminthiasis and dyspepsia. An oil prepared by using the root is useful in scrofula. The powdered leaf is good for skin diseases and is considered a nervine tonic. The flowers are highly esteemed as an alternant, depurative and tonic.

Methods of propagation: Through seeds.

Photo: Vedavyas Kulkarni

SPILANTHES CALVA DC.

Synonyms: *Spilanthes paniculata* Wall.; *Spilanthes acmella* Murr.; *Acmella paniculata* (Wall. ex DC.) R. K. Jansen.; *Acmella clava* (DC.) R. K. Jansen.

Classification:

Kingdom:	Plantae
Subkingdom:	Phanerogamea
Class:	Dicotyledonae
Subclass:	Gamopetalae
Series:	Inferae
Order:	Asterales
Family:	Compositae (Asteraceae)
Genus:	*Spilanthes*
Species:	*Spilanthes calva*

Common names: English: Toothache plant; **Gujarati:** Akkalgaro; **Hindi:** Akarkarha, Pokormul; **Kannada:** Vana mugali, Hem mugulu; **Malayalam:** Kuppa manjet; **Marathi:** Akkalakara, Pipulka; **Telugu:** Maratimogga.

Distribution: Found in the Western and Eastern Ghats and in moist deciduous forests.

Description: An erect annual herb, 50–60 cm in height, stem and branches erect or decumbent, more or less hairy. Leaves opposite, 2.5–5 cm long, ovate, acute or sub-obtuse, irregularly crenate-serrate, base acute. Heads yellow, 0.6–1.3 cm long, ovoid, solitary or sub-panicled; peduncles sometimes reaching 10 cm in length. Flowers bisexual, corolla yellow, 2.5–3 mm, tubular, campanulate above.

Flowering: February–April.

Fruiting: February–April.

Chemical constituents: Aerial parts contain palmitic and stearic acids, tetra-triacontanoic acid, sitosterol, stigmasterol and sitosterol glucoside. Flower heads contain a local anaesthetic, amide, spilanthol and 3 other amides, a sterol and a non-reducing polysaccharide (Ghani, 2003). Acetic, lauric, myristic, linoleic and linolenic acid esters of α- and β-amyrin and myricyl alcohol have also been isolated from the plant (Rastogi and Mehrotra, 1993).

Used in: used as folk medicinal plant.

Parts used: Leaves, buds, flowers and stems.

Uses: Flower head is stimulant and sialagogue; given in toothache, affections of throat and gums and paralysis of the tongue. It is a powerful mosquito larvicide. The decoction of the plant is diuretic and lithontriptic; useful in dysentery and employed as a bath for rheumatism and as a lotion in scabies and psoriasis. Crushed plant is used as fish poison. It is prescribed for a few days to cure glossitis. Juice of the plant and flower head is rubbed to cure scabies. A tincture made from the flower heads is used to treat inflammation of the jaw and dental caries. Roots are used as a purgative.

Methods of cultivation: Conventional method of propagation of this species is through seeds, but viability of seeds is lost within a short period of time. Seeds should be sown immediately after collection.

SPINACIA OLERACEA L.

Synonyms: *Chenopodium oleraceum* (L.) E. H. L. Krause; *Spinacia glabra* Mill.

Classification:

Kingdom:	Plantae
Subkingdom:	Phanerogamea
Class:	Dicotyledonae
Subclass:	Monochlamydeae
Series:	Curvembryae
Order:	Centrospermae
Family:	Chenopodiaceae
Genus:	*Spinacia*
Species:	*Spinacia oleracea*

Common names: Bengali: Palong sak; **English:** Spinach; **Hindi:** Palak; **Kannada:** Palak soppu; **Marathi:** Palak; **Sanskrit:** Chhurika; **Tamil:** Pasalai; **Telugu:** Palakura.

Distribution: It probably originated in North Iran, Afghanistan and Turkmenistan. It is not found wild and is mostly cultivated for its leaves worldwide.

Description: The plant grows to a height of 30 cm. The leaves are simple alternate, dark green, petiole long, about 6–12 cm, leaves succulent, very variable in size, about 2–30 cm long and 1–15 cm broad. The leaves appear at the base of the plant, grow in a rosette, crinkled or flat, angular-ovate or arrow-shaped, with round to pointed basal lobes. The flowers are very small, whitish green, 3–4 mm in diameter, male and female flowers mostly on the same plant, perianth 4-lobed. The fruits are small, hard, dry, and appear in clusters about 10 mm across. Seeds small, several, round, prickly.

Chemical constituents: The leaves contain vitamins such as folate, niacin, vitamin A, vitamin B6, vitamin C and vitamin K, and traces of the other essential vitamins. Other constituents are β-carotene, thiamine, riboflavin, lutein, xanthene and zeaxanthin. Very high in iron.

Used in: Traditional and folk medicine.

Parts used: Leaves.

Uses: The spinach is beneficial for strengthening muscles and improves eyesight, blood pressure, metabolism and the digestive system. It provides neurological benefits and prevents atherosclerosis, osteoporosis, cancer and strokes.

Methods of cultivation: Through seeds.

Photo: Prof Srinath Rao

https://www.organicfacts.net/health-benefits/vegetable/health-benefits-of-spinach.html

STEPHANIA WIGHTII (ARN.) DUNN.

Synonym: *Clypea wightii* Arn.

Classification:

Kingdom:	Plantae
Subkingdom:	Phanerogamea
Class:	Dicotyledonae
Subclass:	Polypetalae
Series:	Thalamiflorae
Order:	Ranales
Family:	Menispermaceae
Genus:	*Stephania*
Species:	*Stephania wightii*

Common names: Malayalam: Mahalikizangu, Malathangi; **Marathi:** Pahadvel; **Sanskrit:** Rajpatha.

Distribution: Indo-Malesia, Kerala (Kannur, Wayanad, Malappuram, Palakkad, Thrissur, Idukki, Kottayam, Kollam and Thiruvananthapuram districts) and Tamil Nadu (Kanyakumari, Nilgiri and Tirunelveli districts).

Description: *Stephania* species are dioecious, i.e. male (staminate) flowers on one plant, and female (pistillate) flowers on another plant. Creeping herbs. Leaves 8 cm across, orbicular, glaucous, margins wavy or coarsely dentate; petiole 8 cm long, slender. Flowers unisexual, subsessile, greenish yellow. Male inflorescence in axillary, capitate cymes, male flowers densely packed in heads; female flowers 3 mm across, in pedunculate cymes; sepals 5, ovate, acute, glabrous; petals ovate, greenish yellow. Carpels 1, globose or ovoid, style short, stigma 3–6-cleft, ovary ovoid. Drupe 8 × 5 mm, oblique, stalked, with 4 rows of tubercles along the margins.

Flowering: April–June.

Fruiting: April–June.

Chemical constituents: The roots, tubers and leaves of this plant contain alkaloids, steroids and fats. The stems contain bisbenzylisoquinoline alkaloids, stephasubine and 3′,4′-dihydro-stephasubine, saponins, steroids and fats. The roots contain the alkaloids fangchinoline, dl-tetrandrine, d-tetrandrine and disochondrodendrine. The aerial parts of the plant contain aknadinine, epistephanine, hernandifoline and magnoflorine. The roots and tubers contain alkaloids such as aknadinine, aknadine and aknadicine. A new alkaloid, 3-O-dimethyl-hernandifoline, was also isolated from the plant Semwal et al., 2010.

Used in: Folk medicine.

Parts used: Aerial parts and tubers.

Uses: *Stephania* has been used to treat dysentery, pyrexia, tuberculosis, diarrhea, dyspepsia, urinary diseases, diabetes, inflammation, infertility, abdominal ills, asthma, ascariasis, wounds, headache, sore breasts and leprosy.

Methods of cultivation: Through seeds.

Photo: Prof Srinath Rao

STRYCHNOS NUX-VOMICA L.

Synonym: *Strychnos spireana* Dop.

Classification:

Kingdom:	Plantae
Subkingdom:	Phanerogamea
Class:	Dicotyledonae
Subclass:	Gamopetalae
Series:	Bicarpellatae
Order:	Gentianales
Family:	Loganiaceae
Genus:	*Strychnos*
Species:	*S. nux-vomica*

Common names: English: Strychnine tree, Nux-vomica, Poison nut; **Hindi:** Kuchla, Kajra; **Kannada:** Kanjira, Hemmust, Karaskarah; **Malayalam:** Kanniram; **Marathi:** Kajra; **Tamil:** Etti, Kancirai; **Telugu:** Musti.

Distribution: A native of India, Sri Lanka, Southeast-Asia and Australia. Grows in open habitats. In India the plant is distributed in the semi-evergreen and moist deciduous forests of Maharashtra spreading across Konkan and the Sahyadri foothills of Western Ghats up to altitudes of 360 m.

Description: This is a deciduous tree growing up to 40 m, armed with stiff spines. The branches are irregular and covered with a smooth ashen bark. The young shoots are deep green in colour with a shiny coat. Leaves: simple, opposite, ovate and shining, about 10 cm long and 8 cm wide, broadly elliptic or elliptic-ovate, base acute to attenuate, apex acute or obtuse, margin entire, glabrous, shiny, coriaceous; 3–5-nerved from base, intercostal reticulate, petiole 5–13 mm, slender. Flowers: small, white, borne in terminal compound cymes; stamens 4, anthers sessile inserted at the throat of the corolla tube; ovary 2-celled, ovules many in each cell. Seeds 3 or 4, orbicular, flat, shiny, greenish white, coin-like, 2–2.5 cm. Fruit: a berry, 3–5 cm in diameter, hard-shelled, orange-red in colour, glabrous. Seeds: 3–5 in number, covered with a soft woolly substance, hard, discoid and grey in colour. They have the shape of a flattened disk.

Flowering: March–December.

Fruiting: March–December.

Chemical constituents: The seeds contain 2.5–5 % alkaloids, the prominent ones being strychnine and brucine. The minor alkaloids present are α- and β-colubrines, icajine, vomicine, novacine, N-oxystrychnine and pseudostrychnine. The seeds also contain chlorogenic acid, iridoids and 4–5 % fixed oil.

Used in: Ayurveda and folk medicine; homeopathy.

Parts used: Fruits, seeds, leaves and bark.

Uses: This is a highly toxic plant, but in small doses it is a cure for certain forms of paralysis and nervous disorders. The seeds are also used in chronic dysentery, rheumatism, insomnia and as an anthelmintic. Strychnine is highly toxic and a central stimulant; it increases appetite and is used as a tonic. The tree's bark is febrifuge and tonic. In Ayurvedic preparation it is used in *mundi churna*, *mundi panchang swarasa* and *mundi kavatha*. In-vitro *Strychnos nux-vomica* inhibited the growth of AGS human gastric carcinoma cells.

Methods of propagation: This tree can be propagated through seed and stem cuttings. It prefers well-drained, humus-rich soil and partial shade.

- Collect the seeds during the period extending from December to the end of April.
- Wash the seeds to remove the adhering pulp of the fruit and dry the seeds in the sun.
- Store the seeds in gunny bags. They can be stored for 1 year.
- Before sowing, the seeds are treated with warm water and soaked for 48 hours. Alternately, they may be dipped in cow-dung slurry for 24 hours.
- During early summer, the seeds are sown on raised beds. They can be planted in poly bags during March. A thin layer of hay is spread over the beds and the bags to maintain the temperature.
- The seeds germinate after 45–50 days. Seedling growth is very slow but the roots grow very fast. Soaking of seeds for 72 hours in cow-dung slurry gives maximum germination.
- In 60–70 days, they will be seen sprouting with a success rate of 25 %.
- Semi-hardwood cuttings are used for propagating the plant in early summer.

Photo: https://en.wikipedia.org/wiki/Strychnos_nux-vomica

http://www.stuartxchange.org/StrychninePlant.html

STRYCHNOS POTATORUM L. F.

Synonyms: *Strychnos heterodoxa* Gilg; *Strychnos stuhlmannii* Gilg.

Classification:

Kingdom:	Plantae
Subkingdom:	Phanerogamea
Class:	Dicotyledonae
Subclass:	Gamopetalae
Series:	Bicarpellatae
Order:	Gentianales
Family:	Loganiaceae
Genus:	*Strychnos*
Species:	*Strychnos potatorum*

Common names: Bengali: Nirmali; **English:** Clearing nut tree; **Gujarati:** Nirmali; **Hindi:** Nirmali; **Kannada:** Chilla; **Malayalam:** Thettamaram, Chillam; **Tamil:** Tattan, Sirungamara, Kottei; **Telgu:** Chilla ginga.

Distribution: Found in Central and Southern India as well as in Sri Lanka and Myanmar.

Description: Trees up to 12 m high, bark 10–12 mm thick, brownish black, rough, corky with deep and narrow vertical cracks; blaze creamy yellow. Leaves simple, opposite, estipulate; petiole 2–8 mm, glabrous; lamina 5–15 × 3.5–7.5 cm, elliptic, ovate, base obtuse or round, apex acute, margin entire, glabrous, shiny, chartaceous; nerves 3–5 from the base, lateral nerves from the midrib 4 pairs, glabrous; intercostae reticulate, prominent. Flowers bisexual, white, 0.85 cm long, in short axillary cymes, 6–8 mm long; pedicel 2 mm; calyx lobes 5, small; corolla 3.5 mm across, salver-shaped, hairy within, tube cylindric, lobes 5, longer than tube; stamens 5, inserted on the corolla tube; anthers 2-celled, introrse; ovary 1 mm, superior, glabrous, ovules 1 in each cell; style glabrous; stigma obscurely 2-lobed. Fruit a berry, 16–18 mm across, globose, black, pericarp hard; seeds 1–2, 10–12 mm across, turgid, round, compressed, yellowish.

Flowering: March–May.

Fruiting: March–May.

Chemical constituents: Phytochemical studies revealed the presence of diaboline (major alkaloid) and its acetate brucine, loganin, mannose, sucrose, arachidonic, lignoceric, linoleic, oleic, palmitic, and stearic acids. On saponification of the oil: β-sitosterol, stigmasterol (also in leaves and bark along with campesterol); oleanolic acid and its 3β acetate, saponins containing acid oleanic, galactose and mannose (seeds) and triterpenes and sterols mannogalactans. The successive extracts of root, stem bark, and seeds of *Strychnos potatorum* have revealed the presence of alkaloids, flavonoids, glycosides, lignins, phenols, saponins, sterols, and tannins (Yadav Kavita et al., 2014).

Used in: Ayurveda and folk medicine.

Parts used: Seeds.

Uses: The fruit is antidiabetic, anti-dysenteric, emetic. The fruit pulp is an expectorant. The seeds have been used as a demulcent, diuretic, emetic, stomachic and tonic. They are effective in the treatment of eye infections, inflammation, diabetes, diarrhea, gonorrhea, gastropathy, leucorrhea and renal and vesicle calculi.

Mode of propagation:

The plant is propagated through seeds, stem cuttings and root suckers. It germinates only in soil and does not tolerate water logging.

Steps to follow:

- Collect the fruits between February and March. Remove the pulp to obtain the seeds. Dry the seeds in shade before use. Seeds have poor viability. Since they do not tolerate water logging, seedbeds are preferably made with sand.
- To ensure good germination, seeds are treated with warm water. They are soaked in water for 48 hours.
- Seeds are sown in poly bags during April. After 45 days, the shoot emerges with the cotyledons retained on the tip resembling a round cap.
- Root growth is faster than shoot growth and hence a longer polythene bag is required for raising the seedlings.
- 30–40 days is the time it takes for germination to occur, with germination success being 40 %.
- Maintain the seedling in the nursery for 60 days.
- For vegetative propagation, pencil-thick and pencil-long hardwood cuttings with 2–6 nodes are used.

Photo: Dr R.S. Modi

SYZYGIUM CUMINI (L.) SKEEL

Synonyms: *Eugenia cumini* (L.) Druce; *Eugenia jambolana* Lam.; *Myrtus cumini* L.

Classification:

Kingdom:	Plantae
Subkingdom:	Phanerogamea
Class:	Dicotyledonae
Subclass:	Polypetalae
Series:	Calyciflorae
Order:	Myrtales
Family:	Myrtaceae
Genus:	*Syzygium*
Species:	*Syzygium cumini*

Common names: English: Java plum; **Hindi:** Jamun; **Kannada:** Neeladahannu; **Marathi:** Jamun; **Sanskrit:** Jambul; **Tamil:** Nagai; **Telugu:** Neralepandu.

Distribution: Native to Bangladesh, India, Nepal, Pakistan and Indonesia. Historically, the tree was exclusive to the Indian Subcontinent, and so widespread across the region that one of the old names of India (or the Indian region) is Jambudvipa (the island of jambul fruit).

Description: Jambul is a large evergreen and densely foliaceous tree with greyish-brown thick bark, exfoliating in woody scales. The leaves are leathery, oblong-ovate to elliptic, 6–12 cm long. Flowers are scented, greenish white, in clusters of just a few or 10–40, round or oblong in shape and found in dichotomous paniculate cymes. The calyx is funnel-shaped, about 4 mm long, and toothed. The plant produces small purple plums, which have a very sweet flavor, turning slightly astringent on the edges of the pulp as the fruit becomes mature. The dark-violet ripe fruits resemble the fruit of the olive tree and have an astringent taste.

Flowering: April–May.

Fruiting: June–July.

Chemical constituents: This plant is a rich source of anthocyanins, glucoside, ellagic acid, isoquercetin, kaempferol and myricetin. The seeds are claimed to contain the alkaloid jambosine, and the glycoside jambolin. The leaves are rich in acylated flavonol glycosides, quercetin, myricetin, myricetin 3-O-4-acetyl-L-rhamnopyranoside, triterpenoids, esterase, galloyl carboxylase and tannin. The stem bark is rich in betulinic acid, friedelin, epi-friedelanol, β-sitosterol, ugenin, fatty acid ester of epi-friedelanol, β-sitosterol, quercetin kaempferol, myricetin, gallic and ellagic acids, bergenins, flavonoids and tannins. The flowers are rich in kaempferol, quercetin, myricetin, isoquercetin (quercetin-3-glucoside), myricetin-3-L-arabinoside, quercetin-3-D-galactoside, dihydromyricetin, oleanolic acid, acetyl oleanolic acid, eugenol-triterpenoid A and eugenol-triterpenoid. The roots are rich in flavonoid glycosides and isorhamnetin 3-O-rutinoside (Ayyanar and Subash-Babu 2012).

Used in: Ayurveda and Unani medicine.

Parts used: Fruits.

Uses: Anti-cancer, anti-inflammatory, cardio-protective, hepatoprotective, anti-diabetic and anti-diarrheal properties. The bark is acrid, sweet, digestive, astringent to the bowels, anthelmintic and used for the treatment of sore throat, bronchitis, asthma, thirst, biliousness, dysentery and ulcers.

Methods of propagation: This plant is propagated through both seeds and vegetative methods.

Photo: Dr R.S. Modi

TERMINALIA ARJUNA (ROXB. EX DC.) WT. & ARN.

Synonyms: *Pentaptera cuneata* (Roth) Kuntze; *Pentaptera arjuna* Roxb.; *Terminalia berryi* Wight & Arn.; *Terminalia ovalifolia* Rottl. ex C. B. Clarke.

Classification:

Kingdom:	Plantae
Subkingdom:	Phanerogamea
Class:	Dicotyledonae
Subclass:	Polypetalae
Series:	Calyciflorae
Order:	Myrtales
Family:	Combretaceae
Genus:	*Terminalia*
Species:	*T. arjuna*

Common names: English: Arjuna; **Hindi:** Arjun; **Kannada:** Maddi; **Malayalam:** Nirmaruta; **Marathi:** Srdhaval; **Sanskrit:** Arjunah; **Tamil:** Attumarutu; **Telugu:** Erramaddi.

Distribution: Very common in the Western Ghats of India.

Description: This is a large tree with a thick trunk, horizontally spreading branches and greenish-white bark, flaking off in large flat pieces. Leaves: large, subopposite, oblong, serrate with glands at the tip of the petiole. Flowers: yellowish green, born in axillary spikes. Fruit: a 5-winged drupe.

Flowering: May–June.

Fruiting: October–November.

Chemical constituents: The bark is found to contain tannins, arjunic acid, arjunetin, β-sitosterol, friedelin and gallic acid. Fruits contain tannins, arjunic acid, β-sitosterol and mannitol.

Used in: Ayurveda and traditional system medicine.

Parts used: Bark and fruit.

Uses: The bark of arjuna is a well-known cardiotonic, used in all types of cardiac failure and dropsy. It is also a good diuretic, febrifuge and anti-dysenteric.

Methods of propagation: Through seeds.

Photo: Dr R.S. Modi

TERMINALIA BELLIRICA (GAERTN.) ROXB.

Synonyms: *Myrobalanus bellirica* Gaertn.; *Myrobalanus laurinioides* Kuntze; *Terminalia attenuata* Edgew; *Terminalia punctata* Roth.

Classification:

Kingdom:	Plantae
Subkingdom:	Phanerogamea
Class:	Dicotyledonae
Subclass:	Polypetalae
Series:	Calyciflorae
Order:	Myrtales
Family:	Combretaceae
Genus:	*Terminalia*
Species:	*T. bellirica*

Common names: English: Belliric myrobalan, Beleric, Baehra; **Hindi:** Baheda, Behara; **Kannada:** Santikayi, Aakaashamara; **Malayalam:** Tanni, Tannikai; **Marathi:** Baheda; **Sanskrit:** Aksaka, Aksha, Anilaghnaka; **Tamil:** Kalanduri, Kalitturumam; **Telugu:** Bhutavasamu tadi, Kattu-olupoe, Tadi.

Distribution: This tree is seen throughout India.

Description: This is a large deciduous tree. Leaves: opposite, large, elliptical and crowded at the ends of branches. Flowers: unisexual and bisexual, yellowish green, borne in axillary spikes, male flowers towards top. Fruit: a globose drupe, tomentose, containing a single seed.

Flowering: April–May.

Fruiting: November–February.

Chemical constituents: Fruits contain tannins, β-sitosterol, ellagic acid, gallic acid, chebulagic acid, ethyl gallate and sugars.

Used in: Ayurveda, Siddha, and folk medicine; homeopathy.

Parts used: Fruit and leaves.

Uses: The rind of the fruit is a component of *triphala churna* (a group of three fruits – *T. chebula, T. bellirica* and *Emblica officinalis*) which is used as a potent laxative and as a *rasayana*. This plant is also used to treat anaemia, cough, leprosy, tuberculosis, inflammations and problems of eye, nose and throat. This is also a source of tannin and dye.

Methods of propagation: Through seeds.

Photo: Prof Srinath Rao

TERMINALIA CHEBULA RETZ.

Synonym: *Myrobalanus chebula* (Retz.) Gaertn., Fruct.

Classification:

Kingdom:	Plantae
Subkingdom:	Phanerogamea
Class:	Dicotyledonae
Subclass:	Polypetalae
Series:	Calyciflorae
Order:	Myrtales
Family:	Combretaceae
Genus:	*Terminalia*
Species:	*T. chebula*

Common names: English: Chebulic myrobalan; **Hindi:** Harad, Haritaki, Harada; **Kannada:** Alalekaayi, Harra; **Malayalam:** Kadukka; **Marathi:** Hirada; **Sanskrit:** Haritaki; **Tamil:** Kadukkai; **Telugu:** Karakkaya.

Distribution: This tree is abundant in the Western Ghats of India.

Description: This is a moderate-sized, much-branched tree. Leaves: large, opposite, elliptical-oblong, pubescent when young; 2 glands are present at the top of the petiole. Flowers: pale yellow, borne in axillary spikes, spicate inflorescence, dioecious, stamens 10, ovules 2. Fruit: ovoid, smooth.

Chemical constituents: Tannins are the major constituent of fruits. Some of the tannins reported are chebulagic acid, chebulinic acid and corilagin. Also present are 3,6-digalloyl glucose, gallic acid, ellagic acid and terchebin.

Used in: Ayurveda and traditional medicine,

Parts used: Fruit.

Uses: The fruit rind is a component of *triphala churna* (a group of three fruits – *T. chebula, T. bellirica* and *Emblica officinalis*) which is used as a potent laxative and as a *rasayana*. The plant is antiseptic, diuretic and carminative and used in diabetes, fever, anemia and cardiac disorders. It is one of the chief sources of tannin in India. Alterative, astringent, expectorant, anti-inflammatory, anodyne, cardio-tonic, aperients, antiseptic, anti-pyretic, antiemetic, tonic. In Ayurveda, it has been used for *ars, pandu, sotha, jirnajvara, visamjvara, premaha, sirorga, kasa* and *tamak svasa*.

Methods of Propagation:

Through seeds and vegetative methods.

- The fallen fruits are collected and dried thoroughly first.
- Later, the hardened flesh is removed.
- Fermentation of the stones gives the best germinative results, but clipping the broad end of the stone without damaging the embryo, followed by soaking in cold water for 36 hour, gives better results. (Direct sowing is not advisable, because of the risk of predation and because the seeds germinate poorly.)
- Shading is desirable in early stages in the nursery and after transplanting.
- Propagation by cuttings is possible, but less successful than transplanting nursery-raised seedlings into the field.

Photo: Dr Vedavyas Kulkarni

TILIACORA RACEMOSA COLEBR.

Synonyms: *Minispermum acuminatum* Lam.; *Minispermum glabrum* Koen. ex Willd.; *Tiliacora acuminata* (Lam.) Hook. f. & Thoms; *Cocculus acuminatus* DC.

Classification:

Kingdom:	Plantae
Subkingdom:	Phanerogamea
Class:	Dicotyledonae
Subclass:	Polypetalae
Series:	Thalamiflorae
Order:	Ranals
Family:	Minispermaceae
Genus:	*Tiliacora*
Species:	*Tiliacora racemosa*

Common names: Bengali: Tiliacord; **Hindi:** Bagamushada; **Kannada**: Kuriballi; **Malayalam:** Vallikkanniram; **Tamil:** Nakamutti, Perunkattukkoti; **Telugu:** Kappatige, Pataveru, Nagamusadi, Tigamushti, Verricitramulaum.

Distribution: Throughout India along streams and rivers up to an altitude of 2000 m; also cultivated in gardens as an ornamental.

Description: Large woody climber, about 15–20 feet tall, stem woody, branches striated. Leaves simple, alternate, ovate to lanceolate, subpalmate 8 × 15 × 3–8 cm across, chartaceous, glabrous, 3–5 nerved at base, petiole 2–3 cm long, leaves of the male plant slightly larger than female, plant base cordate, acuminate. The plants are dioecious, i.e. male and female flowers are on different plants. Inflorescence axillary, panicle 4–10 cm long, pubescent. Flowers yellow, male flowers 2–7 at apex of inflorescence, sepals 6 in 2 rows, petals 6, obovate, stamens 6, cylindric. Female flowers solitary, sepals and petals as in male flowers, carpels 8–12, apocarpous, glabrous on stalked gynophor, style curved, stigma simple. Fruits drupes, ovoid, compressed, 10–15 × 5–6 mm, red when ripe.

Flowering: April–December.

Fruiting: April–December.

Chemical constituents: Leaves contain alkaloids tilioresine and N-methyl tiliamosine. The root contains an alkaloid tiliacorine.

Used in: Ayurveda, folk and Siddha medicine.

Parts used: Leaves, roots.

Uses: Paste of leaves and roots is applied externally for treating cuts and wounds. Leaves have anti-dandruff properties and kill lice and nits. They possess choleretic, antioxidant and anti-cancer properties. Roots are ground in water and given for snake bite.

Methods of cultivation: Through seeds.

TINOSPORA CORDIFOLIA (WILLD.) HOOK. F. & TH.

Synonyms: *Cocculus cordifolius* (Willd.) DC.; *Menispermum cordifolium* Willd.

Classification:

Kingdom:	Plantae
Subkingdom:	Phanerogamea
Class:	Dicotyledonae
Subclass:	Polypetalae
Series:	Thalamiflorae
Order:	Ranales
Family:	Menispermaceae
Genus:	*Tinospora*
Species:	*T. cordifolia*

Common names: Bengali: Gulancha; **English:** Tinospora; **Hindi:** Giloya, Gilo; **Kannada:** Amrutaballi, Anebule; **Malayalam:** Anrytu; **Marathi:** Gulvel; **Sanskrit:** Amritavalli; **Tamil:** Amritavalli, Thippa Thega; **Telugu:** Guduchi.

Distribution: The plant occurs throughout tropical regions of India.

Description: A large woody climber with succulent stems and filiform green aerial roots. The stem has a warty surface (due to lenticels). Leaves: thick, heart-shaped and acuminate. Flowers: small, unisexual, borne on racemes in the axils of leaves. Fruit: a drupe.

Flowering: January–June.

Fruiting: January–June.

Chemical constituents: The stem, which forms the drug, contains clerodane furano diterpenes, viz. columbine and tinosporaside, alkaloids such as jatrorrhizine, palmatin, berberine and timbeterine.

Used in: Ayureda, Siddha, Unani, folk and traditional medicine.

Parts used: The whole plant.

Uses: The stems of the plant, collected in the hot season and dried, with the bark intact, constitute the drug. The drug is believed to be useful as a tonic and antiperiodic; it is also considered aphrodisiac. It is one of the rejuvenating drugs useful in general debility, rheumatism and diabetes, and improves the intellect. It also exhibits immunosuppressive, anti-inflammatory, analgesic and antihepatotoxic properties. The starch obtained from the roots and stems of the plant is useful in diarrhea and dysentery; it is also a nutrient Saha and Ghosh 2012; Upadhyay et al., 2010.

Methods of propagation: The plant is propagated either through seeds or through cuttings.

- Soak the seeds in cold water for 24 hours for better germination.
- Sow these in poly bags during May–July. It takes 10–12 days for the seeds to germinate, with the germination success being 80–90 %. The seedlings are kept in the nursery for about 1 and ½ months.
- For vegetative propagation, 4–6 in long, pencil-thick, semi-hard wood cuttings with 4–8 nodes are planted in poly bags. June–July is the most favourable time for planting the cuttings.
- The cuttings root in about 20–30 days. About 90 % of the cuttings sprout in 45 days.

Photo: Prof Srinath Rao

Photo: Dr Kotresh

TRIBULUS TERRESTRIS L.

Synonyms: *Tribulus lanuginosus* L.; *Tribulus maximus* var. Roseus Kuntze.

Classification:

Kingdom:	Plantae
Subkingdom:	Phanerogamea
Class:	Dicotyledonae
Subclass:	Polypetalae
Series:	Disciflorae
Order:	Geraniales
Family:	Zygophyllaceae
Genus:	*Tribulus*
Species:	*Tribulus terrestris*

Common name: Bengali: Gokhru kanta; **English:** Puncture vine, Caltrop, Yellow vine, Goathead; **Hindi:** Gokharu; **Malayalam:** Nerinnil; **Marathi:** Goghari; **Sanskrit:** Goshurah; **Telugu:** Cinnpalleru; **Tamil:** Palleru-mullu.

Distribution: Native to the warm, temperate and tropical regions of the world: Southern Europe, Southern Asia, throughout Africa, Australia and India.

Description: It is a herbaceous perennial plant that grows as a summer annual in colder climates. The stems radiate from the crown to a diameter of about 10 cm to over 1 m, often branching. They are usually prostrate, forming flat patches, though they may grow more upwards in shade or among taller plants. The leaves are pinnately compound with leaflets 3 mm long. The flowers are bisexual, solitary, small, 4–10 mm wide; sepals 5, 3–5 mm wide, pale green and lanceolate in shape. Petals 5, lemon-yellow, obovate, 0.5–1 cm. Stamens 10, free, in 2 whorls. Ovary 5-carpled, sessile, with 1 ovule in each locule. A week after each flower blooms, it is followed by a fruit that easily falls apart into 4–5 single-seeded nutlets. The nutlets or 'seeds' are hard and bear 2 sharp spines, 10 mm long and 4–6 mm broad point-to-point. These nutlets strikingly resemble a goat's or a bull's head.

Flowering: August–September.

Fruiting: September–November.

Chemical constituents: It contains protodioscin, a steroidal saponin that forms 45 % (dry weight) of the extract, β-carboline alkaloids harman (harmane) and norharman (norharmane).

Used in: Ayurveda, Folk medicine, Homoeopathy, Sowa-Rigpa, Unani, Siddha, Traditional chinese medicine.

Parts used: Fruit, leaf and root.

Uses: *Tribulus* is mentioned in ancient Indian Ayurvedic medical texts dating back thousands of years. Tribulus has been widely used in Ayurveda for the treatment of erectile dysfunctions and various urinary disorders. It is diuretic. It has been used in the treatment of postpartum hemorrhage, epistaxis and gastrointestinal bleeding. It is being promoted as a testosterone booster for the purpose of building muscle and increasing sex drive. It is used for treating gonorrhea, liver disease (hepatitis), inflammation, joint pain (rheumatism), leprosy, coughs, headache, dizziness (vertigo) and chronic fatigue syndrome (CFS), as well as to enhance athletic performance. It is also used for stimulating appetite and as an astringent, tonic and mood enhancer. *Tribulus* is employed for kidney problems, including kidney stones, painful urination and a kidney disorder called Bright's disease, as well as for skin disorders including eczema (atopic dermatitis), psoriasis and scabies. The bark is astringent, tonic and anthelmintic. It is useful in gallbladder disorders (Mamdouh et al., 2013).

Photo: Prof Srinath Rao

TRIDAX PROCUMBENS L.

Synonyms: *Balbisia canescens* Rich.; *Balbisia divaricata* Cass.; *Balbisia elongata* Willd.

Classification:

Kingdom:	Plantae
Subkingdom:	Phanerogamea
Class:	Dicotyledonae
Subclass:	Gamopetalae
Series:	Inferae
Order:	Asterals
Family:	Compositae (Asteraceae)
Genus:	*Tridax*
Species:	*Tridax procumbens*

Common names: Bengali: Tridhara; **English:** Coat button; **Gujarati:** Ghaburi; **Hindi:** Ghamra; **Kannada:** Jayanthi; **Malayalam:** Chiravanka; **Marathi:** Kambarmodi, Jakhamjudi; **Sanskrit:** Jayanthi veda; **Tamil:** Vettukaaya poondu, Kinatruppasan; **Telugu:** Gayaparaku, Gaddi chamanthi.

Distribution: Native to tropical America, but introduced to tropical, sub-tropical and mild temperate regions worldwide; found in fields, road sides and in meadows.

Description: It is a small perennial, semi-prostrate herb and grows to up to 15–30 cm. The leaves are opposite, fleshy, ovate with serrate margins 3–6 × 2–4 cm, hirsute on abaxial and adaxial sides. The inflorescence is a capitulum, with 3-toothed white or yellow ligulate ray florets. Disc florets present on inner side, flowers yellow, tubular, bisexual; calyx represented by scales; corolla about 6 mm long. Fruits are achenes, covered with stiff hairs, with a feathery plume like white pappus.

Flowering: Throughout the year.

Fruiting: Throughout the year.

Chemical constituents: A flavonoid, procumbenetin, has been isolated from the aerial parts of the plant. Other compounds isolated include esters, Please club sterols here.

sterols, pentacyclic triterpenes, fatty acids and polysaccharides. The leaves also contain arachidic, behenic, lauric, linoleic, linolenic, myristic, palmitic, palmitoleic and stearic acids.

Used in: Ayurveda, folk and traditional medicine (Ramesh et al., 2013).

Parts used: Leaves.

Uses: Traditionally, this plant has been used in India for wound healing and as an anticoagulant, antifungal and insect repellent. The juice extracted from the leaves is directly applied on wounds and to cure infectious skin diseases. In Ayurveda, it is used to cure liver disorders, gastritis and heartburn. It is also used to treat boils, blisters and cuts and in the treatment of bronchial catarrh, diarrhea and dysentery. Moreover, it is an immunomodulator and anti-hypertensive. The leaf powder combined with *Cicer arietinum* in a 2: 1 ratio is taken orally to treat diabetes (Sreeramalu et al., 2013; Gamboa-Leon et al., 2014; Beck et al., 2018).

Methods of cultivation: Root stock and seeds.

Photo: Prof Srinath Rao

TRIGONELLA FOENUM GRAECUM L.

Synonym: *Trigonella tibetana.*

Classification:

Kingdom:	Plantae
Subkingdom:	Phanerogamea
Class:	Dicotyledonae
Subclass:	Polypetalae
Series:	Calyciflorae
Order:	Rosales (Fabales)
Family:	Leguminosae (Fabaceae)
Subfamily:	Papilionidae
Genus:	*Trigonella*
Species:	*Trigonella foenum graecum*

Common names: Hindi: Methi; **Marathi:** Methi; **Sanskrit:** Methika; **Tamil:** Vendayam

Distribution: Fenugreek originated in the region of the Middle East (especially Lebanon and Syria) and Southeast Europe, India and China.

Description:

The leaves are trifoliate and obovate; the flowers are pale yellow and pea-like, in leaf axils. The pods are curved and sickle-shaped, and there are 10–20 angular seeds per pod which are aromatic, yellow-brown, rhomboidal and divided into 2 lobes by a furrow.

Flowering: June.

Chemical constituents: Fenugreek seeds contain 45–60 % carbohydrates, mainly galactomannan, a mucilage and 20–60 % proteins. They are especially high in the amino acids lysine and tryptophan; they also contain 4-hydroxyisoleucine, a free amino acid, pyridine alkaloids (trigonelline), steroidal saponins (diosgenin) and flavonoids such as flavone glycosides (apigenin, luteolin, quercetin).

Used in: Ayurveda and folk medicine.

Parts used: Leaves.

Uses: It contains a broad spectrum of therapeutic properties. The seed is extremely nutritious and is recommended to improve body-weight gain, particularly in anorexia nervosa. Moreover, the seeds inhibit cancerous cells of the liver and decrease blood cholesterol levels. They also have anti-diabetic properties. Fenugreek seed and leaf showed anticholesterolemic, anti-inflammatory, antitumor, carminative, demulcent, deobstruent, emollient, expectorant, febrifuge, galactagogue, hypoglycemic, laxative, parasiticide, restorative and uterine tonic effects.

Methods of cultivation: Through seeds.

Photo: Prof Srinath Rao.

TYLOPHORA INDICA (BURM. F.) MERRILL

Synonym: *Tylophora asthmatica* (L. f.) Wight & Arn.

Classification:

Kingdom:	Plantae
Subkingdom:	Phanerogamea
Class:	Dicotyledoane
Subclass:	Gamopetalae
Series:	Bicarpellatae
Order:	Gentianales
Family:	Asclepiadaceae
Genus:	*Tylophora*
Species:	*Tylophora indica*

Common names: Bengali: Antomnl ananthamiil; **Hindi:** Jangli-pikvam; **Kannada:** Adumutadhagida; **Malayalam:** Vallipaala; **Marathi:** Khodiki raasna, Pilkari; **Sanskrit:** Lataksiri; **Tamil:** Nach-cltiiruppam; **Telugu:** Yerripaala.

Distribution: Found in Assam, West Bengal, Orissa and Peninsular India, ascending to an altitude of 900 m. Also occurs in the Car Nicobar Islands.

Description: A branching climber, 1.5 m high. Root stocks short, knotty, 3.2 mm. thick; roots numerous, fine, 15 cm long, very brittle, consisting of fibers or fibrils; leaves simple, ovate to orbicular, cordate-based, 3–10 cm. long, thick; flowers greenish yellow out, purplish within, in many-flowered umbels; follicles fusiform, divaricate, 5 cm. corona 5-lobed, striate. Pollinia horizontal or ascending; seeds ovate, elongated into 2.0–2.5 cm-long coma.

Chemical constituents: The powdered leaves, stem and root contain alkaloids (0.42–0.46 %) The alkaloids present are tylophorine, tylocebrine and tylophorinine. A new potential antitumour alkaloid, tyiophorinidine, has recently been isolated from the roots of 3-year-old plants in 0.009 % yield (Sassya sampada). 3 minor alkaloids, designated alkaloid A, alkaloid B and alkaloid C, have also been reported from the plant. Alkaloid C showed significant activity in murine leukemia. Other constituents reported in the plant (leaves, stem and root) include cetyl alcohol, a phytosterol, a substance of an alcoholic nature, resin and tannin. The leaves also contain c-amyrin, quercetin, kaempferol, and an unidentified flavonoid.

Used in: Ayurveda, folk and traditional medicine.

Parts used: Leaves and roots.

Uses: They possess stimulant, emetic, cathartic, expectorant, stomachic and diaphoretic properties and are used for the treatment of asthma, bronchitis, whooping cough, dysentery and diarrhea.

- They are reported to be given in rheumatic and gouty pains.
- The roots are said to possess bacteriostatic properties and have been suggested as a good natural preservative of food.
- All the 3 alkaloids possess anti-cancerous properties.
- Leaves are used as an emetic.
- Roots are used for jaundice.

Methods of propagation:

Tylophora can be propagated through seeds and vegetatively through cuttings.

- However, as the seeds have poor germination capacity, it is preferred to propagate it by cuttings for commercial cultivation.
- To propagate it vegetatively, cuttings of finger thickness with 6" length stem having two nodes are used.
- The cuttings are planted in raised beds or in poly bags to get greater success.
- The rooting of cuttings takes about 8–10 weeks and cuttings will be ready for planting in the main field after about 10–12 weeks.
- The usual planting time is during the onset of monsoon in May–June.

Photo: Dr. Kotresh

Photo: Dr. R.S. Modi

VITEX NEGUNDO L.

Synonyms: *Vitex cannabifolia* Siebold & Zucc.; *Vitex incise* Lam.

Classification:

Kingdom:	Plantae
Subkingdom:	Phanerogamea
Class:	Dicotyledonae
Subclass:	Gamopetalae
Series:	Bicarpellatae
Order:	Lamiales
Family:	Verbanaceae
Genus:	*Vitex*
Species:	*Vitex negundo*

Common names: Bengali: Nishinda; **English:** Five-leaved chaste tree; **Hindi:** Samhalu, Nirgandi; **Kannada:** Lakkigida; **Malayalam:** Vennocchi, Karunocci; **Sanskrit:** Sinduvara, Indrani, Nilanirgundi; **Tamil:** Nocchi, Nirkundi; **Telugu:** Nallavavlli.

Distribution: This tree is common in India, Afghanistan, Philippines and Sri Lanka.

Description: This is a small tree with grey pubescent and quadrangular branches. Leaves: opposite, 3–5 foliate, exstipulate, long-petioled; leaflets lanceolate, white, tomentose beneath, lateral leaflets small and terminal leaflet larger. Inflorescence a paniculate cyme. Flowers: pentamerous, small, blue and bilabiate. Corolla blue; stamens 4, didynamous. Ovary pilose at apex, bilocular. Fruit: a drupe containing 4 seeds, globose or ovoid.

Flowering: February–July.

Fruiting: February–July.

Chemical constituents: The leaves contain aucubin, agnuside and alkaloids nishindine and glycoflavonoides orientin and isoorientin.

Used in: Ayurveda, Siddha, traditional and folk medicine.

Parts used: The whole plant.

Uses: The leaves in the form of paste are used for inflammatory swellings of the joints, hydrocele and splenic enlargement. They are also used in skin diseases, nervous disorders, leprosy and as a vermifuge. Oil prepared with leaves is useful for hair growth and in increasing the function of brain. The roots are also used in rheumatism, dyspepsia, piles and as anthelmintic. It is an ingredient of the following preparations: Himcolin, Pilex, Rumalaya gel, V-Gel, Acne-n-Pimple Cream, Pimple Clear Cream, Complete Care Toothpaste, Diaper Rash Cream, Muscle & Joint Rub.

Methods of propagation: Through seeds and vegetative method.

Photo: Prof Srinath Rao

WITHANIA SOMNIFERA (L.) DUNAL

Classification:

Kingdom:	Plantae
Subkingdom:	Phanerogamea
Class:	Dicotyledonae
Subclass:	Gamopetalae
Series:	Bicarpellatae
Order:	Polemoniales
Family:	Solanaceae
Genus:	*Withania*
Species:	*Withania somnifera*

Common names: English: Winter cherry; **Hindi:** Asgandh, Punir; **Kannada:** Hiremaddinagida; **Malayalam:** Amukkuram; **Marathi:** Asakndha, Bhuikohala; **Sanskrit:** Asvagandha, Varahakarni; **Tamil:** Amukkira; **Telugu:** Vajigandha, pannirugadda.

Distribution: This plant is distributed throughout India.

Description: This is a green herbaceous perennial with long, hard, tuberous roots. The stem is dichotomously branched. Leaves: simple, ovate. Flowers: Small, sessile, greenish yellow, crowded in the axils of leaves. Fruit: a small berry, enclosed in a green persistent calyx.

Flowering: January.

Fruiting: March.

Chemical constituents: The roots form the drug. The majority of the constituents are a group of steroidal lactones, the withanolides. The known withanolides are more than 20: withanone, Withaferin A, withanolides I, II, III, A– L, WS-1, P and S and withasomnidienone. About a dozen biochemically heterogeneous alkaloids are reported. The major compounds are tropane, pseudotropine, hygrine, piperidine, anaferine, withasomnine and anahygrine. Also present is a large amount of starch.

Used in: Ayurveda, traditional and folk medicine.

Parts used: Roots, leaves and seeds.

Uses: *Ashwagandha* is a rejuvenating drug used for rheumatism, hypertension and as a tonic. It shows anti-tumour, bradycardic, respiratory-stimulant and antispasmodic properties. The tuberous roots are astringent, bitter, acrid, somniferous, thermogenic, stimulant, aphrodisiac, diuretic and tonic. They are useful in vitiated conditions of *vata,* leucoderma, constipation, insomnia, tissue-building and nervous breakdown. The root paste is applied to swellings and to kill pain.

Methods of propagation: This plant is propagated through seeds. It prefers sandy loam or light red silts. Seeds are collected during November–December and soaked in water for 24 hours before sowing. Germination takes place within 8–10 days; 80 % of the seeds germinate and seedlings are transplanted when they are 6 weeks old.

Photo: Prof Srinath Rao

WOODFORDIA FRUTICOSA (L.) KURZ.

Synonyms: *W. floribunda* Salisb.

Classification:

Kingdom: Plantae
Subkingdom: Phanerogamea
Class: Dicotyledonae
Subclass: Polypetalae
Series: Calyciflorae
Order: Myrtales
Family: Lythraceae
Genus: *Woodfordia*
Species: *Woodfordia fructicosa*

Common names: Bengali: Dhai; **English:** Fire flame bush, Woodfordia; **Hindi:** Dawi, Thawi; **Kannada:** Bela, Tamrapushpi; **Malayalam:** Tatiri, Tatrippu; **Sanskrit:** Dhataki; **Tamil:** Dhattari, Jargi, Vellakai; **Telugu:** Dhataki, Jargi.

Distribution: Endemic to India, found throughout North India, in dry deciduous forests; less common in South India, mostly cultivated in gardens for its flowers.

Description: Evergreen or sub-deciduous shrubs, with long spreading branches; bark is reddish brown, peeling off into thin fibrous strips. Leaves opposite, shortly petiolate, lanceolate or oblong-lanceolate, entire, base rounded, apex acuminate; flowers: numerous, orange-red, long, clustered, in short-panicled axillary cymes; calyx: tubular, scarlet, slightly curved and widened above the mouth; petals: scarlet or red, nearly equal to calyx lobes; fruits: capsules, ellipsoid, long, enclosed within the calyx tube; seeds: many, minute, ovate and brown in colour.

Flowering: January–June.

Fruiting: January–June.

Chemical constituents: Woodfordins A–I identified in flowers; quercetin-3-o-(6"-o-galloyl)-β-D galactopyranoside, Quercertain-3-o-α-larabinoside, quercetin-3-o-oxylopyranoside and myricetin-3-o-arabinopyranoside were isolated from dried flowers. Alcoholic extract showed significant abortifacient activity in female albino rats (Das et al., 2006).

Used in: Ayurveda, Siddha and traditional medicine.

Parts used: Flowers.

Uses:

The flowers have astringent and stimulant properties; they are styptic, depurative, constipating, antibacterial, febrifuge and alexiteric. They are useful in conditions of *kapha* and *pitta*, leprosy, burning sensation, skin diseases, diarrhea, dysentery, fever, headache, hemorrhoids, herpes, internal hemorrhage, leukorrhea, liver disorders, menorrhagia, ulcers and wounds. The juice from the leaves is used in bilious sickness. It is also valued as a stimulant in pregnancy. Useful in diabetes. (Khushalani et al., 2006).

Methods of Propagation: By seeds and vegetative method.

Photo: Dr Kotresh

WRIGHTIA TINCTORIA (ROXB.) R. BR.

Synonyms: *Allamanda verticillata* Desf.; *Allamanda oleandrifolia* Lodd.; *Wrightia lacinata* A. DC.; *Wrightia timorensis* Miq.

Classification:

Kingdom:	Plantae
Subkingdom:	Phanerogamea
Class:	Dicotyledonae
Subclass:	Gamopetalae
Series:	Bicarpellatae
Order:	Gentianales
Family:	Apocynaceae
Genus:	*Wrightia*
Species:	*Wrightia tinctoria*

Common names: English: Sweet Indrajao, Ivory wood; **Gujarati:** Dudhalo; **Hindi:** Dudhi, Kapar, Mitaindarjau; **Kannada**: Aale, Beppale, Ajamara, Kondamurki; **Marathi:** Bhurevadi, Kalakuda; **Malayalam:** Ayyappala, Dandappala, Vettupala; **Sanskrit:** Hayamaraka, Shwetha kutaja; **Tamil:** Paalai, Vet-palai, Irum-palai; **Telugu:** Amkuda, Anakuduchettu, Chiti-anikudu, Kondajemudu.

Distribution: Native to Burma and India. Also found in Australia, Myanmar, Nepal and Vietnam. In India it is distributed throughout up to an altitude of 1200 m. It is common in moist deciduous forests.

Description: It is a deciduous tree reaching a height of 25–30 feet. The bark of the stem is light grey with a scaly bark, and produces milky-white latex. Leaves are simple, opposite, 10–15 cm long and 5 cm broad, ovate, obtusely acuminate and glabrous. Inflorescence: corymb-like cymes, 5–15 cm across. Flowers: bisexual, white, fragrant, seen at the tip of the branches, whitish. Calyx: 5-lobed, ovate, obtuse, ciliate, glandular inside. Corolla: salver-shaped lobes 5, oblong, rounded at the tip, white, 12 × 5 mm throughout with 1–2 series of erect, fimbriate corona scales. Stamens 5, inserted on the mouth of the tube, anthers sagittate, adhering to fruits, pendulous, cylindrical, greenish black, about 40–50 cm long, paired follicles joined at their tips. Seeds brown, flat with a bunch of white hairs.

Flowering: March–May.

Fruiting: August.

Chemical constituents: The seed pods contain wrightial, cycloartenone, cycloeucalenol, β-amyrin and β-sitosterol (Srivastava 2014).

Used in: Ayurveda, Siddha, Unani and traditional medicine.

Parts used: Leaves and stem bark.

Uses: The bark is useful in treating fever, gastrointestinal disorders and skin diseases, especially psoriasis. It has anti-inflammatory and anti-dandruff properties, hence it is used in hair oil preparations. The sap added to milk has been reported to have preservative properties.

Methods of cultivation: The tree responds well to coppicing, and also produces root suckers. Furthermore, the plant may be propagated through seeds; however, the germination percent is low. Seeds can be collected in the month of February.

Photo: Milind Giridhari

WRIGHTIA TOMENTOSA ROEM. & SMITH

Synonyms: *Wrightia mollissima*; *Wrightia wallichii*; *Wrightia arborea.*

Classification:

Kingdom:	Plantae
Subkingdom:	Phanerogamea
Class:	Dicotyledonae
Subclass:	Gamopetalae
Series:	Bicarpellatae
Order:	Gentianales
Family:	Apocynaceae
Genus:	*Wrightia*
Species:	*W. tomentosa*

Common names: Bengali: Dudhkoraiya; **English**: Kutaja; **Hindi:** Dudhi; **Kannada:** Bilikudegidda; **Malayalam:** Mailam pala; **Marathi:** Kala inderjau; **Tamil:** Pala; **Telugu:** Tellapaala.

Distribution: Found throughout the (tropical) warmer parts of India, ascending to an altitude of about 2000 feet in the Himalayas and to 4000 feet in the Nilgiris. The plant is placed in the category I of the Red Data Book.

Description: A small, deciduous tree, up to 12 m high, abounding in yellow milky juice, with opposite divaricate scabrous branches; branchlets pubescent; young parts densely tomentose. Bark grayish yellow to rust-coloured, corky, with light-coloured specks; leaves: elliptic, often tomentose, 7.5–15.0 cm long; flowers pale yellow with orange coronal scales, born in corymbose cymes; calyx pubescent outside, glandular inside; segments 3 mm long, broadly ovate, obtuse, with ciliate membranous margins. Corolla tube 5–6 mm long; lobes 1.3–1.6 cm long, oblong, rounded at the apex; corona orange, of 5–10 oblong, often laciniate scales. Follicles 15–30 cm. Long, connate into a cylinder, with a deep groove on each side; seeds with white silky coma.

Chemical constituents: The bark contains alkaloids, conessine, holarrhine, kurchicine and conkurchine. Also present are β-sitosterol, lupeol and α-amyrin. The seeds yield an oil similar to *Holarrhena*. Leaves yield quercetin, kaempferol, procyanidin, iso-orientin and phenolic acids such as vanillic, syringic and sinapic acids.

Used in: Ayurveda, Unani and Siddha medicine.

Parts used: Leaves, seeds, bark, fruit, oil from seed.

Uses: Inflammatory and antipyretic properties. The root bark and seeds are often used for chronic diarrhea and dysentery. They are also used in curing piles, hemorrhage, fever and cardiac diseases. Stem bark and root bark are believed to be useful in snake bite and scorpion stings. In the Siddha system, this plant is used to cure psoriasis and other skin diseases. The leaves show analgesic.

Methods of propagation: Through seeds.

Photo: Dr Kotresh

ZIZYPHUS MAURITIANA LAM.

Synonyms: *Zizyphus jujuba* (L.) Gaertn. non Mill.

Classification:

Kingdom:	Plantae
Subkingdom:	Phanerogamea
Class:	Dicotyledones
Subclass:	Polypetalae
Series:	Disciflorae
Order:	Celastrales
Family:	Rhamnaceae
Genus:	*Zizypus*
Species:	*Zizypus mauritiana.*

Common names: Bengali: Ber, Boroi; **English:** Indian plum, Indian jujube; **Hindi:** Baer, Ber; **Kannada:** Yalachi, Elanji, Hanji, Jelachi; **Malayalam:** Ilentha, Lantappalam, Cherumali; **Sanskrit:** Badarah, Karkandhu, Madhuraphala; **Tamil:** Elandai, Elladu, Ilanti; **Telugu:** Regu.

Distribution: Plant native to South Asia. It grows wild, particularly in sub-Himalayan hill country. Dry, tropical, medium rainfall areas of the Indo-Gangetic Plains and dry areas of Rajasthan. It is now found in many parts of Asia, Australia, the West Indies, tropical America and Africa.

Description: A small much-branched tree with oval to spreading irregular canopy. Main trunk is strong and branching is profuse. Bark is light brown to dark brown and scaly. Height of the tree may go up to 6–12 m. Root system is deep and extensive. Young branchlets softly pubescent. Leaves bifarious, ovate-elliptic, rounded at both ends, faintly and irregularly denticulate, glabrous above, covered beneath with a dense whitish or buff tomentum, strongly 3-nerved from the base. Prickles solitary or in pairs.

Flowers greenish yellow, in small axillary clusters or short-peduncled axillary cymes. Calyx pubescent or woolly outside, glabrous within, cleft about halfway down; lobes triangular, prominently keeled on the inner face for about ½ their length. Petals cucullate, rounded at the apex, deflexed with the stamens between the calyx lobes. Pedicels short, hairy. Stamens enclosed in the petals, but exserted, filaments slightly dilated at the base. Disk with 10 grooved lobes. Ovary about half sunk in the disk, ovoid; styles 2, connate for about ½ their length.

Fruit is a drupe, round to elliptical or oblong, edible, globose, fleshy, smooth, yellow or orange when ripe. Epicarp is smooth, shining, and does not separate from the mesocarp and endocarp. Mesocarp is thick, juicy, sweet and varies in thickness. Both are edible. Seeds: 1 in each fruit, round or oblong to oval. Seed coat is hard and light brown to yellow in colour, smooth or rough, non-endospermic, pitted.

Flowering: July–October.

Fruiting: November.

Chemical constituents: The seeds, which form the drug, contain 2 saponins: jujubosides A and B. Betulinic acid is also reported. This acid cures melanoma cells completely and does not show toxic effects on normal cells. Leucocyanidin in contained in the bark. Leucopelargonidin, betulinic and ceabothic acids are found in the wood. Rutin comes from the leaves. Mauritines A, B, C, D, E and F, frangufoline and amphbines B, D and F were also found from leaves.

Used in: Ayurveda and folk medicine.

Parts used: The whole plant.

Uses: The roots are bitter, cooling, anodyne and tonic and are useful in vitiated conditions of *pitta*, fever, wounds, ulcers and cephalalgia. The bark is astringent, constipating and tonic and is useful in dysentery, diarrhea, gingivitis and boils. The leaves are bitter, cooling, astringent, anthelmintic, diaphoretic, antipyretic and are used in stomatitis, ulorrhea, wounds, syphilitic ulcers, asthma, typhoid fever, diarrhea and obesity. The fruits are sweet, cooling, anodyne, purgative, mucilaginous, pectoral, styptic, aphrodisiac, invigorating, depurative, appetizer and tonic. The seeds are acrid, sweet, astringent, soporific and tonic. They are useful in encephalopathy, ophthalmopathy, dipsia, cough, asthma, wounds, vitiated conditions of *pitta*, burning sensations, diarrhea, vomiting, leucorrhoea, general debility and insomnia.

Methods of propagation:

Seeds are used for raising seedlings.

- Seeds are sometimes soaked to quicken the germination or may be cracked to induce early germination. They are sown in February or in early spring. They complete germination in 8–10 weeks. Seedlings are budded or grafted.
- Transplanting of 1–1.5 year-old trees is done during early winter or at the onset of the monsoon season. Spacing is usually 4–8 m.

Photo: Prof Srinath Rao

References

Adinarayana D and Syamsundar KV. 1982. Occurence of a rare diholosylflavone. 2–0-glucosylvitexin in *Desmodium triflorum. Current Science* 51: 936–937.

Agbor GA, Vinsonb JA, Obenc JE and Ngogangd JY. 2006. Comparative analysis of the in vitro antioxidant activity of white and black pepper. *Nutrition Research* 26(12): 659–663.

Ahmad A, et al. 2013. A review on therapeutic potential of *Nigella sativa*: A miracle herb. *Asian Pacific Journal of Tropical Biomedicine* 3(5): 337–352.

Akihisa T, Hayashi Y, Patterson GW, Shimizo N and Tamura T. 1992. 4α-methylvernosterol and other sterols from *Vernonia anthelmintica* seeds. *PhytoChemistry* 31(5): 1759–1763.

Al-Adhroey Abdulelah H, Nor Zurainee M, Al-Mekhlafi Hesham M and Mahmud Rohela. 2010. Median Lethal Dose, antimalarial Activity, Phytochemical Screening and radical Scavenging of methanolic Languas galanga Rhizome Extract. *Molecules* 15(11): 8366–8376.).

Al-Snafi AE. 2018. The chemical constituents and pharmacological effects of *Foeniculum vulgare* - A review. *ISRO Journal of Pharmacy* 8(5): 81–96.

Amadi BA, Duru MKC and Agomuo EN. 2012. Chemical profilesof leaf, stem, root and flower of *Ageratum conyzoides. Asian Journal of Plant Science and Research* 2(4): 428–432.

Ambavade SD, Mhetre NA, Tate VD and Bodhankar SL. 2006. Pharmacological evaluation of the extracts of *Sphaeranthus indicus* flowers on anxiolytic activity in mice. *Indian Journal of Pharmacology* 38(4): 254–259.

Amin GR. 2005. *Poular Medicinal Plants of Iran*. Tehran, Iran: Vice-Chancellorship of Research, Tehran University of Medical Science Press.

Anonymous. 2001. *The Wealth of India: A Dictionary of Indian Raw Materials and Industrial Products*, vol. 2. New Delhi: National Institute of Science Communication, ISBN: 81-7236-2137.

Arayne S, Sultana MS and Nand Bahadur AS. 2007. The Berberis story: *Berberis vulgarisi* Therapeutics Pak. *Journal of Pharmaceutical Sciences* 20(1): 83–92.

Arya A, et al. 2012. Chloroform fraction of *Centratherum anthelminticum* (L.) seed inhibits tumor necrosis factor alpha and exhibits pleotropic bioactivities: Inhibitory role in human tumor cells. Evidence-based Complementary and Alternative Medicine. doi: 10.1155/2012/627256.

Asgarpanah J and Ramezanloo F. 2012. Chemistry, pharmacology and medicinal properties of *Peganum harmala* L. *African Journal of Pharmacy and Pharmacology* 6(22): 1573–1580.

Aswathy TR, Gayathri E, Praveen J, Achuthsankar S Nair and Sugunan VS. 2019. Phytoprofiling of medicinal plant Cayratia pedata by qualitative and quantitative method. *Journal of Pharmacognosy and Phytochemistry* 8(2): 1637–1642.

Atal C, Girotra R and Dhar K. 1977. Occrance of sesamine in Piper longum. *Indian Journal of Chemistry* 15B: 583.

Avani Kothari, Harish Padh and Neeta Shrivastava. 2006. Ex situ conservation method for *Clerodendrum inerme*: A medicinal plant of India. *African Journal of Biotechnology* 5(5): 415–418.

Ayurvedic Pharmacopoeia of India. 2001. *Vol. IV, Govt. of India Ministry of Health and Family Welfare, Indian system of Medicine & Homeopathy*. New Delhi: The Controller of Publication Civil Lines, 66.

Ayyanar M and Subash-Babu P. 2012. *Syzygium cumini* (L.) Skeels: A review of its phytochemical constituents and traditional uses. *Asian Pacific Journal of Tropical Biomedicine* 3(3): 240–246.

Babu BH, Shylesh BS and Padikkala J. 2001. Antioxidant and hepatoprotective effect of *Acanthus ilicifolius. Fitoterapia* 72(3): 272–277.

Babu BH, Shylesh BS and Padikkala J. 2002. Tumour reducing and anti-carcinogenic activity of *Acanthus ilicifolius. Journal of Ethnopharmacology* 79(1): 27–33.

Badola KC, Mohinder P, Bhanderi HCS and Pal M. 1993. Vegetative propagation of ranbaigan (*Solanum torvum* Sw.) by rooting branch cuttings. *Indian Forester* 119(12): 1–27.

Bargal K and Bargali SS. 2009. *Acacia nilotica*: A multipurpose leguminous plant. *Nature and Science* 7(4): 11–19.

Basu K and Ghosh BK. 1972. Chemical investigation of *Coccinia indica. Transactions of the Bose Research Institute* (Calcutta) 35(2): 43–44.

Bean MF, Antoun M, Abramson D, Chang CJ, Mc Laughin J and Cassady J. 1985. Cucurbitacin B and Isocucurbitacin B: Cytotoxic compounds of *H.ixora. Journal of Nature Products* 48(3): 500.

Beck S, Heather Mathison H, Todorov T, Esli-Armando CJ and Kopp Olga. 2018. A review of medicinal uses and pharmacological activities of *Tridax procumbens* (L.). *Journal of Plant Studies* 7(1): 19–34.

Beena Jose and Joji Reddy L. 2009. Aanalysisofthe essential oils of the stems, leaves and rhizomes of the medicinal plant *Costu spictus* from Southern India International. *Journal of Pharmacy and Pharmaceutical Sciences* 2(2): 100–101.

Bhandary MJ, Chandrashekhar KR and Kaveriappa KM. 1995. Ethnobotany of Gowlis of Uttara district, Karnataka, India. *Journal of Ethnopharmacology* 47(3): 149–158.

Bhattacharjee D, Das A, Das SK and Chakraborthy GS. 2011. *Clerodendrum Infortunatum Linn.*: A Review. *JAPHR* 1(3): 81–85.

Bouaziz MA, Rassaoui R and Besbes S. 2014. Chemical composition, functional properties, and effect of inulin from Tunisian Agave americana L. Leaves on textural qualities of pectin gel. *Journal of Chemistry* 2014: 1–11.

Bukar A, Danfillo IS, Adeleke OA and Ogunbodede EO. 2004. Traditional oral health practices among Kanuri women of Brono State Nigeria. *Odonto-Stomatologie Tropicale = Tropical Dental Journal* 27(107): 25–31.

Burkil HM. *The Useful Plants of West Tropical Africa 1985–2004*. Kew: Royal Botanical Gardens.

Chakrabortty S and Choudhary R. 2014. *Hemidesmus indicus* (Anantmool): A rare herb of Chhattisgarh. *Indian Journal of Scientific Research* 4(1): 89–93.

Chatterjee A and Pakrashi SC. 2001. *The Treatise of Indian Medicinal Plants*, vol. 5. New Delhi: Council of Industrial and Scientific Research, 8–9.

Chau C, Chen C and Lin C. 2004. Insoluble fiber-rich fractions derived from *Averrhoa carambola*: Hypoglycemic effects determined by *in vitro* methods. *Lebensmittel-Wissenschaft und -Technologie* 37(3): 331–335.

Chen LJ, Games DE and Jones J. 2003. Isolation and identification of four flavonoid constituents from the seeds of *Oroxylum indicum* by high-speed counter-current chromatography. *Journal of Chromatography: Part A* 988(1): 95–105.

Chetty KM and Rao KN. 1989. Ethnobotany of Sarakallu and adjacent areas of Chittoor district. *Vegetos* 2: 51–58.

Chithra V and Leelamma S. 1999a. *Coriandrum sativum* changes the levels of lipid peroxides and activity of antioxidant enzymes in experimental animals. *Indian Journal of Biochemistry and Biophysics* 36(1): 59–61.

Chithra V and Leelamma S. 1999b. Hypolipidemic effect of coriander seeds (*Coriandrum sativum*): Mechanism of action. *Plant Foods for Human Nutrition* 51(2): 167–172.

Cuervo C, Blunden G and Patel AV. 1991. Chlorogenone and neochlorogenone from the unripe fruits of *Solanum torvum*. *Phytochemistry* 30(4): 1339–1341.

Das PK, Sahu NP, Banerjee S, Sett S, Goswami S and Bhattacharya S. 2006. Anti peptic ulcer activity of an extract of a plant flower *Woodfordia fruticosa*. United States Patent No.20060040005.

Desha M, Binaibabu N and Doss J. 2012. Futer prospect for the critically endangered medicinally important species, Canarium strictum Roxb. A review. *International Journal of Conservation Science* 3(3): 231–237.

Devi Priya M. 2013. Review on pharmacological activity of *Hemigraphis colorata* (Blume) H.G. Hallier. *International Journal of Herbal Medicine* 1(3): 120–121.

Forouzanfar F, Fazly Bazzaz BS and Hosseinzadeh H. 2014. Black cumin (*Nigella sativa*) and its constituent (thymoquinone): A review on antimicrobial effects. *Iranian Journal of Basic Medical Sciences* 17(12): 929–938.

Gaitonde BB, Raiker KP, Shroff FN and Patel JR. 1957. Pharmacological studies with Malkangunin indigenous tranquilizing drug. *Current Medical Practice* 1: 619–621.

Gamboa-Leon R, Vera-Ku M, Peraza-Sanchez SR, Ku-Chulim C, Horta-Baas A and Rosado-Vallado M. 2014. Antileishmanial activity of a mixture of Tridax procumbensand Allium sativumin mice. *Parasite* 21: 15. doi: 10.1051/parasite/2014016. PMCID 3980668. PMID 24717526.

Garazd MM, Garazd YL and Khilya VP. 2003. Neoflavones. 1. Natural distribution and spectral and biological properties. *Chemistry of Natural Compounds* 39(1): 54–121.

Gauthaman KK, et al. 2006. Cardioprotective effect of the *Hibiscus rosa-sinensis* flowers in an oxidative stress model of myocardial ischemic reperfusion injury in rat. *BMC Complementary and Alternative Medicine* 6: 32.

Ghosal RS, et al. 1971. Alkaloids of *Desmodium triflorum*. *Phytochemistry* 10: 3313–3314.

Ghosh AK, Bhattacharyya FK and Ghosh DK. 1990. Leishmania donovani: A mastigote Inhibition and mode of action of berberine. Exp. *Parasitd* 60(3): 404–413.

Govindappa M, Anil Kumar NV and Santoyo G. 2011. *Crotalaria pallida* extracts as a putative HIV-protease inhibitors. *Journal of Research in Biology* 1(4): 285–291.

Gowthami M, Tamil Selvi S, Senthil Kumar G. and Panneerselvam A. 2012. Phytochemical analysis and antibacterial properties of leaf extract of *Azima tetracantha* (Lam.). *Asian Journal of Plant Science and Research* 2(2): 110–114.

Gray AM and Flatt PR. 1999. Insulin-releasing and insulin-like activity of the traditional anti-diabetic plant Coriandrum sativum (coriander). *British Journal of Nutrition* 81(3): 203–209.

Grover JK and Yadav SP. 2004. Pharmacological actions and potential uses of *Momordica charantia*: A review. *Journal of Ethnopharmacology* 93(1): 123–132.

Guhabakshi DN, Sensarma P and Pal DC. 1999–2001. *A Lexicon of Medicinal Plants in India*, vol. 1&2. Calcutta: Naya Prokash.

Gulcin I, Oktay M, Kirecci E and Kufrevioglu OI. 2003. Screening of antioxidant and antimicrobial activities of anise (*Pimpinella anisum* L.) seed extracts. *Food Chemistry* 83(3): 371–382.

Gupta B, Nayak S and Solanki Sarita. 2011. *Abroma augusta* Linn f: A review. *Pharmacia Sinica* 2(4): 253–261.

Gupta M, Bagchi A and Ray AB. 1991. Additional withanolides of *Datura metel*. *Journal of Natural Products* 54(2): 599–602.

Gupta M, Singh A and Joshi HC. 2015. Berberis Lycium multipotential medicinal application: An overview. *International Journal of Chemical Studies* 3(4): 10–13.

Gupta R and Gupta GD. 2015. A review on plant *Cordia obliqua* Willd. (Clammy cherry). *Pharmacognosy Reviews* 9(18): 127–131.

Hayashi T, Kawaski M, Miwa Y, Taga T and Morita N. 1990. Antiviral agents of plant origin III. Scopadulin, a novel tetracyuclic diterpene from *Scoparia dulcis* L. *Chemical and Pharmaceutical Bulletin* 38(4): 945–994.

Hemshekhar M, et al. 2011. An overview on genus *Garcinia*: Phytochemical and therapeutical aspects. *Phytochemistry Reviews* 10(3): 325–351.

Hossain MZ, Shibib, BA and Rahman R. 1992. Hypoglycemic effects of *Coccinia indica*: Inhibition of key gluconeogenic enzyme, glucose-6-phosphatase. *Indian Journal of Experimental Biology* 30(5): 418–420.

Hua L, Peng Z, Chia LS, Goh NK and Tan SN. 2001. Separation of kaempferols in *Impatiens balsamina* flowers by capillary electrophoresis with electrochemical detection. *Journal of Chromatography: Part A* 909(2): 297–303.

Hussain AI, Rathore HA, Sattar MZA, Chatha SAS, Sarker SD and Gilani AH. 2014. *Citrullus colocynthis* (L.) Schrad (bitter apple fruit): A review of its phytochemistry, pharmacology, traditional uses and nutritional potential. *Journal of Ethnopharmacology* 155(1): 54–66.

Igoli JO, Ogaji OG, Tor-Anyiin TA and Igoli NP. 2005. Traditional Medicine Practice Amongst the Igede People of Nigeria. Part II. *Journal of Traditional CAM* 2(2): 134–152.

Ito C, Kondo Y, Rao KS, Tokuda H, Nishino H and Furukawa H. 1999. Chemical constituents of *Glycosmis pentaphylla*. Isolation of a novel naphthoquinone and a new acridone alkaloid. *Chemical and Pharmaceutical Bulletin* 47(11): 1579–1581.

Itoh A, Tanaghashi T and Nagakura N. 1995. Tetrahydroisoquinoline monoterpene glucosides and tetrahydro-beta-carbolinemonoterpene glucoside from *Alangium lamarckii*. *Journal of Natural Product* 58(8, 14): 1228–1239.

Itoh T, Nayashiro H. 1994. Tetrahydroisoquinoline monoterpene glucosides from Alangium lamarckii and Cephaelis ipecauanha. *Phytochemistry* 36(2): 383–387.

Jadhav VM, Thorat RM, Kadam VJ and Sathe NS. 2009. Traditional medicinal uses of Hibiscus rosa-sinensis. *Journal of Pharmacy Research* 2(8): 1220–1222.

Jayasri MA, Mathew L and Radha A. A report on the antioxidant activity of leaves and rhizomes of *Costus pictus* D.Don. *International Journal of Integrative Biology* 5: 1–20.

Jeeshna MV, Manorama S and Paulsamy S. 2009. Antimicrobial properties of the medicinal shrub *Glycosmis pentaphylla* CORREA. *Journal of Applied Biology* 3: 25–27.

Jha RK, Garud N. and Nema RK. 2009. Excision and incision wound healing activity of flower head alcoholic extract of Sphaeranthus indicus L. in Albino rats. *Global Journal of Pharmacology* 3: 32–37.

Jha RK, Mangilal A, Bhandari and Nema RK. 2010. Antidiabetic activity of flower head petroleum ether extracts of *Sphaeranthus indicus* Linn. *Asian Journal of Pharmaceutical and Clinical Research* 3: 16–19.

Joseph B and Jini D. 2013. Antidiabetic effects of *Momordica charantia*(bitter melon) and its medicinal potency. *Asian Pacific Journal of Tropical Disease* 3(2): 93–102.

Kalidass C, Abragam Amish D and Mohan VR. 2009. Pharmacognostic studies of the whole plant of *Asclepias curassavica* Linn. *Journal of Pharmacy Research* 2(7): 1214–1217.

Kamboj A and Saluja AK. 2008. A review on Its phytochemical and pharmacological profile. *International Journal of Green Pharmacy* 2(2): 59–68.

Kaur R, Kaur H and Dhindsa AS. 2018. Glycyrrhiza glabra: A phytopharmacological review. *The International Journal of Pharmaceutical Research* 4: 2470–2477.

Khan I, Najeebullah S, Ali M and Shinwari ZK. 2016. Phytopharmacological and ethnomedicinal uses of the Genus Berberis (Berberidaceae): A review. *Tropical Journal of Pharmaceutical Research*2016 15(9): 2047–2057.

Khan MR. 1999. Chemical composition and medicinal properties of *Nigella sativa* Linn. *Inflammo-Pharmacology* 7: 13–35.

Khushalani H, Tatke P and Singh KK. 2006. Antifertility activity of dried flowers of *Woodfordia fruticosa* kurz. *Indian Journal of Pharmaceutical Sciences* 68(4): 528–529.

Kirtikar KR and Basu BD. 1998. Dehradun: Shiva offset press. *Indian Medicinal Plants*: 1674–1681.

Krishna PVG, Venkata Rao E and Venkata Rao D. 1975. Crystalline principles from the leaves and twigs of *Leptadenia reticulata*. *Planta Medica* 27: 395–400.

Kumar DS, et al. 2010. A medicinal potency of *Momordica charantia*. *International Journal of Pharmaceutical Sciences Review and Research* 1(2): 95–99.

Kumar P, Dhakad Sharma PK and Kumar S. 2017. A review on phytochemical studies and biological potential of *Citrullus colocynthis* (L.) Schrad. *Bioengineering and Bioscience* 5(4): 55–64.

Kumar Sunil, Malhotra Rashmi and Kumar Dinesh. 2010. *Euphorbia hirta*: Its chemistry, traditional and medicinal uses, and pharmacological activities. *Pharmacognosy Reviews* 4(7): 58–61.

Kumari Sangeeta Singh DC and Khatoon S. 2017. A precious herb - Gmelina arborea Linn W.S.R to its medicinal uses and parmacological activity. *International Journal of Ayurveda and Pharma Research* 5: 47–53.

Kundu S and Ray AB. 1987. Chemical examination of *Coccinia indica* fruits. *Journal of the Indian Chemical Society* 64(12): 776–777.

Latha Mand Pari L. 2005. Effect of an aqueous extract of *Scoparia dulcis* on plasma and tissue glycoproteins in steptozotocin induced diabetic rats. *Pharmazie* 60(2): 151–154.

Lawal OA, Ogunwande IA, Ibirogba AE, Layod OM and Opoku AR. 2015. Chemical constituents of essential oils from *Catharanthus roseus* (L.) G. Don Grown in Nigeria. *Journal of Essential Oil Bearing Plants* 18(1): 57–63.

Li Y, Chen X, Satake M, Oshima Y and Ohizumi Y. 2004. Acetylated flavonoid glycosides potentiating NGF action from Scoparia dulcis. *Journal of Natural Products* 67(4): 725–727.

Liu J, Chen J, Wang C and Qui M. 2009. New cucurbitane triterpenoids and steroidal glycoside from *Momordica charantia*. *Molecules* 14(12): 4804–4813.

Lu Y, Jianguang L, Xuefeng H and Lingyi K. 2009. Four new steroidal glycosides from Solanum torvum and their cytotoxic activities. *Steroids* 7(4): 95–101.

Maganha EG, Da costa HR, Rosa RM, Pegas HOA, Lia de paula RAL and Saffi J. 2009. Pharmacological evidences for the extracts and secondary metabolites from plants of the genus *Hibiscus*. *Food Chemistry* 118: 1–10.

Mallika J, Devi MC and Shyamala S. 2005. In vitro and in vivo evaluation of free radical scavenging potential of *Cissus quadrangularis*. *African Journal of Biomedical Research* 8: 95–99.

Malviya S, Rawat S, Kharia A and Verma M. 2011. Medicinal attributes of *Acacia nilotica* Linn. - A comprehensive review on ethnopharmacological claims. *International Journal of Pharmaceutical and Life Sciences* 2(6): 830–837.

Mamdouh NS, Mokhtar M, Ahmed B, Ahmed A, Sayed HM and Mohamed SK. 2013. Pharmacognostical studies on flower of *Tribulus terrestris* L. *Journal of Pharmacognosy and Phytochemistry* 1(5): 19–23.

Mayer R. 1999. Calycopterones and calyflorenones, novel biflavonoids from *Calycopteris floribunda*. *Journal of Natural Products* 62(9): 1274–1278.

Mayer R. 2004. Five biflavonoids from Calycopteris floribunda (Combretaceae). *Phytochemistry* 65(5): 593–601.

Meghwal M and Goswami TKTK. 2012. Chemical composition, nutritional, medicinal and functional properties of black Pepper: A review. *Open Acsess Scientific Reports* 1(2): 2–5.

Mehta BK, Mehta D and Itoriya A. 2004. Structure elucidation by NMR spectroscopy of a new acetylated saponin from *Centratherum anthelminticum*. *Carbohydrate Research* 339(18): 2871–2874.

Meira M, Pereira da Silva E, David JM and David JP. 2012. Review of the genus *Ipomoea*: Traditional uses, chemistry and biological activities. *Revista Brasileira de Farmacognosia* 22(3): 682–713.

Meyre Silva C, Mora TC, Biavatti MW, Santos ARS, Dal-Magro J, Yunes RA, Cechinel-Filho V. 1988. Preliminary phytochemistry and pharmacological studies of *Aleurites moluccana* leaves [L] wild. *Phytomedicine* 5: 109–113.

Monu Milind P, Kadian R and Sharma K. 2014. Phytopharmacology of *Acacia catechu* Willd: A rewiew. *World Journal of Pharmacy and Pharmaceutical Sciences* 3(11): 1380–1389.

Moresco HH, Queiroz GS, Pizzolatti MG and Brighente IMC. 2012. Chemical constituents and evaluation of the toxic and antioxidant activities of *Averrhoa carambola* leaves. *Revista Brasileira de Farmacognosia* 22(2): 319–324.

Muthuswamy R and Senthamara R. 2014. Pharmacognostical studies on stem bark of *Canarium strictum* Roxb. *Pharmacognosy Journal* 6(1): 12–18.

Nadim MM, Malik AM, Ahmad J and Bakshi SK. 2011. The essential oil composition of *Achillea Millefolium* L.cultivated under tropical condition in India. *World Journal of Agricultural Sciences* 7(5): 561–565.

Nadumane VK, et al. 2011. Evaluation of the anti-cancer potential of *Costus pictus* on fibrosarcoma (HT-1080) cell line. *Natural Pharmaceuiticals* 2(2): 72–76.

Nagaratna A, Prakash LH and Harini A. 2014. Pharmacological review on gokhra ganja. *Journal of Pharmacognosy and Phytochemistry* 3: 253–257.

Nayk P and Thirunavoukkarasu M. 2016. A review of the plant *Boerhaavia diffusa*: Its chemistry, pharmacology and therapeutical potential. *Journal of Phytopharmacology* 5(2): 83–92.

Nikitha Devi A and Krishna Mohan G. 2018. A Review on *Chloroxylon swietenia. International Journal of Pharmacy and Biological Sciences* 8(5): 502–506.

Nogueira JHC, Gonçalez E, Galleti SR, Facanali R, Marques MOM. and Felício JD. 2010. *Ageratum conyzoides* essential oil as aflatoxin suppressor of Aspergillus flavus. *International Journal of Food Microbiology* 137(1): 55–60.

Ojo OO, Nadro MS and Tella IO. 2006. Protection of rats by extracts of some common Nigerian trees against acetaminophen-induced hepatotoxicity. *African Journal of Biotechnology* 5: 755–760.

Ozcan MM and Chalchat JC. 2006. Chemical composition and antifungal effect of anise (*Pimpinella anisum* L.) fruit oil at ripening stage. *Annals of Microbiology* 56(4): 353–358.

Panthong A, Kanjanapothi D and Taylor WC. 1986. Ethnobotanical review of medicinal plants from Thai traditional books. Part I. Plants with anti inflammatory, anti-asthmatic and anti-hypertensive properties. *Journal of Ethno Pharmacology* 18(3): 213–228.

Parida R and Nayak S. 2019. Chemical composition of Hedychium coronarium Koen. flowers from eastern India. *Plant Science Today* 6(2): 258–263.

Parmar C and Kaushal MK. 1982. Cordia obliqua. In: Parmar C., Kaushal MK. editors. *Wild Fruits of the Sub-Himalayan Region*. New Delhi: Kalyani Publishers, 136.

Patel JR, Tripathi P, Sharma V, Chauhan NS and Dixit VK. 2011. *Phyllanthus amarus*: Ethnomedicinal uses, phytochemistry and pharmacology: A review. *Journal of Ethnopharmacology* 138(2): 286–313.

Payal C, Gurlaganjeet K, Davinder K, Gagan S, Amit C and Dhawan RK. 2010. A review on phytochemistry and biological activities of *Aerva. Medicinal and Aromatic Plants* 4(2): 2–4.

Payal C, Gurlaganjeet K, Davinder K, Gagan S, Amit C, et al. 2015. A Review on Phytochemistry and Biological Activities of *Aerva. Medicinal and Aromatic Plants* 4: 187. doi:10.4172/2167-0412.1000187

Peana AT, Moretti MDL, Manconi V, Desole G and Pippia P. 1997. Anti-inflammatory activity of aqueous extracts and steroidal sapogenins of *Agave americana. Planta Medica* 63(3): 199–202.

Pearline D, Kamat N and Padma T. 2015. Rosa sinensis (Hibiscus) -a versatile Indian origin plant. *Journal of Chemical and Pharmaceutical Sciences* 8(4): 970–974.

Peterson DW, et al. 2009. Cinnamon extract inhibits tau aggregation associated with Alzheimer's disease in vitro. *Journal of Alzheimer's Disease* 17(3): 585–597.

Pratima V and Rekha V. 2014. A Review on Limonia acidissima l.: Multipotential Medicinal Plant. *International Journal of Pharmaceutical Sciences Review and Research* 28: 191–195.

Puri D and Bhandari A. 2014. Fagonia: A potential medicinal desert plant. *Journal of NAP* 27(1): 28–33.

Qu WH, Li JG and Wang MS. 1991. Chemical studies on the *Helicteres isora. Zhongguo Yaoke Daxue Xuebao* 22: 203–206.

Quader MA, Nutan MTH and Rashid MA. 1999. Antitumor alkaloid from Glycosmis pentaphylla. *Fitoterapia* 70(3): 305–307.

Ragasa CY, Lorena GS, Mandia EM, Dennis D, Raga DD and Chang She C. 2013. Chemical Constituents of *Abrus precatorius. American Journal of Essential Oil and Natural Products* 1: 7–10.

Rahman AHMM, Biswas MC, Islam AKMR and Zaman ATMN. 2013. Assessment of traditional medicinal plants used by local people of Monirampur Upazilla under Jessore district of Bangladesh. *Wudpecker Journal of Medicinal Plants* 2: 99–109.

Rahmani M, Muhammad R and Ali AM. 2010. Alkaloids and sulphur-containing amides from Glycosmis citrifolia and *Glycosmis elongata*. *Sains Malaysiana* 39: 445–451.

Rajesh R, Chitra K and Parakh PM. 2011. *Aerva lanata* (Linn) Juss. Ex schult-an overview. *Indian Journal of Natural Products and Resources* 2: 5–9.

Ramadan Mohamed and Mörsel J. 2002. Oil composition of coriander (*Coriandrum sativum* L.) fruit-seeds. *European Food Research and Technology* 215(3): 204–209.

Ramesh P, Vijaya C and Parasuraman S. 2013. Anti-arthritic activity of ethanolic extract of Tridax procumbens (Linn.) in Sprague Dawley rats. *Pharmacognosy Research* 5(2): 113–117.

Rao V, et al. 2013. Preliminary phytochemical screening and anthelmintic activity of *Desmodium triflorum* (L.) DC leaf and root extracts. *International Journal of Pharmaceutical Sciences* 3(1): 156–158.

Rastogi RP and Mehrotra BN. 1960–1969. *Compendium of Indian Medicinal Plants*, vol. I. New Delhi: PID, 237.

Rastogi RP and Mehrotra BN. 1970–1979. *Compendium of Indian Medicinal Plants*, vol. II. New Delhi: PID, 403.

Rastogi RP and Mehrotra BN. 1980. *Compendium of Indian Medicinal plants, Drug Research Perspectives*, vol. 2. New Delhi: Publication and information, Directorate, 196.

Rastogi RP and Mehrotra BN. 1980–1984. *Compendium of Indian Medicinal Plants*, vol. III. New Delhi: PID, 378.

Rastogi RP and Mehrotra BN. 1985–1989. *Compendium of Indian Medicinal Plants*, vol. IV. New Delhi: PID, 422.

Rastogi RP and Mehrotra BN. 1990–1994. *Compendium of Indian Medicinal Plants*, vol. V. New Delhi: PID, 58.

Rastogi RP and Mehrotra BN. 1990–1994. *Compendium of Indian Medicinal Plants*, vol. V. New Delhi: PID, 473.

Ratra M and Gupta R. 2015. Comprehensive review on pharmacological profile of *Alangium salvifolium*: A medicinal Plant. *UK Journal of Pharmaceutical and Biosciences* 3(3): 22–28.

Ravikiran S, Reddy Shiva A, Devi Sita P and Janardhan Reddy K. 2006. Insecticidal, antifeedant and oviposition deterrent effects of the essential oil and individual compounds from leaves of *Chloroxylon swietenia* DC. *Pest Management Science* 1121: 62–1116.

Rehana Banu, H and Nagarajan N. 2012. Phytochemical screening for active compounds in *Gloriosa superba* Leaves and Tubers. *International* Journal of Pharmacognosy and Phytochemistry Research 4(1): 20.

Reshma RK, Josepp S, Neethu EK, Kavya V. Anjali KM and Suga Bharathi M. 2018. Preliminary phytochemical and biochemical analysis of Hemigraphis colorata H.G. Hallier. *International Journal of Research in Pharmacy and Pharmaceutical Sciences* 3(2): 05–09.

Riaz M, Khalid MR and Chaudhary FM. 1995. Essential oil composition of Pakistani *Ageratum conyzoides* L. *Journal of Essential Oil Research* 7(5): 551–553.

Rodriguez E, Vander VG and Marby TJ. The structure of calycopterin. *Phytochemistry* 11(7): 2311–2312.

Saha S and Ghosh S. 2012. *Tinospora cordifolia*: One plant, many roles. *Ancient Science of Life* 31(4): 151–159.

Sasya Sahyadri- A CD on Plants of Western Ghats, 2003, K.N. Ganeshaiah.

Sain M and Sharma V. 2013. *Catharanthus roseus* (an anti-cancerous drug yielding plant) - A review of Potential Therapeutic Properties. *Indian Journal of Pure & Applied Biosciences* 1(6): 139–142.

Sakunphueak A and Panichayupakaranant P. 2010. Simultaneous determination of three naphthoquinones in the leaves of *Impatiens balsamina* L. by reversed-phase high-performance liquid chromatography. *Phytochemical Analysis* 21(5): 444–450.

Sannigrahi S. 2004. Biodiversity documentation for Kerala- Flowering Plants, part 6: 81. 2004; Cook, Fl. Bombay 1: 181.1903; Almeida, Fl. Maharashtra 1:207. 1996; Saldanha, Fl. Ka.

Sannigrahi S, Mazumder UK, Dilipkumar P and Mishra SL. 2009. Hepatoprotective potential of methanol extract of clerodendrum infortunatum linn. against CCl4 induced hepatotoxicity in rats Pharmacognosy *Magazine* 5(20): 394–399.

Sastry BS, Vijayalakshami T and Venkata Rao E. 1985. Chemical constituents of the stem bark of *Leptadenia reticulata*. *Indian Drugs* 22: 611–612.

Savanur PV. 2016. Haritamannajari (Acalypha indica Linn.) - A review. *IJAAR* 2(5): 545–550.

Sebastian MK and Bhandari MM. 1984. Medicinal Plant flora of Udaipur district, Rajasthan. *Bulletin of Medicinal Ethanobotany Research* 5: 122–134.

Semwal RB, Semwal DK, Vermaak I and Viljoen A. 2015. A comprehensive scientific overview of *Garcinia cambogia*. *Fitoterapia* 102: 134–148.

Semwal RB, Semwal R, Sudhir Kumar K, Singh G and Rawat U. 2010. The genus *Stephania* (Menispermaceae): Chemical and pharmacological perspectives. *Journal of Ethnopharmacology* 132(2): 369–383.

Shamsuddin T, Rahman W, Khan SA, Shamsuddin KM and Kintzinger JB. 1988. Moluccanin, a coumarino-lignoid from *Aleurites moluccana*. *Phytochemistry* 27(6): 1908–1909.

Sharma N and Vijayvergia R. 2013. Review on *Digera muricata* (l.) Mart. – A Great Versatile Medicinal Plant. *International Journal of Pharmaceutical Sciences Review and Research* 20(1): 114–119.

Sheeja E, Edwin E, Dhanbal S and Suresh B. 2005. Anti-inflammatory activity of the leaves of *Nothapodytes foetida* Miers. *Indian Journal of Pharmaceutical Sciences* 67(2): 251–253.

Shilpa K, Sangeetha KN, Muthusamy VS, Sujatha S and Lakshmi BS. 2009. Probing key targets in insulin signaling and adipogenesis using a methanolic extract of *Costus pictus* and its bioactive molecules, methyl tetracosonoate. *Biotechnology Letters* 31(12): 1837–1841.

Shirwaikar A, Prabhu K, Kumar CD, Rajendran K and Lobo R. 2006. Pharmacognostical evaluation of *Sphaeranthus indicus* L. *Natural Product Sciences* 12: 85–88.

Shirwaikar A and Sreenivasan KK. 1996. Chemical investigation and antihepatotoxic activity of the fruits of *Lagenaria siceraria*. *Indian Journal of Pharmaceutical Sciences* 58(5): 197–202.

Shojaii A and Fard MA. 2012. Review of pharmacological properties and chemical constituents of Pimpinella anisum. *ISRN Pharmacology* 2012: 1–8.

Singh SB, Singh AK and Thakur RS. 1984. Chemical constituents of the leaves of *Helicteres isora*. *Indian Journal of Pharmaceutical Sciences* 46: 148–149.

Sreejith PS, Praseeja RJ and Asha VV. 2012. A review on the pharmacology and phytochemistry of traditional medicinal plant, *Glycosmis pentaphylla* (Retz.). *Correa*. *Journal of Pharmacy Research* 5(5): 2723–2728.

Sreeramulu N, Sateesh S, Ragan, A and Raju, VS. 2013. Ethno-botanico-medicine for common human ailments in Nalgonda and Warangal districts of Telangana, Andhra Pradesh, India. *Annals of Plant Sciences* 2(7): 220–229.

Srivastava M, Srivastava S and Rawat AKS. 2010. Chemical standardization of Cassia aungustifolia Vahl. *Seeds Pharmacognosy Journal* 2(13): 554–560.

Srivastava R. 2014. A review on phytochemical, pharmacological, and pharmacognostical profile of Wrightia tinctoria: Adulterant of kurchi. *Pharmacognosy Reviews* 8(15): 36–44.

Stanley GD. 2001. Black Pepper, Piper nigrum. Medicinal and aromatic Plants—Industrial Profiles, Book reviews. *Phytochemistry* 58(5): 827–829.

Suruse PB, Bodele SB, Duragkarn J and Kale MK. 2008. Anti inflammatory and analgesic activities of isolated compounds from Canarium strictum. *Cell and Tissue Research* 8(3): 1481–1484.

Suruse PB, Duragkar NJ, Shivhare UD, Kale MK and Bodele SB. 2010. Study of antibacterial activity of Canarium strictum gum resin. *Research Journal of Pharmacognosy and Phytochemistry* 2: 435–437.

Tadros SH and Sleem AA. 2004. Pharmacognostical and biological study of the stem and leaf of *Averrhoa carambola* L. grown in Egypt. *Bulletin of the Faculty of Pharmacy* (Cairo University) 42: 225–246.

Tharachand Selvaraj I and Avadhani M. 2013. Medicinal properties of Malabar tamrind [*Garcinia cambogia*. (Gaetrn.) Desr]. *International Journal of Pharmaceutical Sciences Review and Research* 19(2): 101–107.

The Wealth of India. 2005. *A Dictionary of Indian Raw Materials and Industrial Products, First Supplement Series*. New Delhi: National Institute of Sciences Communication and Information Resources and CSIR. 3(D-1): 195–498.

Thirupathi K, Kumar SS, Raju VS, Ravikumar B, Krishna DR and Mohan GK. 2008. A review of medicinal plants of the genus *Cordia*: Their chemistry and pharmacological uses. *Journal of Natural Remedies* 8: 1–10.

Tilebeni HG. 2011. Review to basil medicinal plant. *International Journal of Agronomy and Plant Production* 2(1): 5–7.

Tiwari P, Mishra BN and Sangwan SN. 2014. Phytochemical and pharmacological properties of *Gymnema sylvestre*: An important medicinal plant. *BioMed Research International* 2014: 1–18.

Upadhyay Avnish K, Kumar Kaushal, Kumar Arvind and Mishra Hari S. 2010. *Tinospora cordifolia* (Willd.) Hook. f. and Thoms. (Guduchi) – validation of the ayurvedic pharmacology through experimental and clinical studies. *International Journal of Ayurveda Research* 1(2): 112–121.

Vaishnav MM, Jain P, Jogi SR and Gupta KR. 2001. Coccinioside –K, triterpenoid saponin from *Coccinia indica*. *Oriental Journal of Chemistry* 17(3): 465–468.

Van Kiem P, et al. 2008. Chemical constituents of *Acanthus ilicifolius* L. and effect on osteoblastic MC3T3E1 cells. *Archives of Pharmacal Research* 31(7): 823–829.

Verdoorn IC. 1963. Salvadoraceae. In: Dyer RA., Codd LE. editors. *Flora of Southern Africa*, vol. 26. Pretoria, South Africa: Botanical Research Institute, Department of Agricultural Technical Services, 128–134.

Vorkoc J and Sedmera P. 1977. Evaluation of anti-inflammatory activity and toxicity related studies of Chloroxylon *sweitenia*. *Phytochemistry* 11: 2647.

Wang J, Yang X, Di Y, Wang Y, Shen Y and Hao X. 2006. Isoflavone diglycosides from *Glycosmis pentaphylla*. *Journal of Natural Products* 69(5): 778–782.

Warashina T and Noro T. 2008. Steroidal glycosides from the roots of Asclepias curassavica. *Chemical and Pharmaceutical Bulletin (Tokyo)* 56(3): 315–322.

Wenhao Q, Jingrong L and Mungshi W. 1991. Chemical studies on H. isora Linn. *Zhangguo Yaoke, Dauxe xuebao* 22(4): 203–206, through CA 1992; 116:12485.

Yadav Kavita N, Kadam Prasad V, Patel Jigna A and Patil Manohar J. 2014. *Strychnos potatorum*: Phytochemical and pharmacological review. *Pharmacognosy Reviews* 8(15): 61–66.

Yamunadevi M, Wesely EG and Johnson M. 2011. Phytochemical studies on the terpenoids of medicinally important plant: *Aerva lanata* L. using HPTLC. *Asian Pacific Journal of Tropical Biomedicine* 1: 220–225.

Yeh HF, Luo CY, Lin CY, Cheng SS, Hsu YR and Chang ST. 2013. Methods for thermal stability enhancement of leaf essential oils and their main Constituents from Indigenous Cinnamon (*Cinnamomum osmophloeum*). *Journal of Agricultural and Food Chemistry* 61(26): 6293–6298.

Yousafa Z, Wanga Y and Baydounc E. 2013. Phytochemistry and pharmacological studies on *Solanum torvum* Swartz. *Journal of Applied Pharmaceutical Science* 3(04): 152–160.

Zarga MA. 1986. Three new simple indole alkaloids from *Limonia acidissima*. *Journal of Natural Products* 49(5): 901–904.

Zhen HS, Zhu XY, Lu RM, Liang J, Qiu Q and Meng QM. 2008. Research on chemical constituents from stem of *Gymnema sylvestre*. *Zhong Yao Cai = Zhongyaocai = Journal of Chinese Medicinal Materials* 31(8): 1154–1156.

Zheng C and Qin L. 2007. Chemical components of Centella asiatica and their bioactivities. *Journal of Chinese Integrative Medicine* 5: 348–351.

WEB SOURCE

http://en.wikipedia.org/wiki/File:Phyllanthus_amarus.

http://en.wikipedia.org/wiki/File:Phyllanthus_amarus.

http://en.wikipedia.org/wiki/Terminalia_chebula.

http://herbs.indianmedicinalplants.info/index.php/medicinal-plants-pictures-a-details/1169-bhandira-clero dendrum-infortunatum-medicinal-uses.

http://herbs.indianmedicinalplants.info/index.php/medicinal-plants-pictures-a-details/1169-bhandira-clero dendrum-infortunatum-medicinal-uses.

http://keralaplants.in/keralaplantsdetails.aspx?id=Woodfordia_fruticosa.

http://medicinalplantinindia.blogspot.in/2012/05/melissa-officinalis-lemon-balm.

http://medplants.blogspot.com/search/label/Anthocephalus%20cadamba.

http://shodhganga.inflibnet.ac.in/bitstream/10603/117966/6/chapter%204.pdf.

http://tropical.theferns.info/viewtropical.php?id=Ipomoea+obscura.

http://vikaspedia.in/agriculture/crop-production/package-of-practices/medicinal-and-aromatic-plants/embel ia-ribes-1.

http://www.ayushveda.com/herbs/cassia-auriculata.htm.

http://www.efloraofgandhinagar.in/succulents/bryophyllu.

http://www.fao.org/docrep/x5327e/x5327e0y.htm).

http://www.google.co.in/imgres?imgurl=http://www.ecoplanet.in/herbalextracts/images/Holostemma%2520 ada-kodien.jpg&imgrefurl

http://www.theplantlist.org/tpl/record/kew-2746366.

http://www.toxicologycentre.com/English/plants/Botanical/adapatiyan.html.

http://www.toxicologycentre.com/English/plants/Botanical/adapatiyan.html

http://www.motherherbs.com/inula-racemosa.html.

http://www-ang.kfunigraz.ac.at/~katzer/engl/generic_frame.html?Murr_koe.html

https://sites.google.com/site/efloraofindia/species/a---l/l/lamiaceae/leucas.

https://ayurveda-foryou.com/ayurveda_herb/wrightia.html.
https://ayurveda-foryou.com/ayurveda_herb/wrightia.html.
https://easyayurveda.com/2016/12/13/karonda-carissa-carandas-karamarda.
https://en.wikipedia.org/wiki/Boswellia_serrata.
https://en.wikipedia.org/wiki/Bryophyllum_pinnatum.
https://en.wikipedia.org/wiki/Carissa_carandas.
https://en.wikipedia.org/wiki/Citrullus colocynthis.
https://en.wikipedia.org/wiki/Citrullus_colocynthis.
https://en.wikipedia.org/wiki/Clerodendrum_infortunatum.
http://eol.org/pages/579771/overview.
https://en.wikipedia.org/wiki/Strychnos_nux-vomica.
https://en.wikipedia.org/wiki/Tribulus_terrestris.
https://en.wikipedia.org/wiki/Withania_somnifera.
https://en.wikipedia.org/wiki/Woodfordia_fruticosa.
https://en.wikipedia.org/wiki/Wrightia_tinctoria.
https://herbpathy.com/Uses-and-Benefits-of-Carissa-Carandas-.
https://hort.purdue.edu/newcrop/parmar/02.html.
https://indiabiodiversity.org/observation/show/1812832.
https://indiabiodiversity.org/species/show/268093.
https://indiabiodiversity.org/species/show/31804.
https://indiabiodiversity.org/species/show/31903.
https://ayurveda-foryou.com/ayurveda_herb/wrightia.html.
https://keyserver.lucidcentral.org/weeds/data/media/Html/alternanthera_pungens.htm.
https://sites.google.com/site/efloraofindia/species/a---l/a/apocynaceae/wrightia/wrightia-arborea.
https://sites.google.com/site/efloraofindia/species/a---l/cl/cornaceae/alangium/alangium-salvifolium.
https://www.ayurvitewellness.com/3213/ankola-alangium-salvifolium/.
https://www.bimbima.com/ayurveda/medicinal-plant-sweet-basil-ocimum-basilicum/310/.
https://www.bimbima.com/ayurveda/medicinal-plant-sweet-basil-ocimum-basilicum/310/.
https://www.bimbima.com/ayurveda/medicinal-use-of-aparajita-herb/1165/.
https://www.bimbima.com/ayurveda/medicinal-use-of-aparajita-herb/1165/.
https://www.bimbima.com/ayurveda/medicinal-use-of-harsingarnight-jasmine/20/.
https://www.bimbima.com/ayurveda/medicinal-uses-of-dronapushpi-gomathumbai/334/.
https://www.bimbima.com/herbs/shalaparni-desmodium-gangeticum-information-medicinal-uses-and-caution/61/.
https://www.biotecharticles.com/Biotech-Research-Article/Medicinal-Uses-of-Cassia-Auriculata-2376.html.
https://www.botanical-online.com/english/cumin_cuminum_cyminun.htm.
https://www.flowersofindia.net/catalog/slides/Clearing%20Nut%20Tree.html.
https://www.google.co.in/search?q=feronia+elephantum&tbm.
https://www.hort.purdue.edu/newcrop/morton/karanda.html.
https://www.indianetzone.com/38/akola_plant.htm.
https://www.organicfacts.net/health-benefits/vegetable/health-benefits-of-spinach.html.
https://www.planetayurveda.com/library/dhataki-woodfordia-fruticosa/.
www.pankajoudhia.com.
https://www.indianetzone.com/37/aerva_lanatam.
https://indiabiodiversity.org/species/show/32892.
https://www.stylecraze.com/articles/benefits-and-uses-of-senna-plant/.
http://www.flowersofindia.net/catalog/slides/Gamhar.html.
http://www.flowersofindia.net/catalog/slides/Gurmar.html.
https://www.bimbima.com/hair-care/medicinal-use-of-gudhal-or-hibiscus-rosa/1825/.
https://www.bimbima.com/ayurveda/patol-parwal/231/.
https://www.bimbima.com/ayurveda/medicinal-uses-of-ivy-gourdkunduru/407/.
https://www.bimbima.com/ayurveda/makoi-plant-solanum-nigrum-medicinal-uses-side-effects/375/.
https://www.bimbima.com/herbs/solanum-indicum/2862/.
https://www.bimbima.com/ayurveda/ayurvedic-herb-latakaranja-caesalpinia-bonducella/348/.
https://www.bimbima.com/herbs/viola-odorata/3869/.
https://www.bimbima.com/herbs/shalaparni-desmodium-gangeticum-information-medicinal-uses-and-caution/61/.
https://www.bimbima.com/ayurveda/atibala-indian-mallow-abutilon-indicum-information-uses-and-more/19/.
https://www.bimbima.com/herbs/shivlingi/2943/.

http://www.worldagroforestry.org/treedb/AFTPDFS/Acacia_catechu.PDF.
https://en.wikipedia.org/wiki/Tridax_procumbens.
http://tropical.theferns.info/viewtropical.php?id=Tridax+procumbens.
http://www.flowersofindia.net/catalog/slides/Ceylon%20Satinwood.html.
https://en.wikipedia.org/wiki/Boswellia_serrata.
https://ayushvedah.com/druginfo.php?drugid=575&info=uses.
http://doctorschar.com/dill-anethum-graveolens/.
https://www.motherherbs.com/anethum-graveolens.html.
http://tropical.theferns.info/viewtropical.php?id=Vachellia+farnesiana.
https://herbs.indianmedicinalplants.info/index.php/a/1238-acacia-farnesiana.

Index